Certified Solidworks Professional Advanced Weldments Exam Preparation

To report errors or problems with download files from this Google Drive Location https://drive.google.com/drive/folders/1T0m-nv3eEnsML5Kq4vZkdN90ApWIM2Iw?usp=sharing or the short url form http://bit.ly/CSWPA-WD, please send a note to *cswpasmebook@gmail.com*.

Copyright © 2020 by Matt G Boston.

ISBN: 978-0-620-86825-9

Paperback version.

Acknowledgements

Special thanks to my work colleagues and friends for their encouragement and support in writing this Book. I would also like to thank my wife Judith and kids Charlotte and Malcolm for their support and giving me the space I needed to write this book.

TABLE OF CONTENTS

TABLE OF FIGURES

CHAPTER 1

QUESTION 1 EXAM SCREEN CAPTURES

QUESTION 1 EXAM SCREEN CAPTURE 1

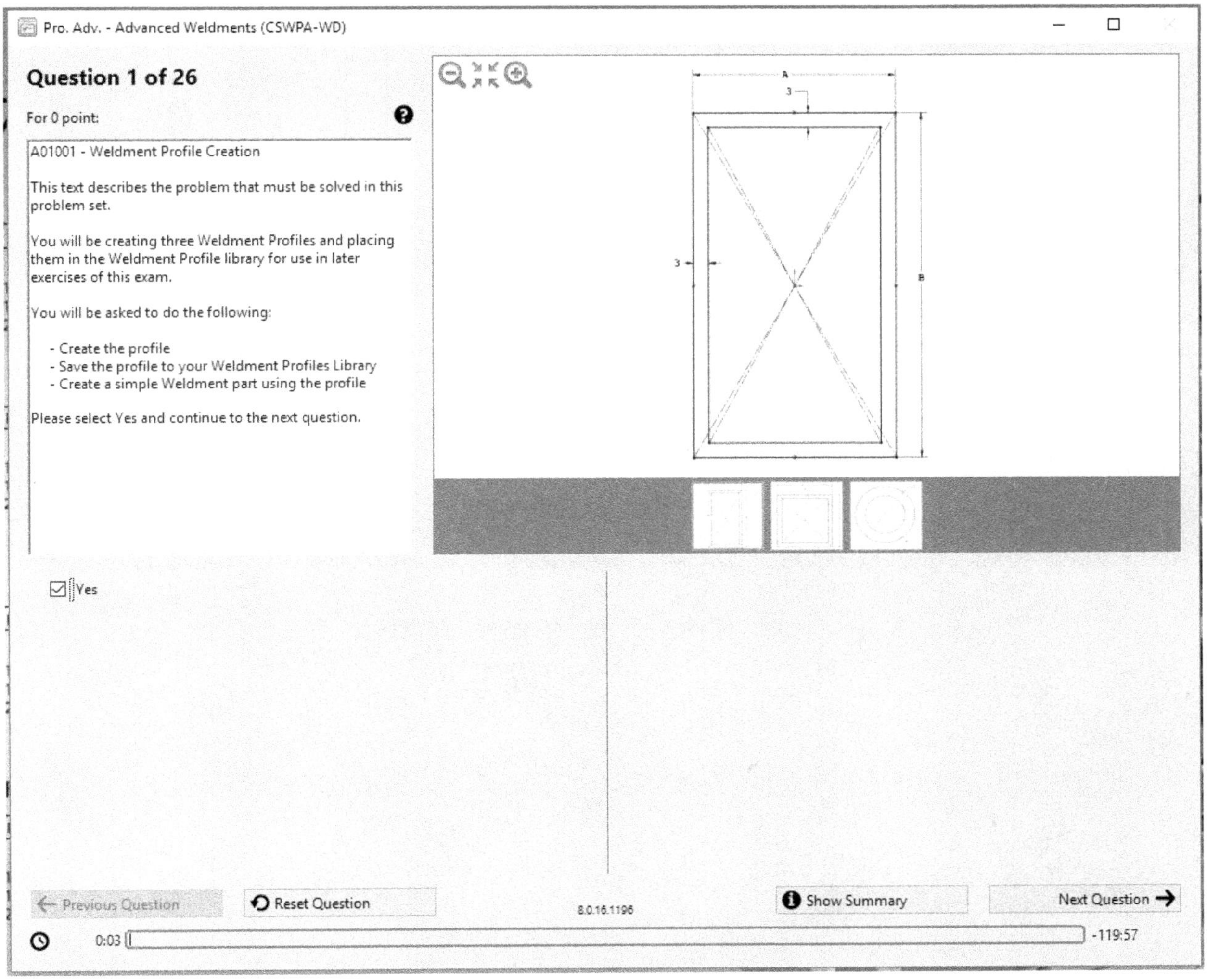

Figure 1 - Question 1 OF 26 - Exam Screen Capture 1

QUESTION 1 EXAM SCREEN CAPTURE 2

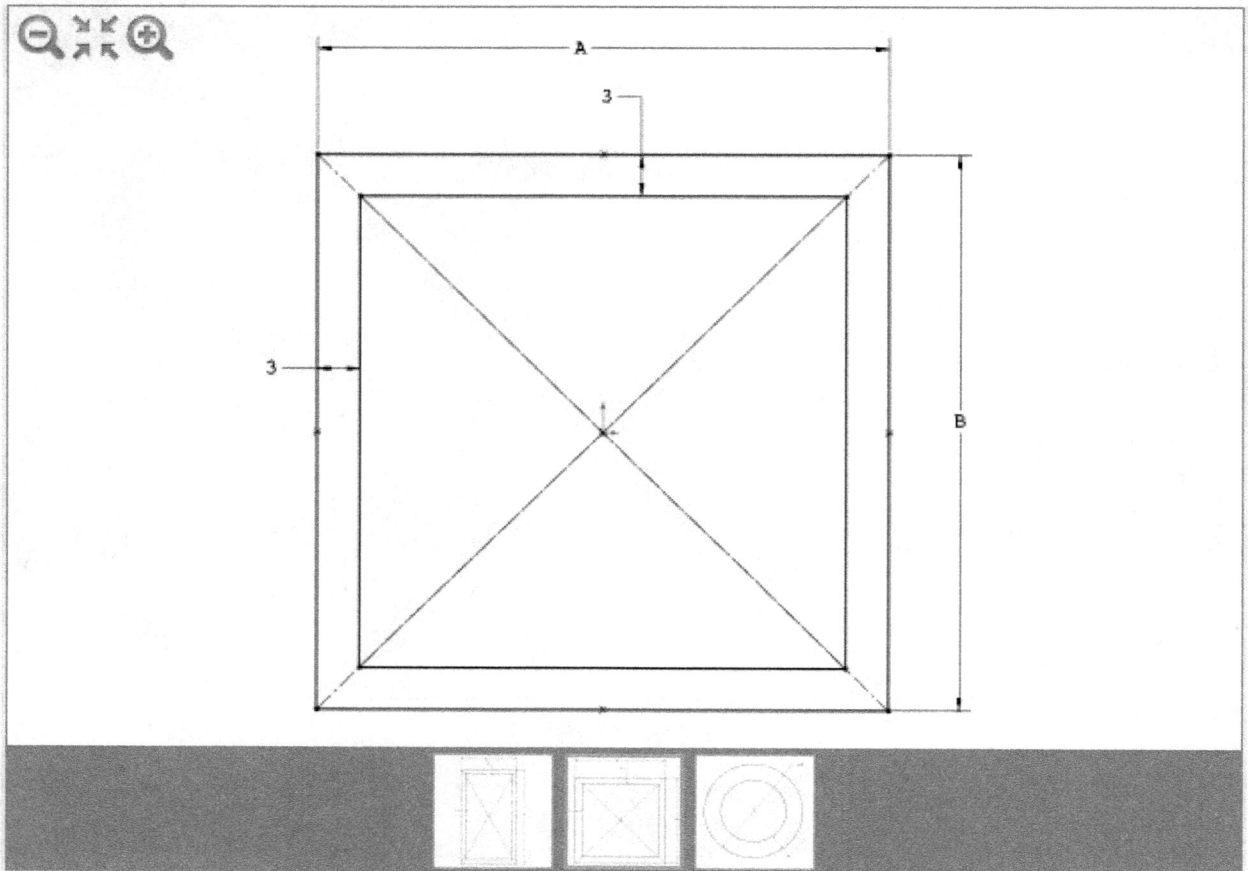

Figure 2 - Question 1 OF 26 - Exam Screen Capture 2

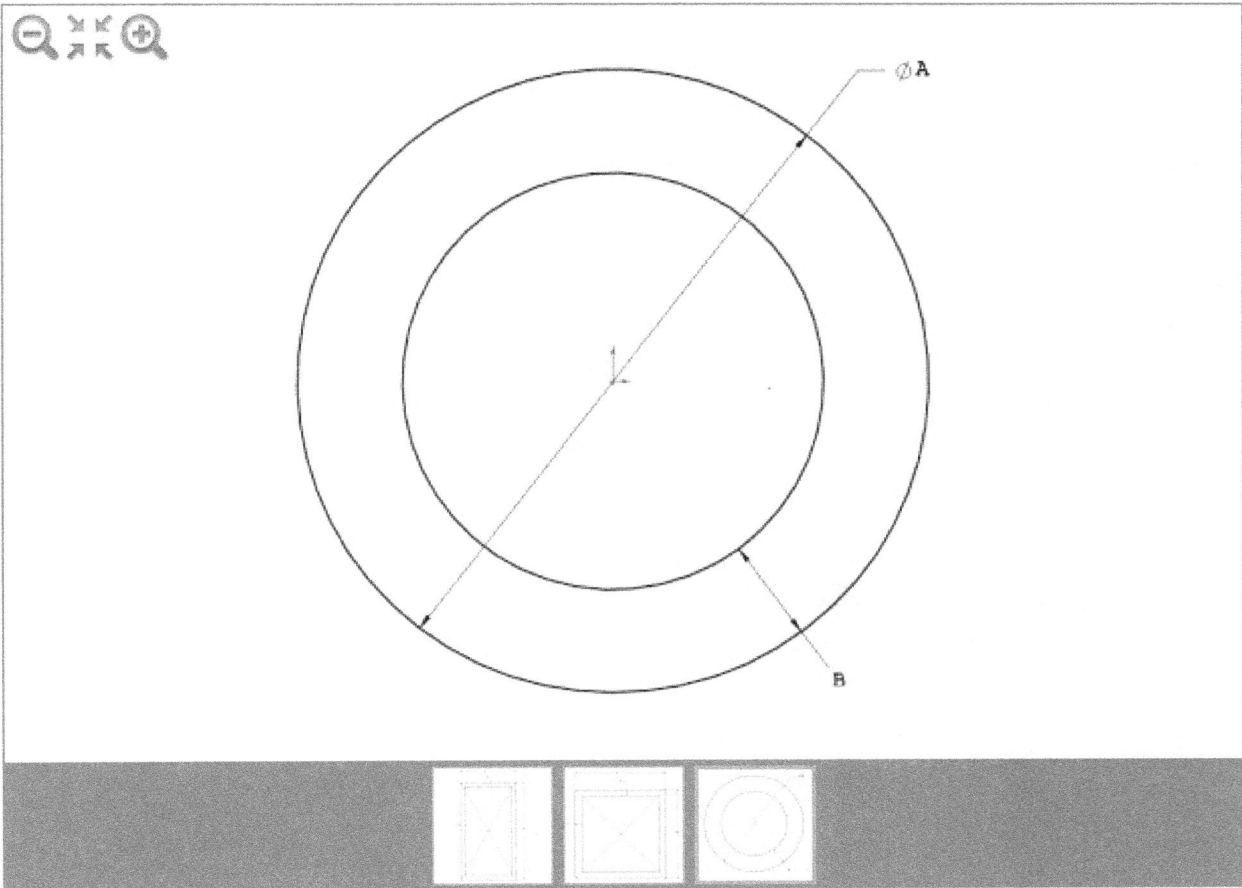

Figure 3 - Question 1 of 26 - Exam Screen Capture 3

Question 1 shows three Weldment Profiles that you will be creating, saving and using to create parts or Weldments as part of this exam in the following questions.

Select Yes and continue to Question 2.

CHAPTER 2

QUESTION 2 EXAM SCREEN CAPTURES

QUESTION 2 EXAM SCREEN CAPTURE 1

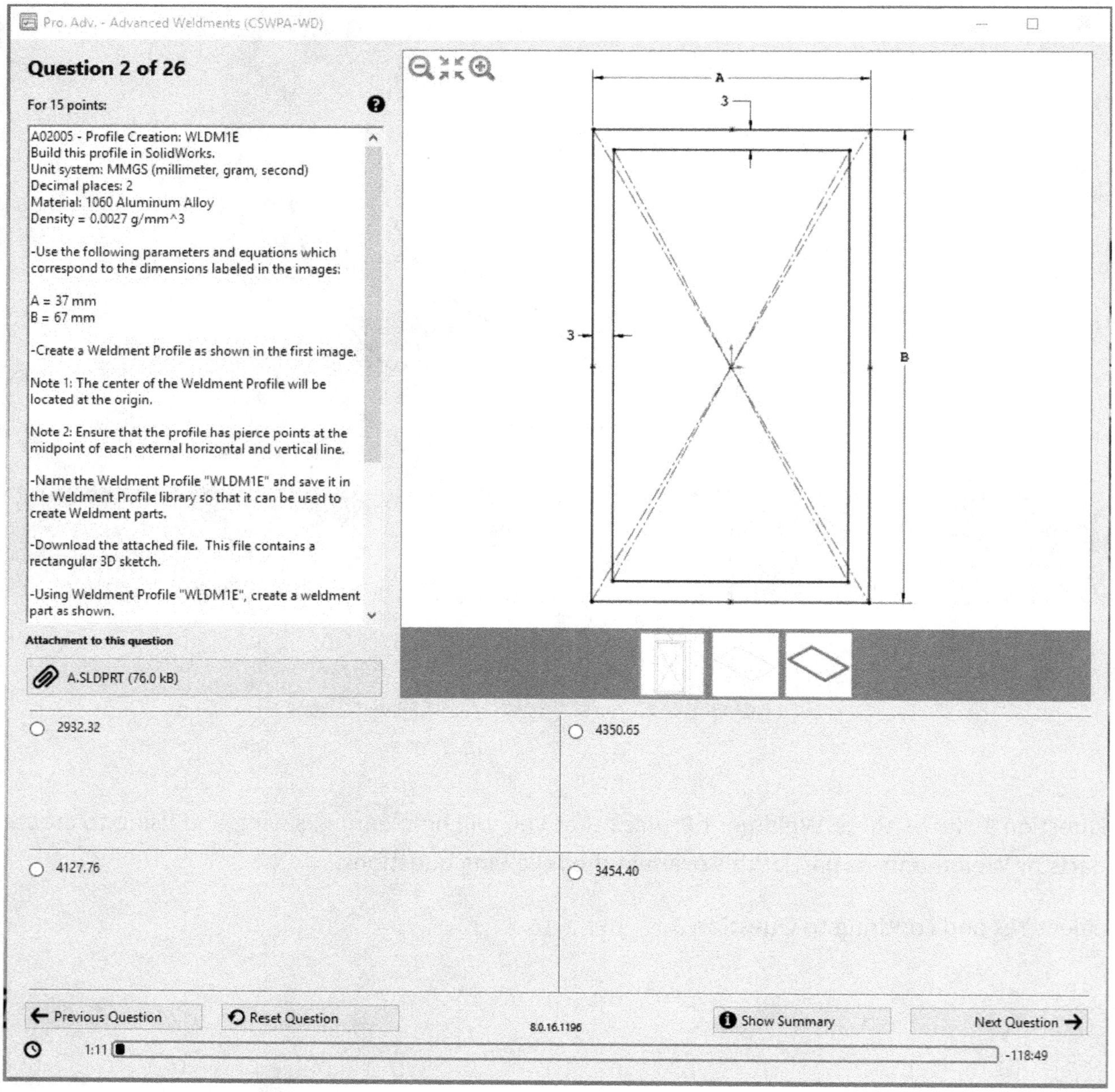

Pro. Adv. - Advanced Weldments (CSWPA-WD)

Question 2 of 26

For 15 points:

A02005 - Profile Creation: WLDM1E
Build this profile in SolidWorks.
Unit system: MMGS (millimeter, gram, second)
Decimal places: 2
Material: 1060 Aluminum Alloy
Density = 0.0027 g/mm^3

-Use the following parameters and equations which correspond to the dimensions labeled in the images:

A = 37 mm
B = 67 mm

-Create a Weldment Profile as shown in the first image.

Note 1: The center of the Weldment Profile will be located at the origin.

Note 2: Ensure that the profile has pierce points at the midpoint of each external horizontal and vertical line.

-Name the Weldment Profile "WLDM1E" and save it in the Weldment Profile library so that it can be used to create Weldment parts.

-Download the attached file. This file contains a rectangular 3D sketch.

-Using Weldment Profile "WLDM1E", create a weldment part as shown.

Attachment to this question

📎 A.SLDPRT (76.0 kB)

○ 2932.32

○ 4350.65

○ 4127.76

○ 3454.40

← Previous Question   ↻ Reset Question   8.0.16.1196   ⓘ Show Summary   Next Question →

🕐 1:11 ▮   -118:49

Figure 4 - Question 2 of 26 - Exam Screen Capture 1

Note 1: The center of the Weldment Profile will be located at the origin.

Note 2: Ensure that the profile has pierce points at the midpoint of each external horizontal and vertical line.

-Name the Weldment Profile "WLDM1E" and save it in the Weldment Profile library so that it can be used to create Weldment parts.

-Download the attached file. This file contains a rectangular 3D sketch.

-Using Weldment Profile "WLDM1E", create a weldment part as shown.

Note 1: Align the center of the Weldment profile to the 3D sketch.

Note 2: Use the "End Miter" option to join all segments to each other.

-Measure the total mass of all four segments created.

Note: Make sure to apply the proper material to the part.

What is the total mass of all four weldment segments(grams)?

Attachment to this question

Figure 5 - Question 2 of 26 - Exam Screen Capture 2

5

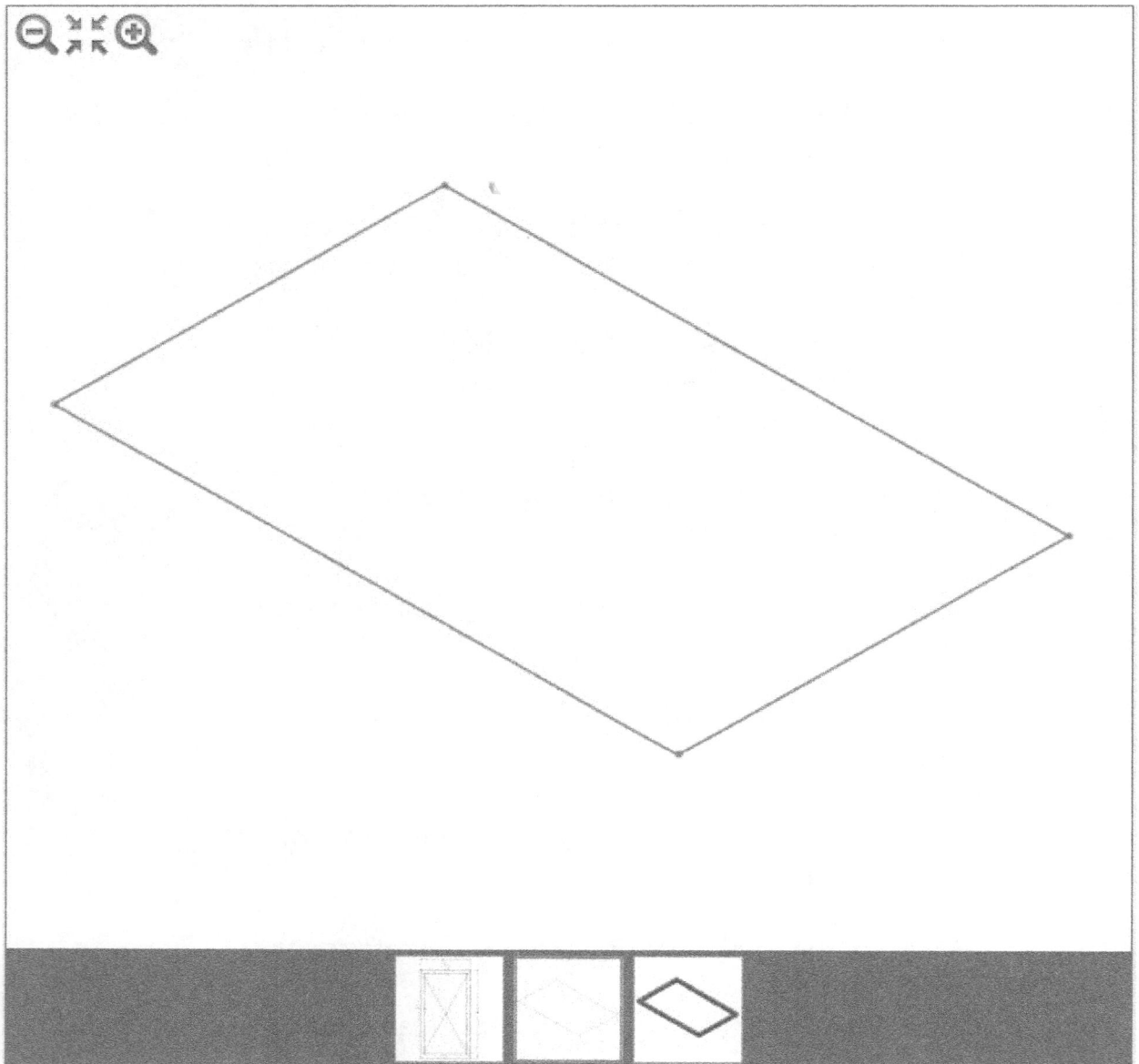

Figure 6 - Question 2 of 26 - Exam Screen Capture 3

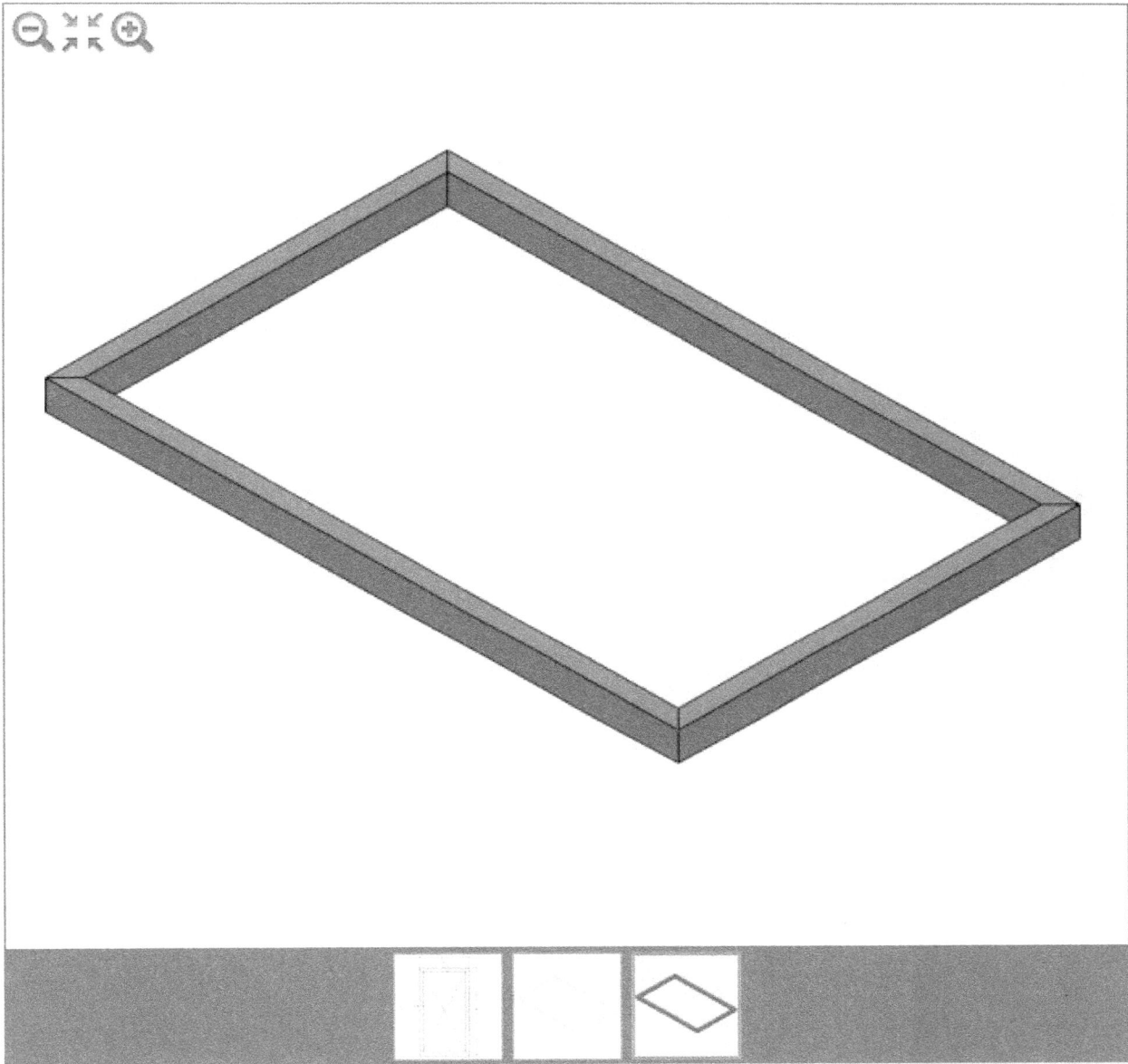

Figure 7 - Question 2 of 26 - Exam Screen Capture 4

Start a new part in Solidworks.

SET A UNIT SYSTEM

Go to Tools > Options > Document Properties > Drafting Standard. From the drop down menu, Select the ISO Standard since our dimensions are given in millimeters.

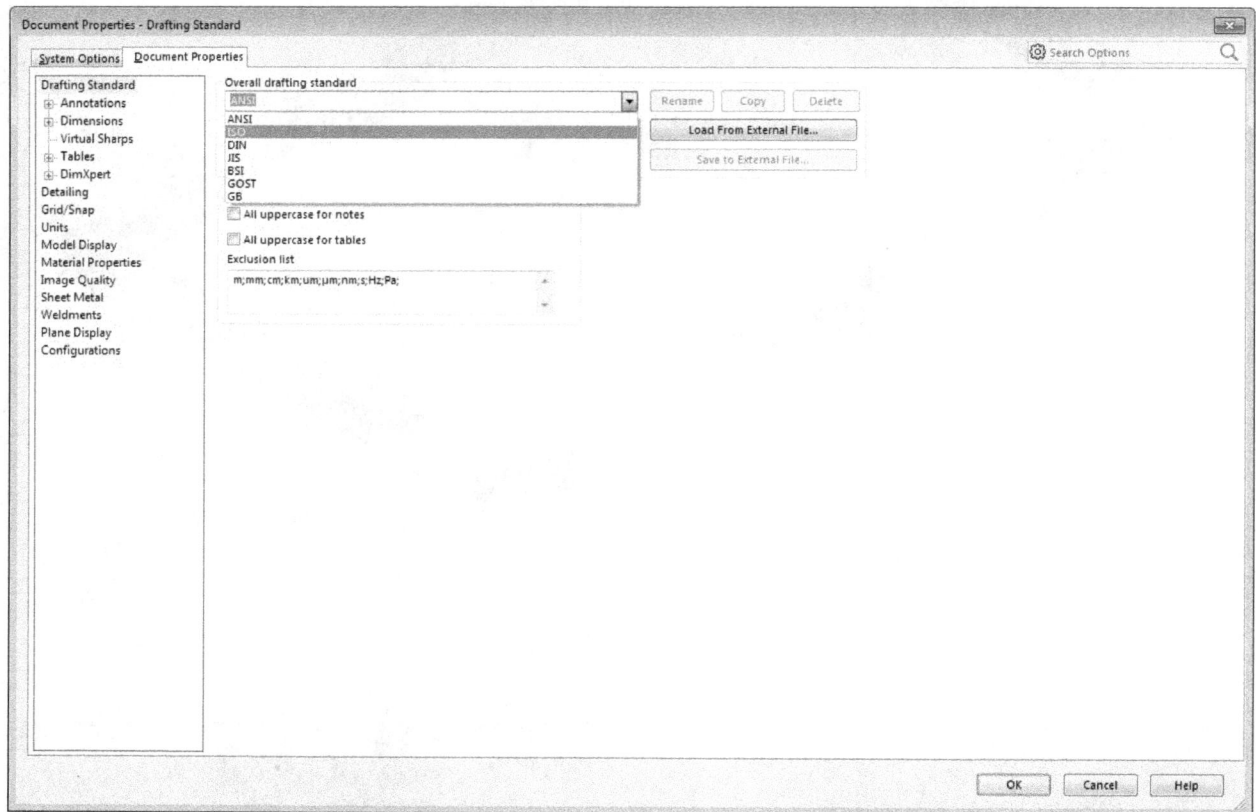

Figure 8 - Question 2 of 26 - Document Properties - Drafting Standard

Go to Tools > Options > Document Properties > Units to change the Unit System to MMGS (millimeter, gram, second) and to also set the number of decimal places to two decimal places.

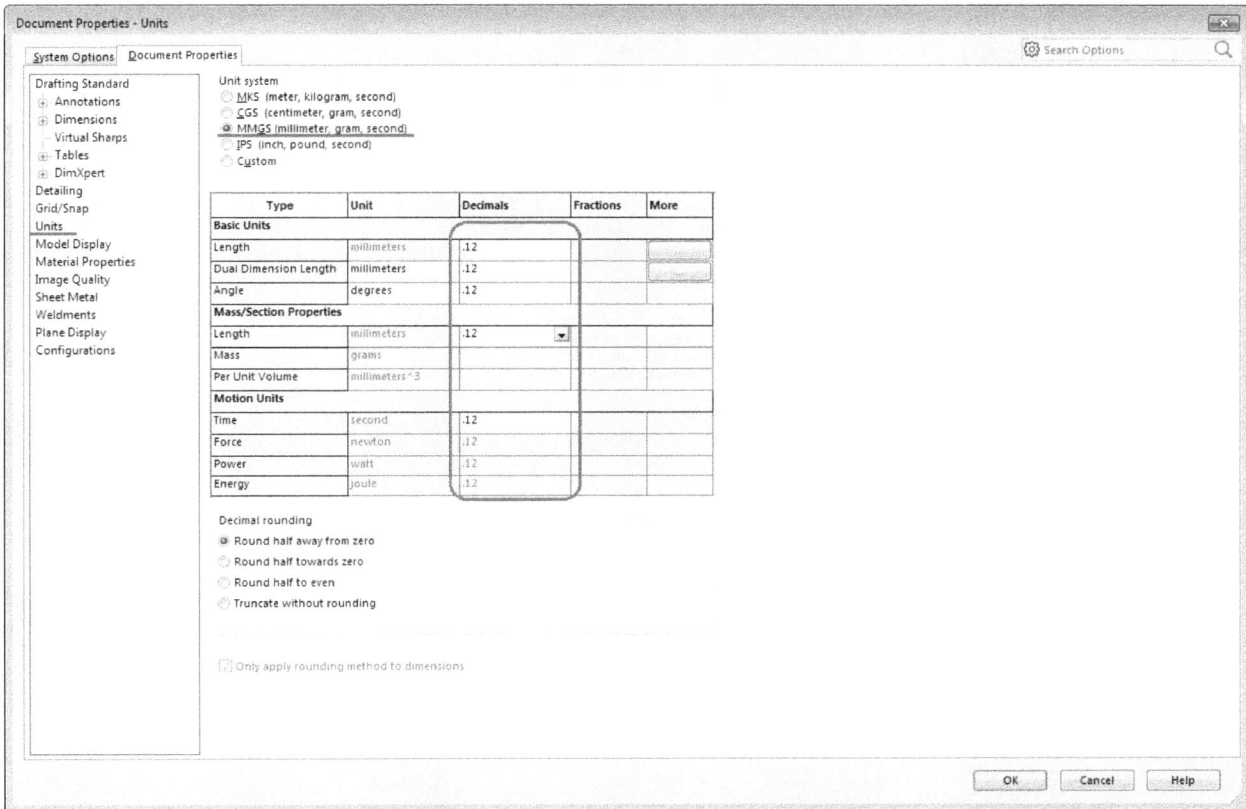

Click OK.

APPLYING A MATERIAL

If the material is not already applied as 1060 Alloy - Right-click Material in the FeatureManager design tree. Click Edit Material, select 1060 Alloy in the material tree under Aluminium Alloys, and click Apply, then Close.

CREATING A WELDMENT PROFILE

Select any plane in the FeatureManager Design Tree and click a sketch entity tool or click Sketch on the Sketch toolbar.

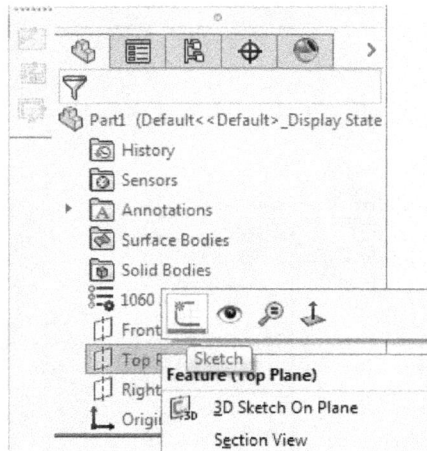

Figure 10 - Question 2 of 26 - Starting A Sketch on a Plane

Under the Sketch Tab, Click Center Rectangle.

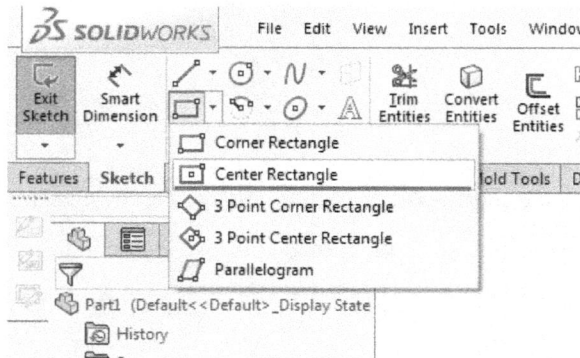

Figure 11 - Question 2 of 26 - Selecting a Center Rectangle

In the Rectangle PropertyManager under Rectangle Type, make sure you have the Center Rectangle selected and under the options select From Midpoints. The From Midpoints option automatically creates the points on the center of the external vertical and horizontal lines which we can use as pierce points later on as required under Note 2 *(Question 2 Screen Capture)*. A pierce point defines the location of the profile, relative to the sketch segment used to create the structural member. The default pierce point is the sketch origin in the profile library feature part. Any vertex or sketch point specified in the profile can also be used as a pierce point.

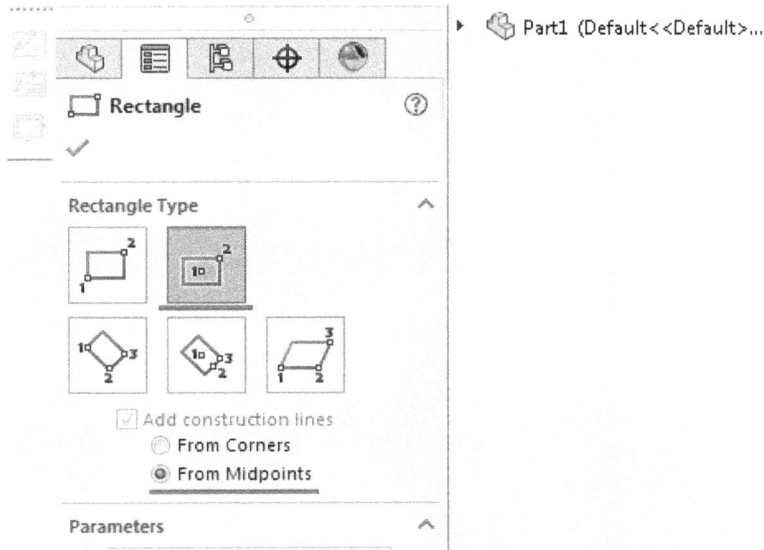

Figure 12 - Question 2 of 26 -Center Rectangle Options

In the graphics area - > Click the Origin to define the center of the rectangle. Drag to sketch the rectangle with centerlines. Release to set the four edges. Your sketch should look as shown in the following image.

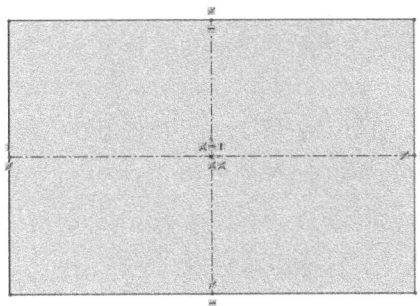

Figure 13 - Question 2 of 26 - Sketching A Center Rectangle

Dimension the sketch as shown in the following image - NB: All dimensions are in millimeters and to two decimal places as set in the beginning of this question.

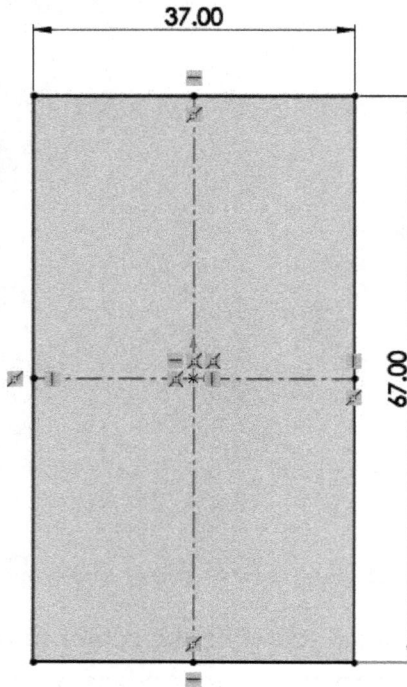

Figure 14 - Question 2 of 26 - Fully Defined Center Rectangle Sketch

While the sketch is still open, select one entity then Click Offset Entities (Sketch toolbar) or Tools > Sketch Tools > Offset Entities.

Set the properties in the Offset Entities PropertyManager as shown in the following image. Select the Reverse Checkbox if necessary to change the offset direction to the inside of the Fully Defined Center Rectangle. When you click in the graphics area, the offset entity is complete. Set the properties before you click in the graphics area.

Click Ok or click in the graphics area.

An Offset Entities relation is created between the new rectangle and our first Fully Defined Center Rectangle. If the original rectangle changes, the new rectangle updates to maintain the offset.

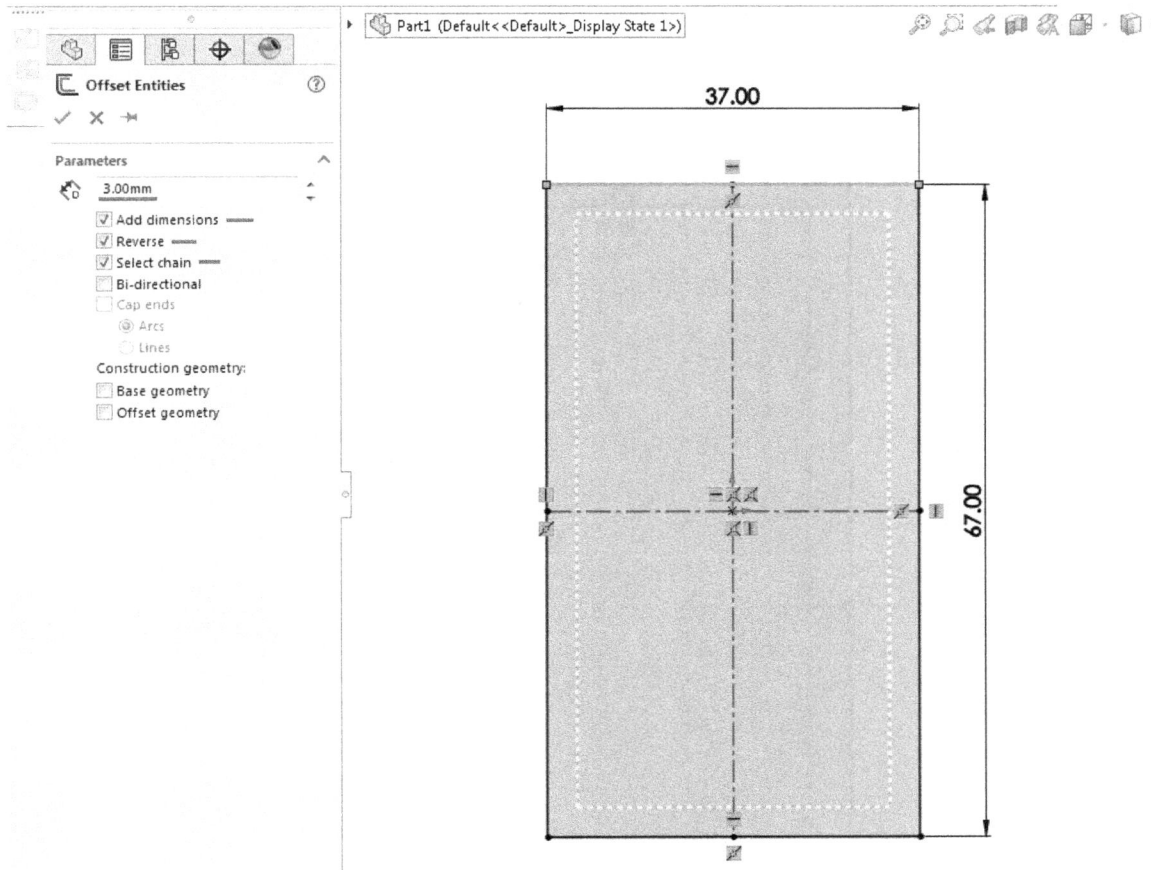

Figure 15 - Question 2 of 26 - Offsetting Sketch Entities

Your sketch should now look as shown in the following image.

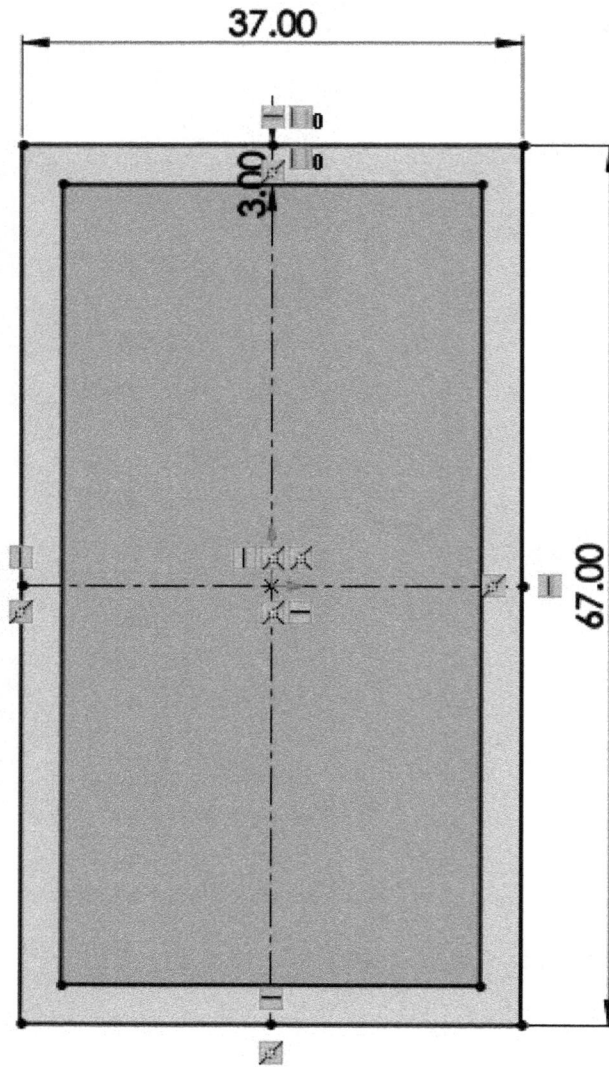

Figure 16 - Question 2 of 26 - Fully Defined Sketch for Weldment Profile Creation

Click ⤶ in the Confirmation Corner or Insert > Exit Sketch to exit the sketch.

CREATING A WELDMENT PROFILE

In the FeatureManager design tree, select Sketch1 *(default name of the sketch we created above)*. Click File > Save As. In the dialog box:

In Save in, browse to the default location for weldment profiles which is <u>install-dir\lang\language\weldment profiles</u> or in my case the location is *C:\Program Files\SOLIDWORKSCorp\SOLIDWORKS(2)\lang\english\weldment profiles* and create **<standard>** and **<type>** subfolders.

We will create a *standard folder* named *EXAM WELDMENT PROFILES* and three *type folders* inside this new standard folder named *RECTANGLE*, *SQUARE* and *PIPE*. See the following images.

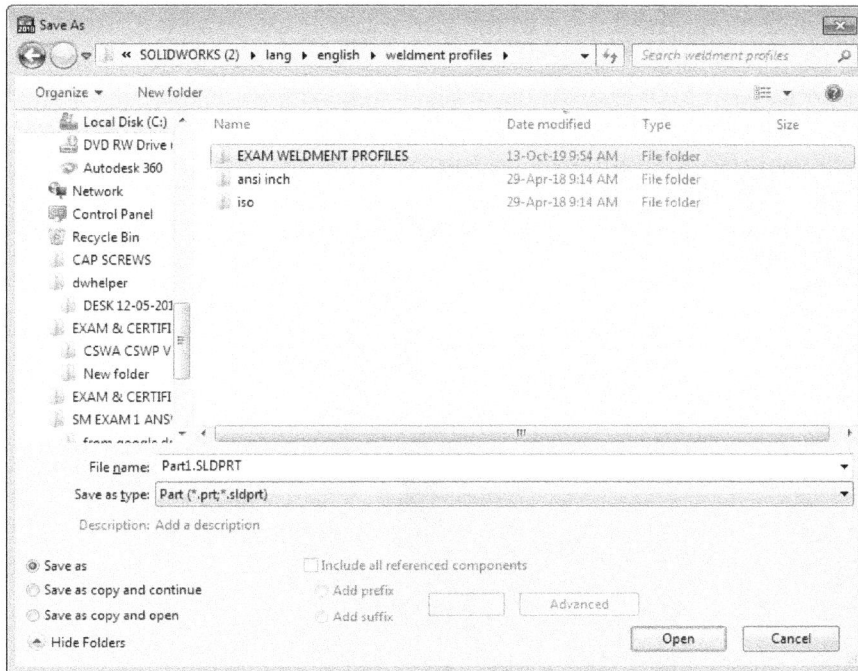

Figure 17 - Question 2 of 26 - Custom Weldment Profile created Standard Folder

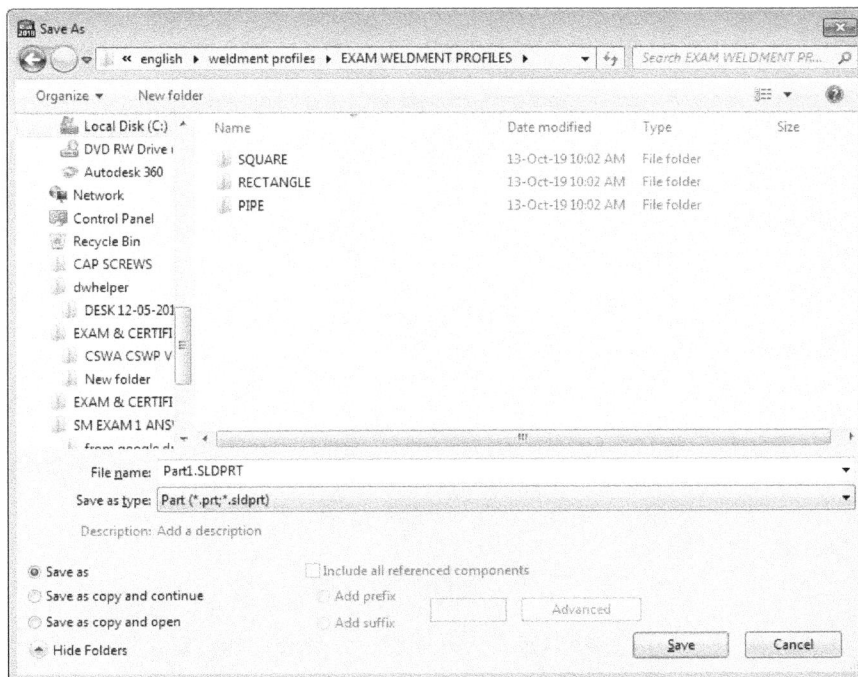

Figure 18 - Question 2 of 26 - Custom Weldment Profile created Type Folders

*NOTE: **STANDARD** FOLDERS* - for example ANSI INCH, ISO e.t.c contain one or more **TYPE** FOLDERS, for example angle iron, c channel, pipe, and so on. In the PropertyManager, after a Standard is selected, the names of each of its *type* sub-folders appear in Type.

We will save this weldment profile into the RECTANGLE type folder.

Under Save as type, select Lib Feat Part (*.sldlfp).

Type a name under Filename that is WLDM1E as requested in question 2 screen capture - see the following image.

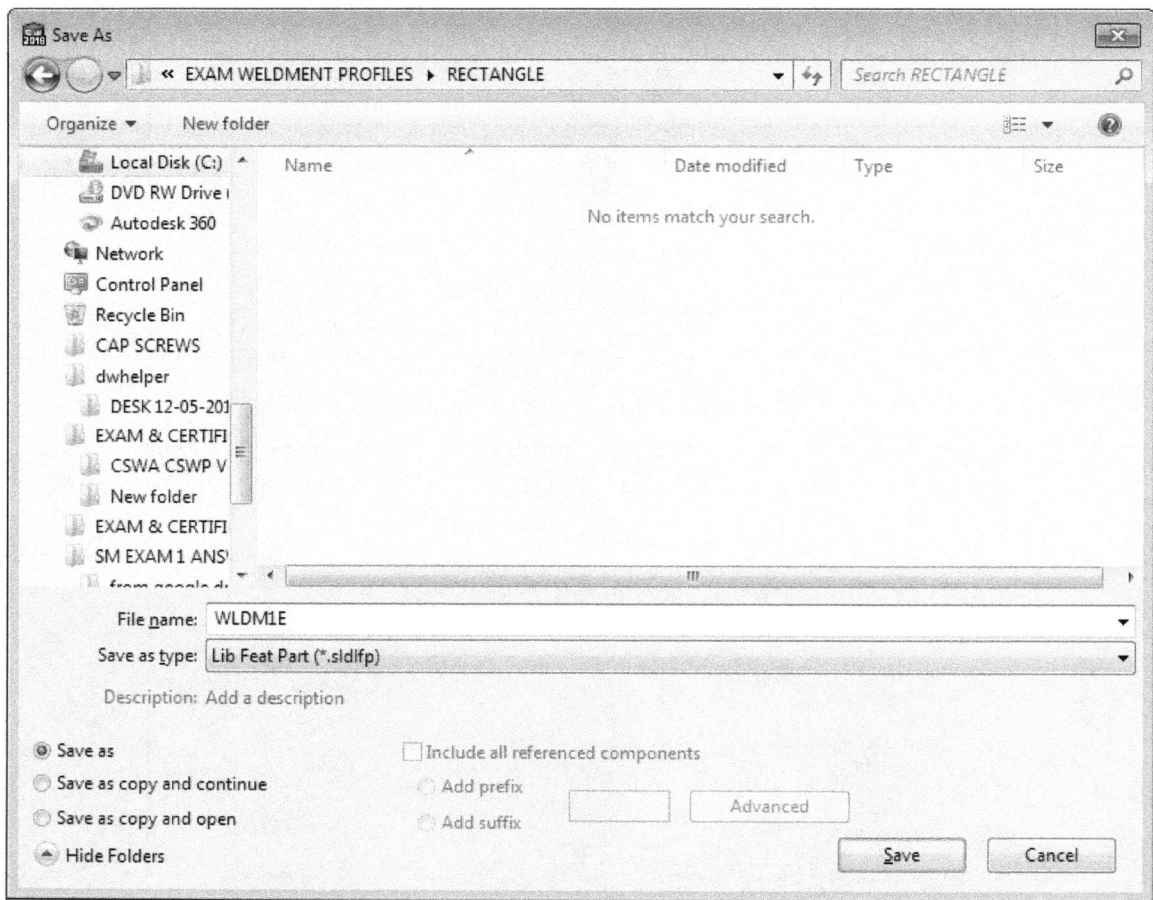

Figure 19 - Question 2 of 26 - Custom Weldment Profile Save In Location

Click Save.

In the FeatureManager design tree you will notice that the Sketch1 icon changes from ⌐ (-) Sketch1 to ⌐ Sketch1 . Close the part, you don't have to save it. Download the part with the file name A.sldprt from this Google Drive location (*http://bit.ly/CSWPA-WD)*. If you experience any problems with downloading any files you may send an email to *cswpasmebook@gmail.com*.

Open the downloaded part - A.sldprt and Save it on your computer as Question 02.

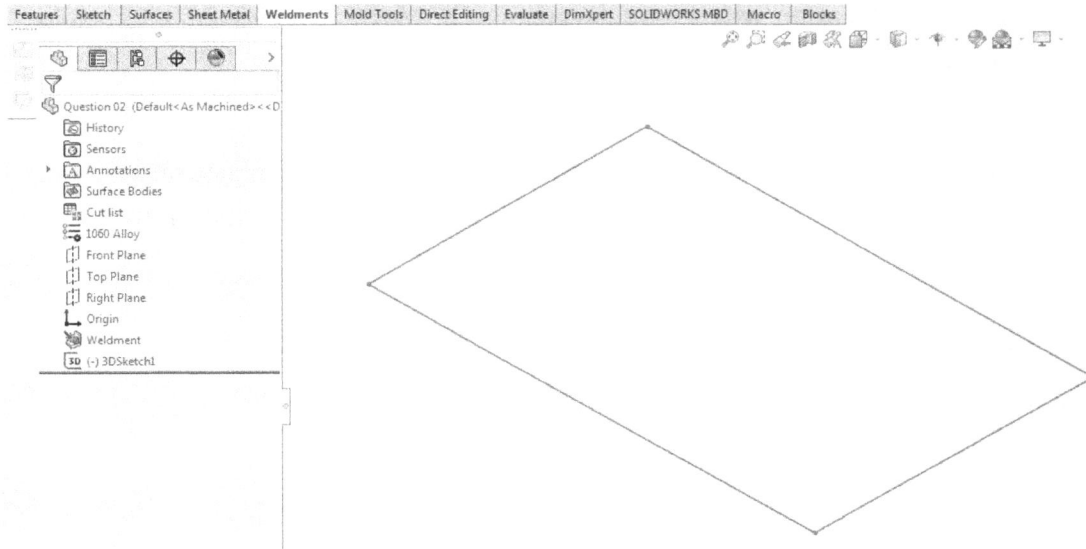

Figure 20 - Question 2 of 26 - Downloaded Part

ADDING STRUCTURAL MEMBERS

Click Structural Member (Weldments toolbar) or Insert > Weldments > Structural Member. Make selections in the PropertyManager to define the profile for the structural member as shown in the following image.

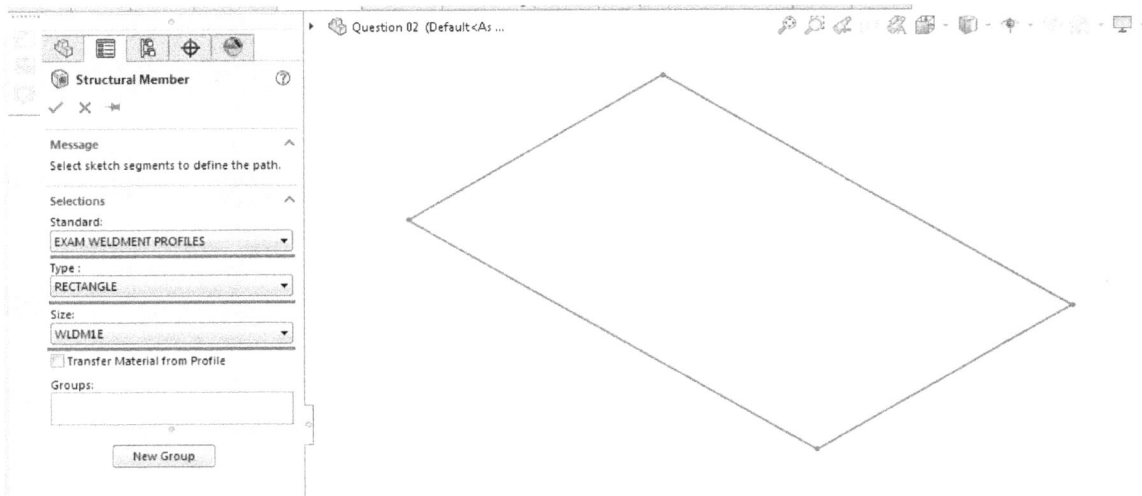

Figure 21 - Question 2 of 26 - Downloaded Part

NOTE: If the EXAM WELDMENT PROFILES folder you created does not appear on the Standard drop down menu of your Structural Member PropertyManager then you need to add the location where you saved it to File Locations . Go to Tools > Options > File Locations and under the Show folders for drop down menu Select Weldment Profiles then Click the Add command button . Browse to the location where you saved the EXAM WELDMENT PROFILES folder and click select folder then click OK. Do not add then EXAM WELDMENT PROFILES folder itself as the location but the folder where it is saved.

In the graphics area, select sketch segments to define the path for the structural member. If the profile you specify has a material assigned, the option Transfer Material from Profile is available.

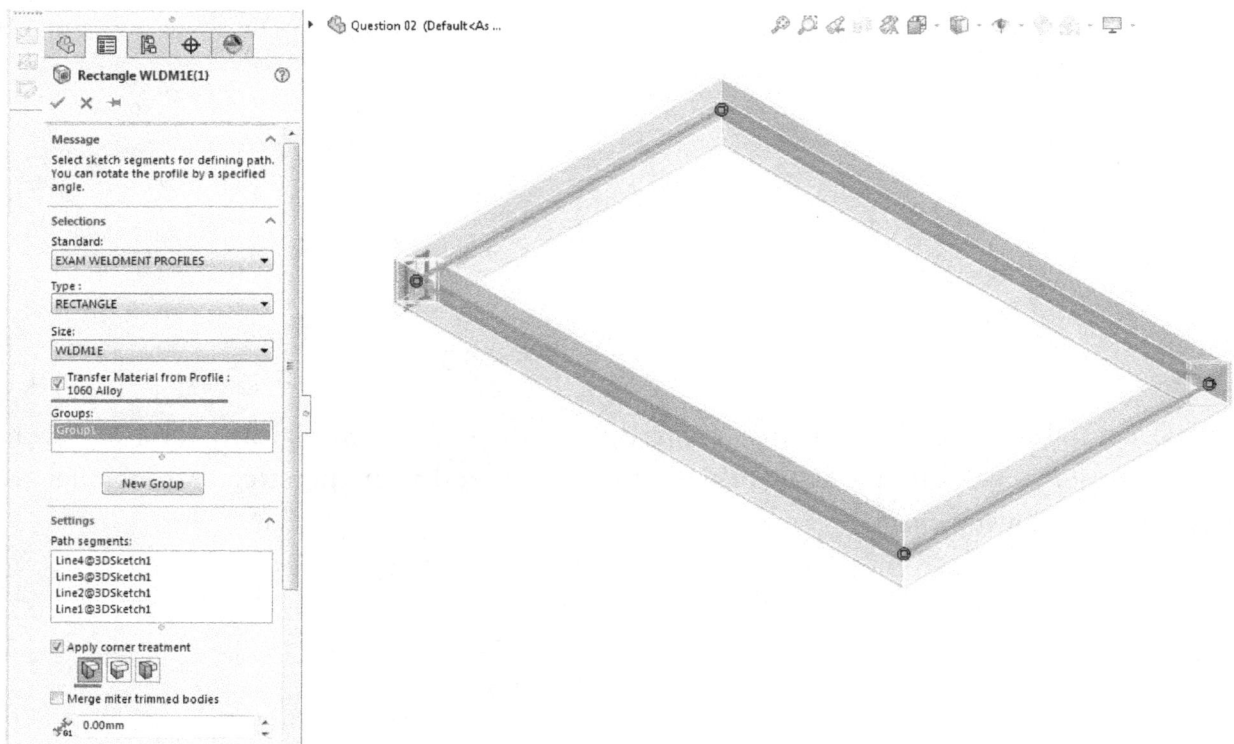

Figure 22 - Question 2 of 26 - Adding Structural Members

Click Ok. Click Save.

MEASURING THE MASS OF THE PART

Under the Evaluate Tab, click the Mass Properties button and the mass properties menu appears as shown below. Type in the mass of 4127.76 grams in the Question 2 textbox and click Next Question to go to Question 3.

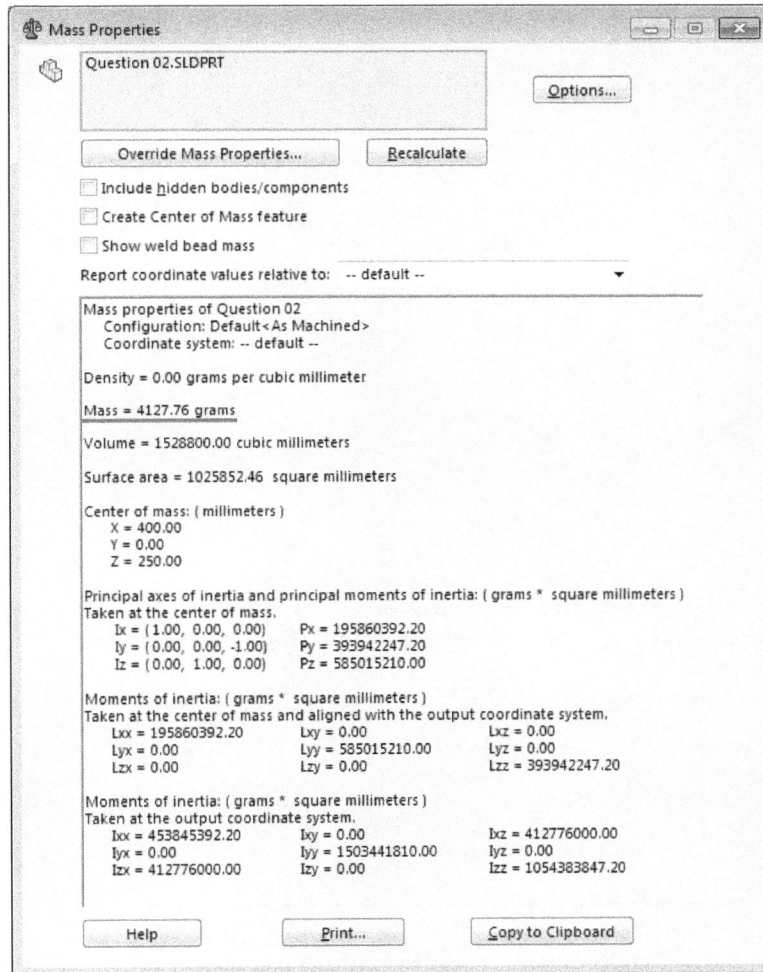

Figure 23 - Question 2 of 26 - Mass Properties

Save and close Question 02 Part.

PREPARING FOR QUESTION 2 AND QUESTION 3

Since we are going to create two more weldment profiles in the standard and type folders created in this question we may create a Desktop Shortcut for quick access to this location to save us time in the following questions since the file path is quite deep.

Open *C:\Program Files\SOLIDWORKSCorp\SOLIDWORKS(2)\lang\english\weldment profiles* or *the location where you saved the standard folder* EXAM WELDMENT PROFILES then right click on the folder and select **Send to** then click on *Desktop (create shortcut)*.

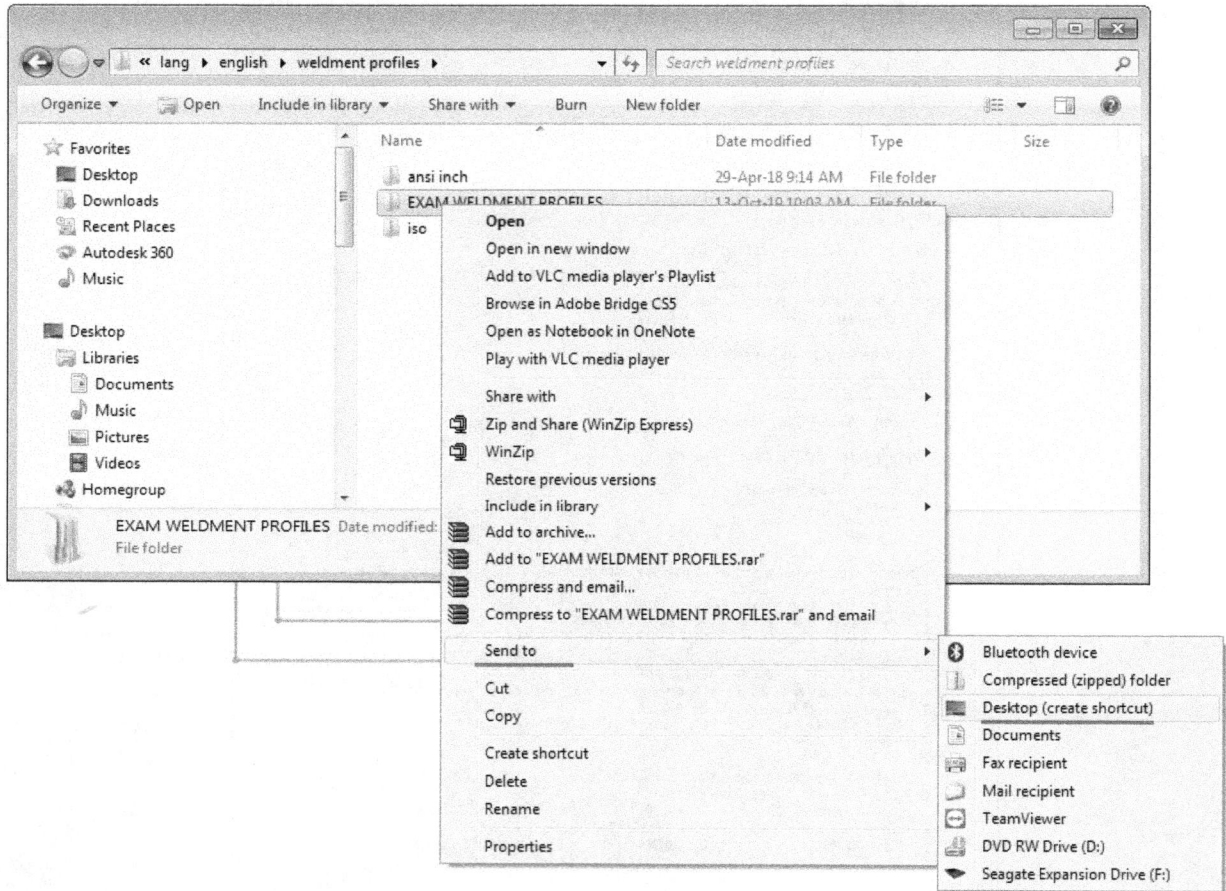

Figure 24 - Question 2 of 26 - Creating A Desktop Shortcut

CHAPTER 3

QUESTION 3 EXAM SCREEN CAPTURES

QUESTION 3 EXAM SCREEN CAPTURE 1

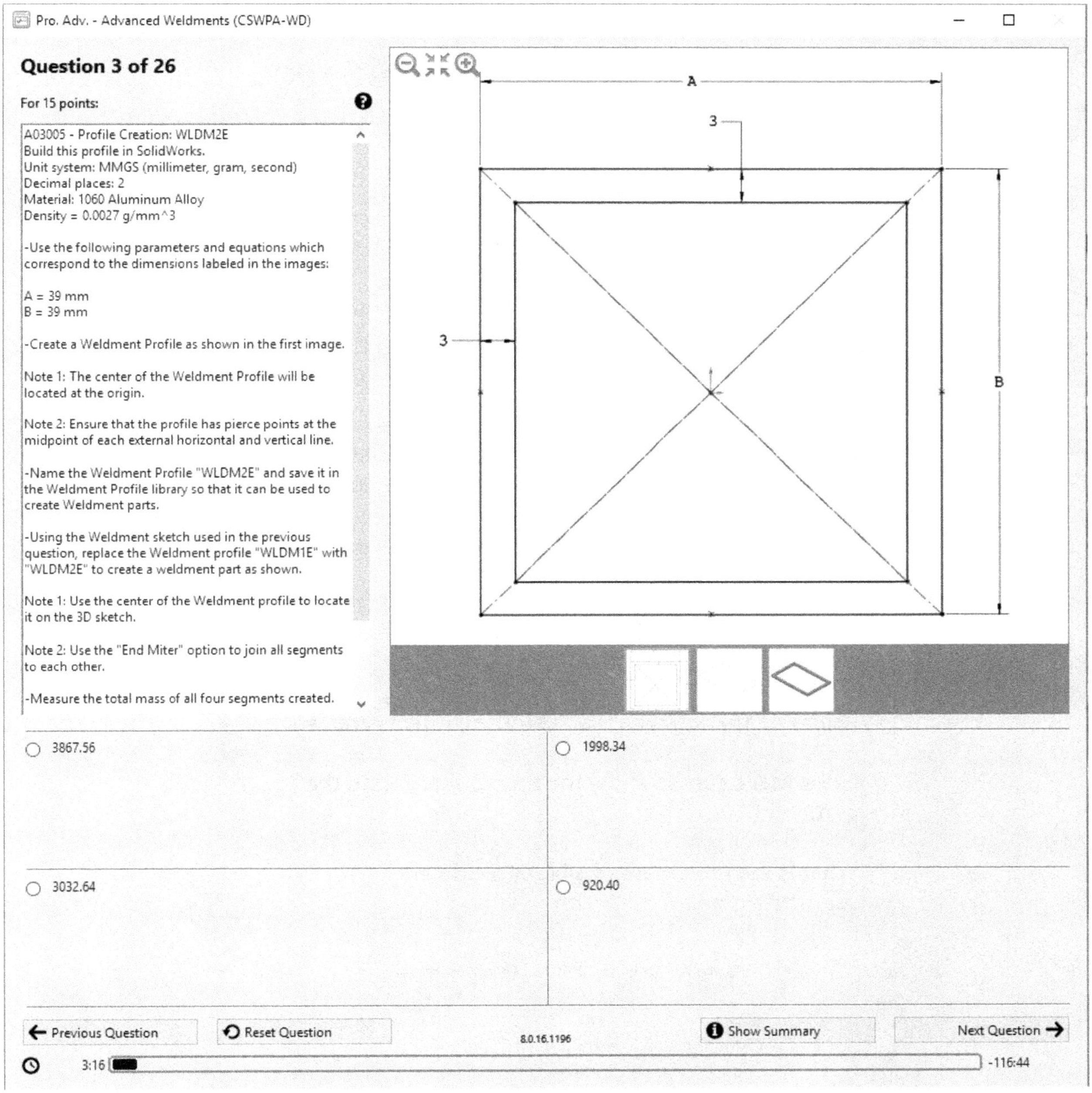

Figure 25 - Question 3 of 26 - Exam Screen Capture 1

QUESTION 3 EXAM SCREEN CAPTURE 2

Density = 0.0027 g/mm^3

-Use the following parameters and equations which correspond to the dimensions labeled in the images:

A = 39 mm
B = 39 mm

-Create a Weldment Profile as shown in the first image.

Note 1: The center of the Weldment Profile will be located at the origin.

Note 2: Ensure that the profile has pierce points at the midpoint of each external horizontal and vertical line.

-Name the Weldment Profile "WLDM2E" and save it in the Weldment Profile library so that it can be used to create Weldment parts.

-Using the Weldment sketch used in the previous question, replace the Weldment profile "WLDM1E" with "WLDM2E" to create a weldment part as shown.

Note 1: Use the center of the Weldment profile to locate it on the 3D sketch.

Note 2: Use the "End Miter" option to join all segments to each other.

-Measure the total mass of all four segments created.

Note: Make sure to apply the proper material to the part.

What is the total mass of all four weldment segments(grams)?

Figure 26 - Question 3 of 26 - Exam Screen Capture 2

QUESTION 3 EXAM SCREEN CAPTURE 3

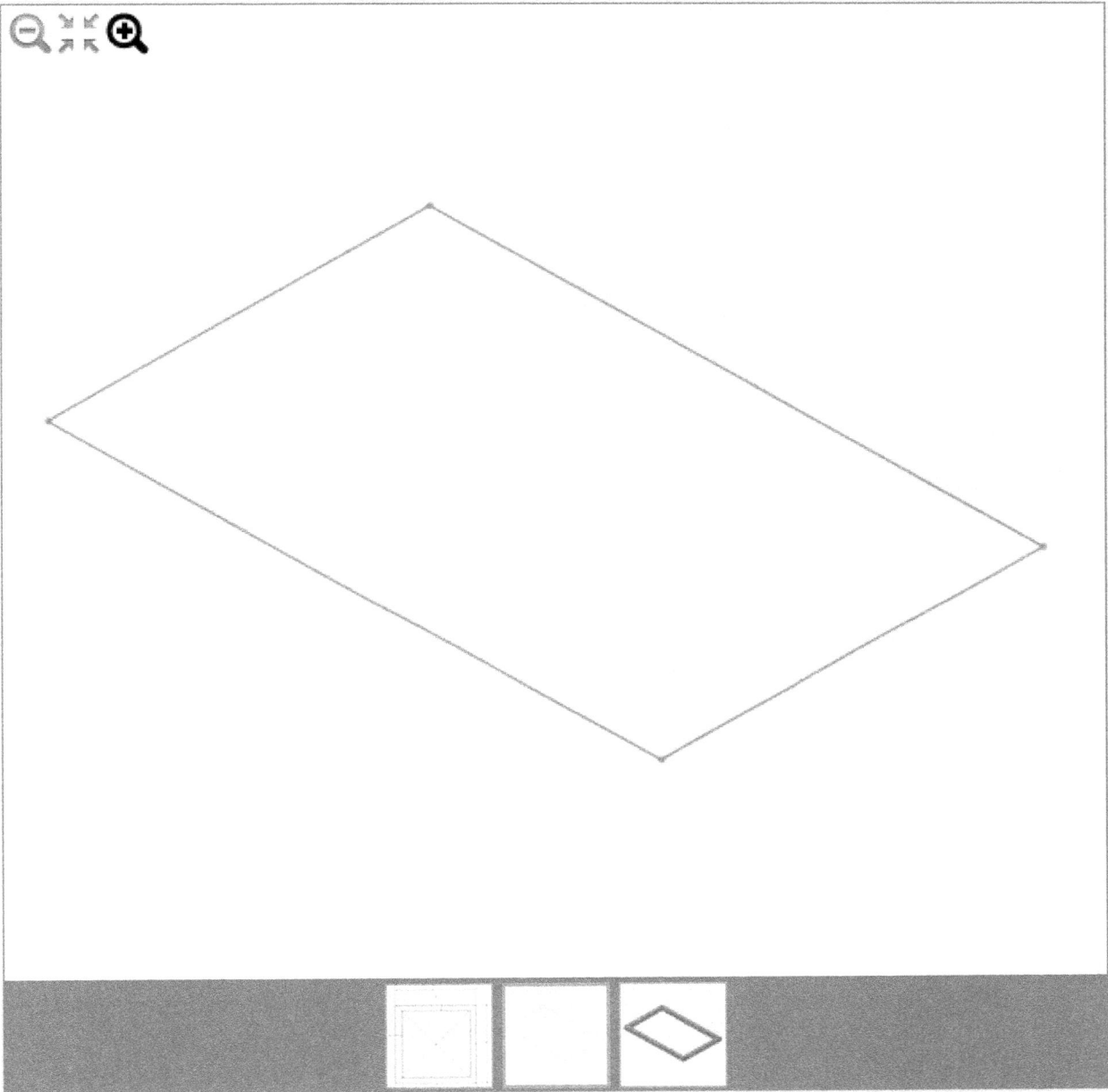

Figure 27 - Question 3 of 26 - Exam Screen Capture 3

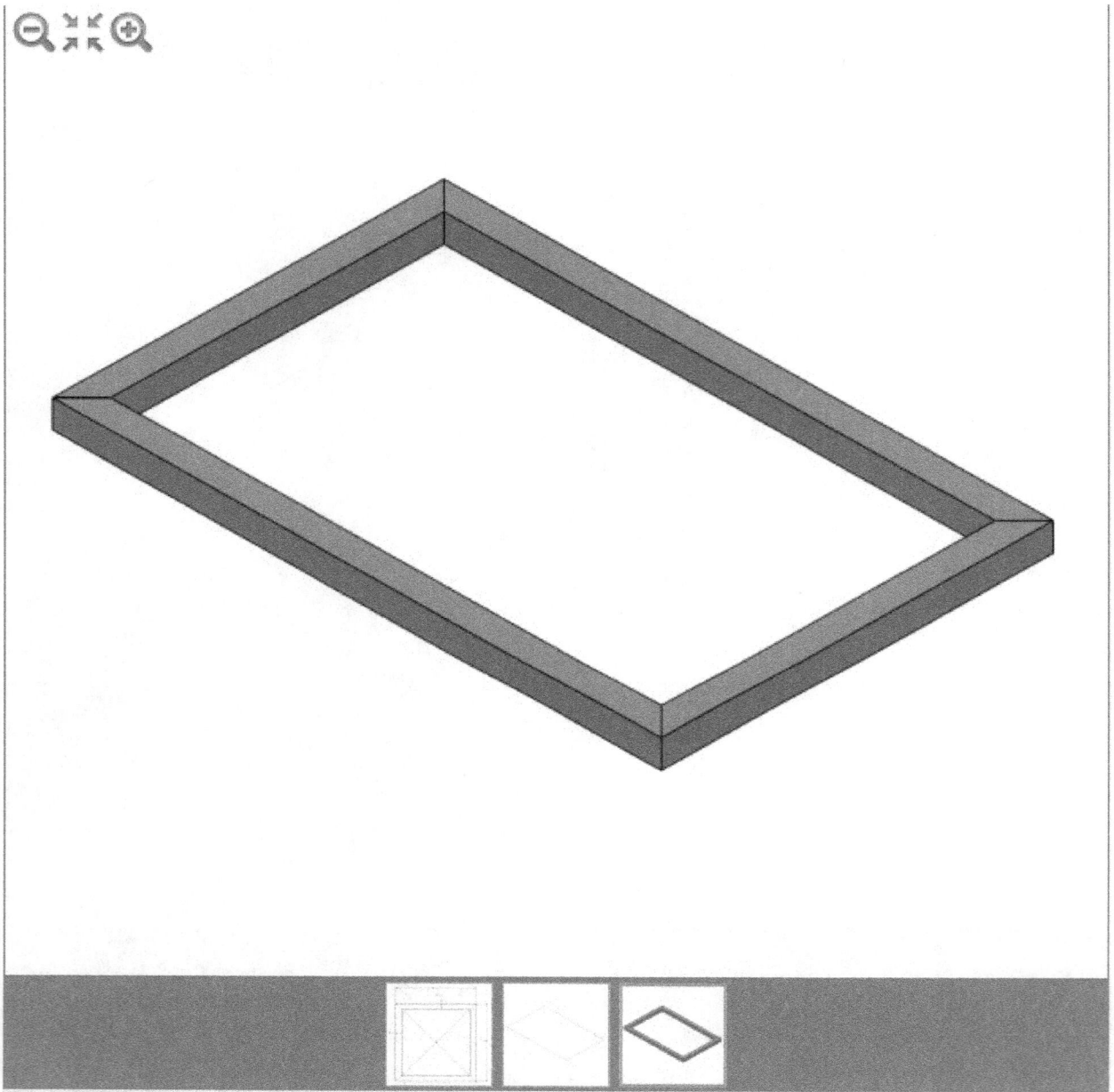

Figure 28 - Question 3 of 26 - Exam Screen Capture 4

Start a new part in Solidworks.

SET A UNIT SYSTEM

Go to Tools > Options > Document Properties > Drafting Standard. From the drop down menu, Select the ISO Standard since our dimensions are given in millimeters.

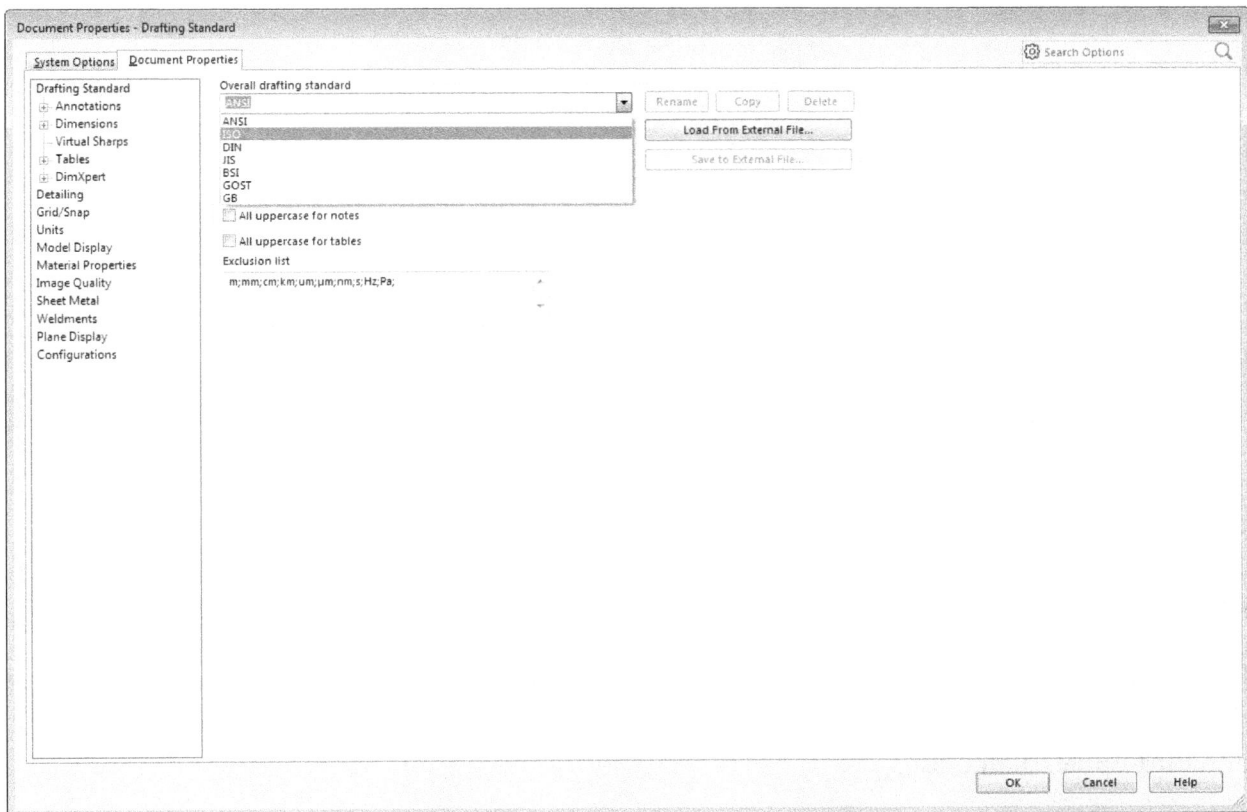

Figure 29 - Question 3 of 26 - Document Properties - Drafting Standard

Go to Tools > Options > Document Properties > Units to change the Unit System to MMGS (millimeter, gram, second) and to also set the number of decimal places.

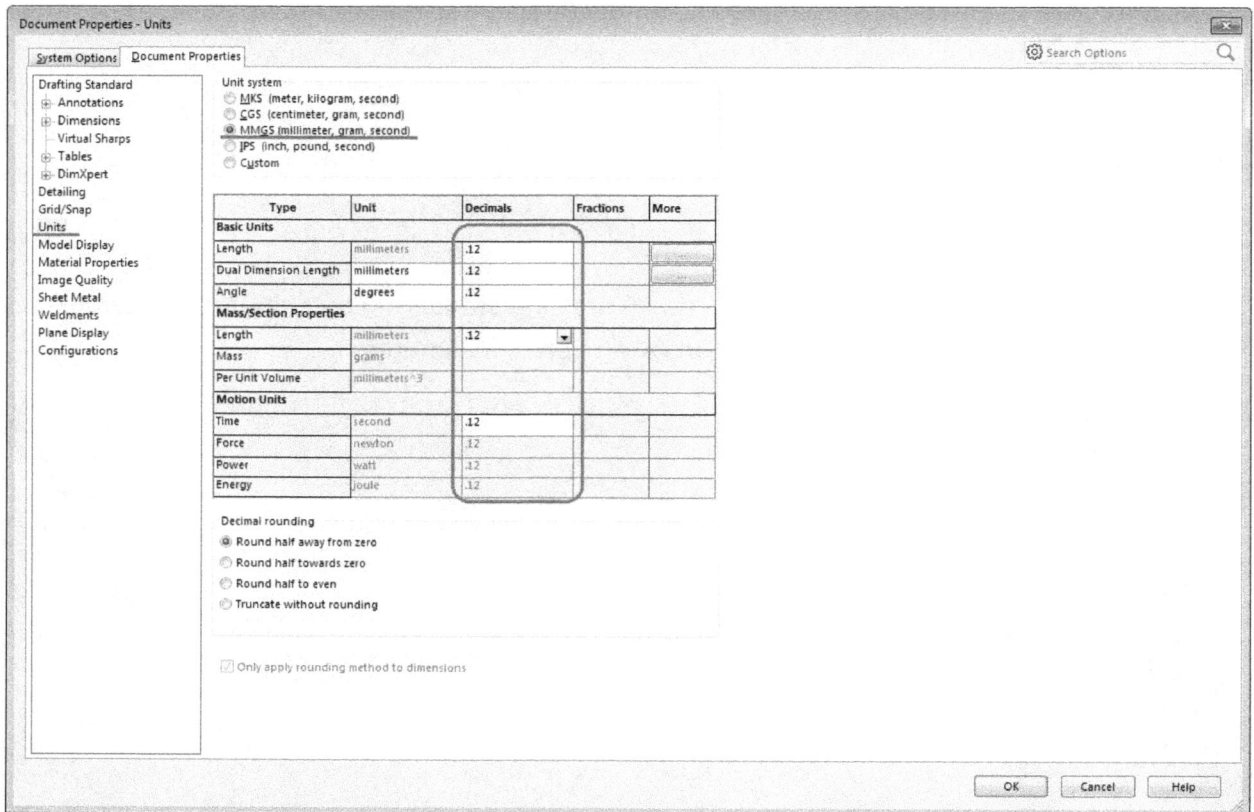

Figure 30 - Question 3 of 26 - Document Properties - Units

Click OK.

APPLYING A MATERIAL

If the material is not already applied as 1060 Alloy - Right-click Material in the FeatureManager design tree. Click Edit Material, select 1060 Alloy in the material tree under Aluminium Alloys, and click Apply, then Close.

CREATING A WELDMENT PROFILE

Select any plane in the FeatureManager design tree and click a sketch entity tool or click Sketch on the Sketch toolbar.

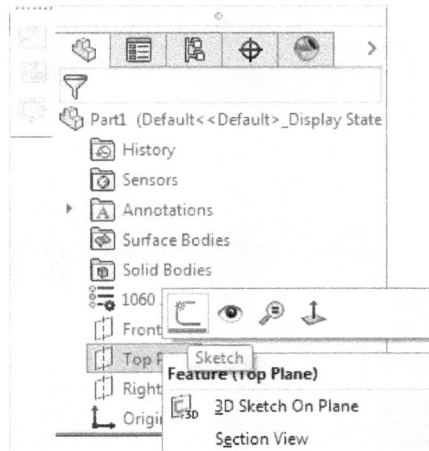

Figure 31 - Question 3 of 26 - Starting A Sketch on a Plane

Under the Sketch Tab, Click Center Rectangle.

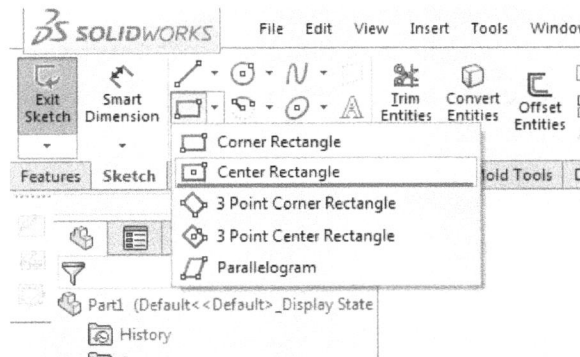

Figure 32 - Question 3 of 26 - Selecting a Center Rectangle

In the Rectangle PropertyManager under Rectangle Type, make sure you have the Center Rectangle selected and under the options select From Midpoints which Adds centerlines from midpoints of line segments.

The From Midpoints option automatically creates the points on the center of the external vertical and horizontal lines which we can use as pierce points later on as required under Note 2. A pierce point defines the location of the profile, relative to the sketch segment used to create the structural member. The default pierce point is the sketch origin in the profile library feature part. Any vertex or sketch point specified in the profile can also be used as a pierce point.

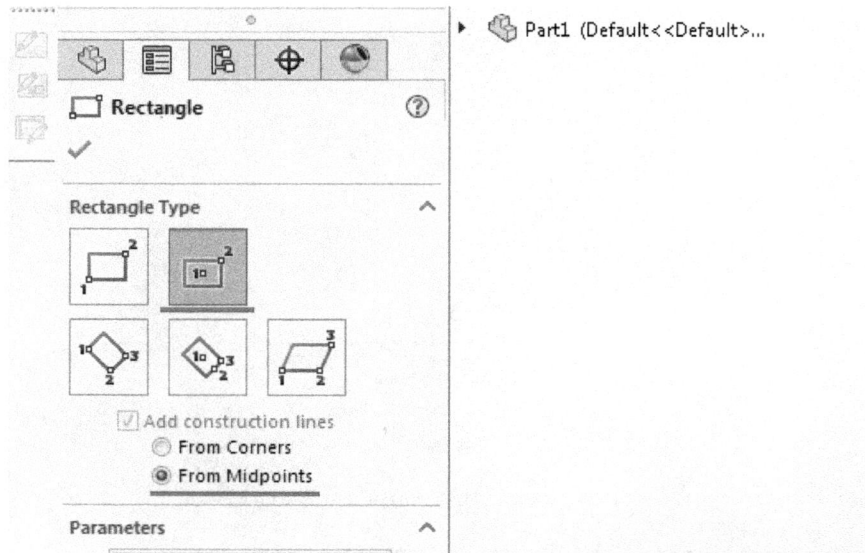

Figure 33 - Question 3 of 26 -Center Rectangle Options

In the graphics area - > Click the Origin to define the center of the rectangle. Drag to sketch the rectangle with centerlines. Release to set the four edges. Your sketch should look as shown in the following image.

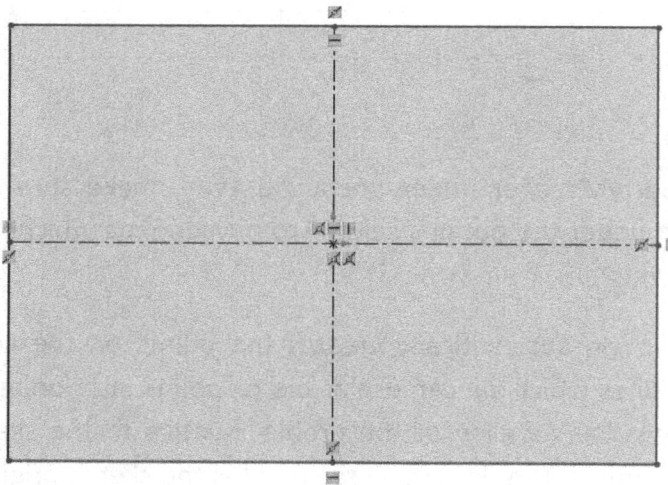

Figure 34 - Question 3 of 26 - Under Defined Center Rectangle

Dimension the sketch as shown in the following image.

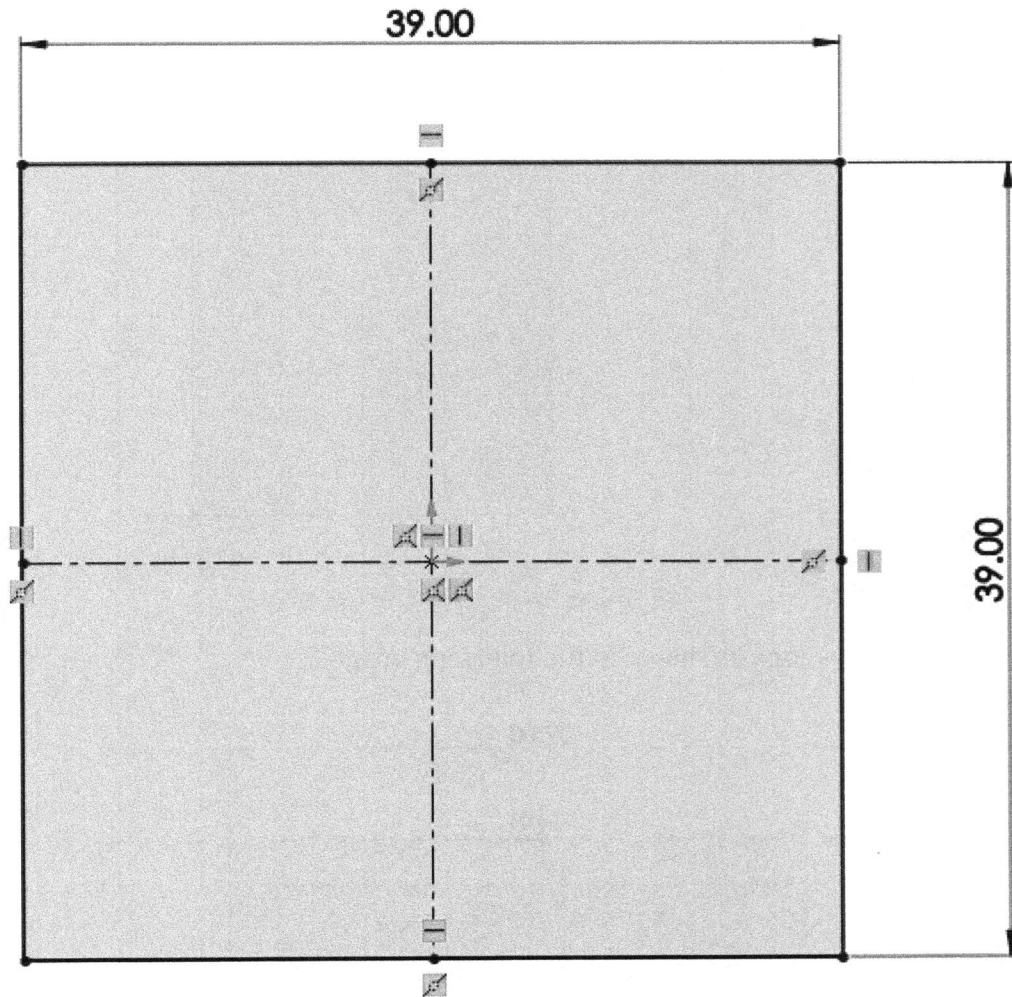

Figure 35 - Question 3 of 26 - Fully Defined Center Rectangle Sketch - SQUARE

While the sketch is still open, select one entity then Click Offset Entities (Sketch toolbar) or Tools > Sketch Tools > Offset Entities.

Set the properties in the Offset Entities PropertyManager as shown in the following image. Select the Reverse Checkbox if necessary to change the offset direction to the inside of the Fully Defined Center Rectangle. When you click in the graphics area, the offset entity is complete. Set the properties before you click in the graphics area.

Click Ok or click in the graphics area.

An Offset Entities relation is created between the new rectangle and our first Fully Defined Center Rectangle. If the original rectangle changes, the new rectangle updates to maintain the offset.

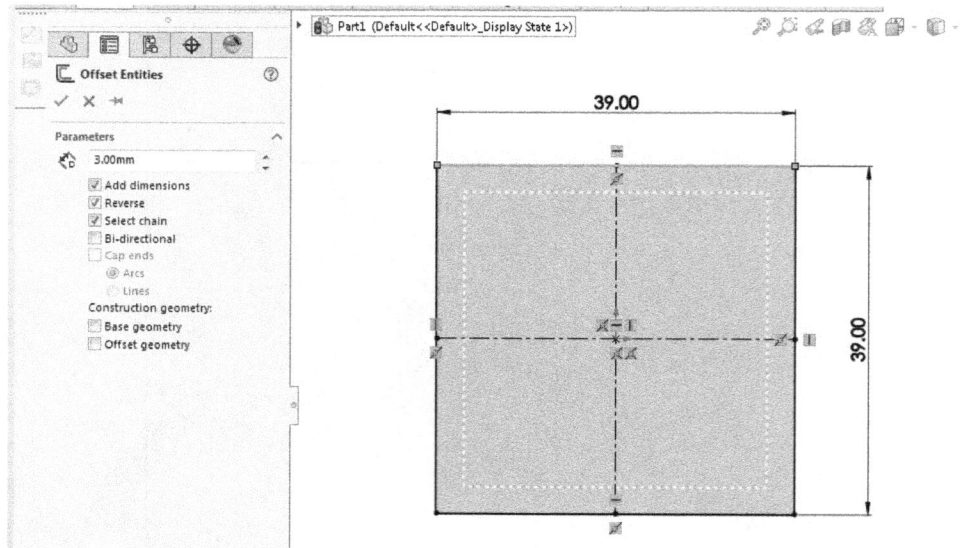

Figure 36 - Question 3 of 26 - Offsetting Sketch Entities

Your sketch should now look as shown in the following image.

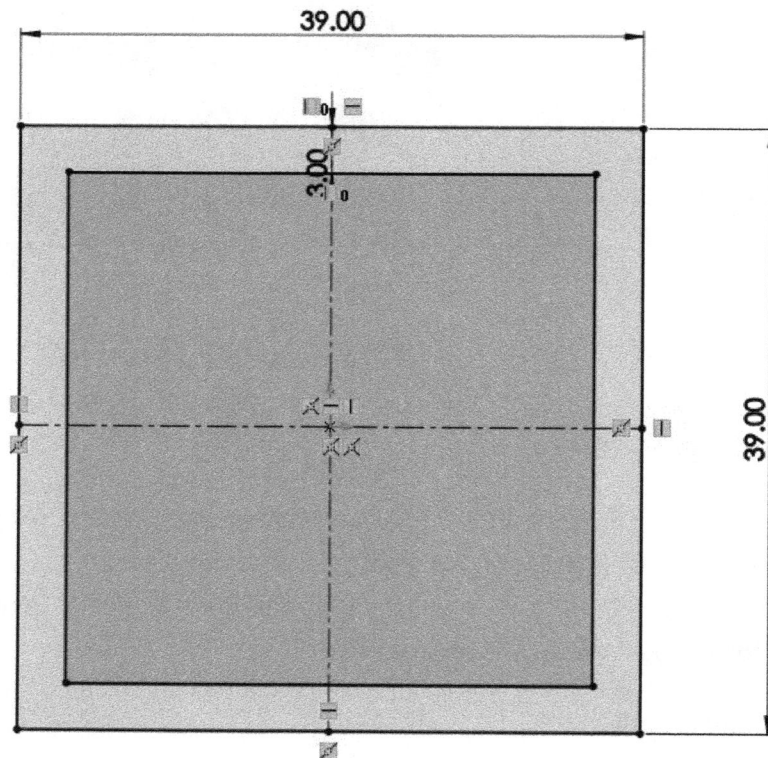

Figure 37 - Question 3 of 26 - Fully Defined Sketch for Weldment Profile Creation

Click ⤶ in the Confirmation Corner or Insert > Exit Sketch to exit the sketch.

CREATING A WELDMENT PROFILE

In the FeatureManager design tree, select Sketch1 *(default name of the sketch we created above)*. Click File > Save As. In the dialog box:

Under Save in, browse to the **standard folder** named **EXAM WELDMENT PROFILES** which we created in Question 02 or simply click on the Desktop Shortcut also created in Question 02 and select the **type folder** named **SQUARE** also created in Question 02. See the following images.

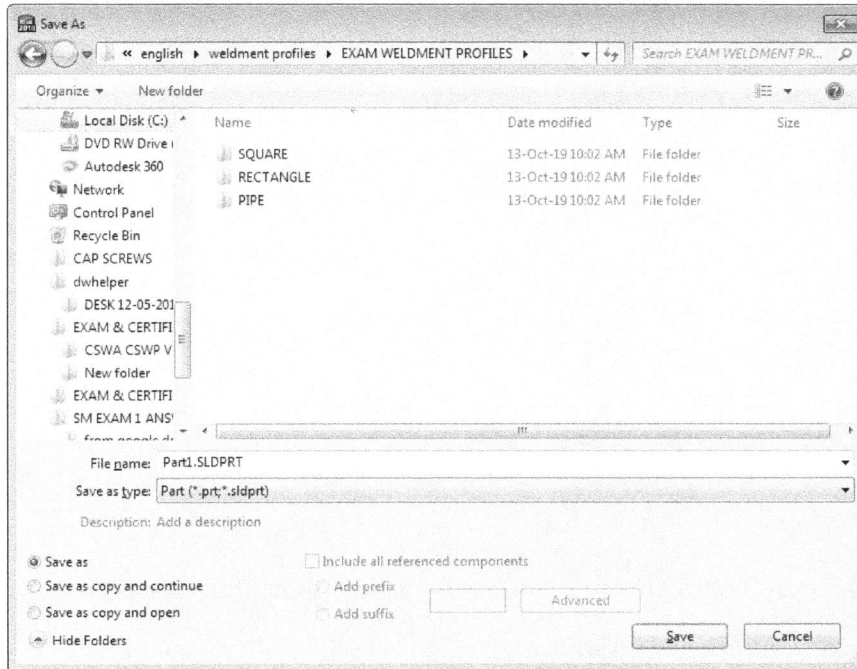

Figure 38 - Question 3 of 26 - Custom Weldment Profile created Type Folders

We will save this weldment profile into the SQUARE type folder.

Under Save as type, select Lib Feat Part (*.sldlfp).

Type a name under Filename that is WLDM2E as requested in question 3 screen capture - see the following image.

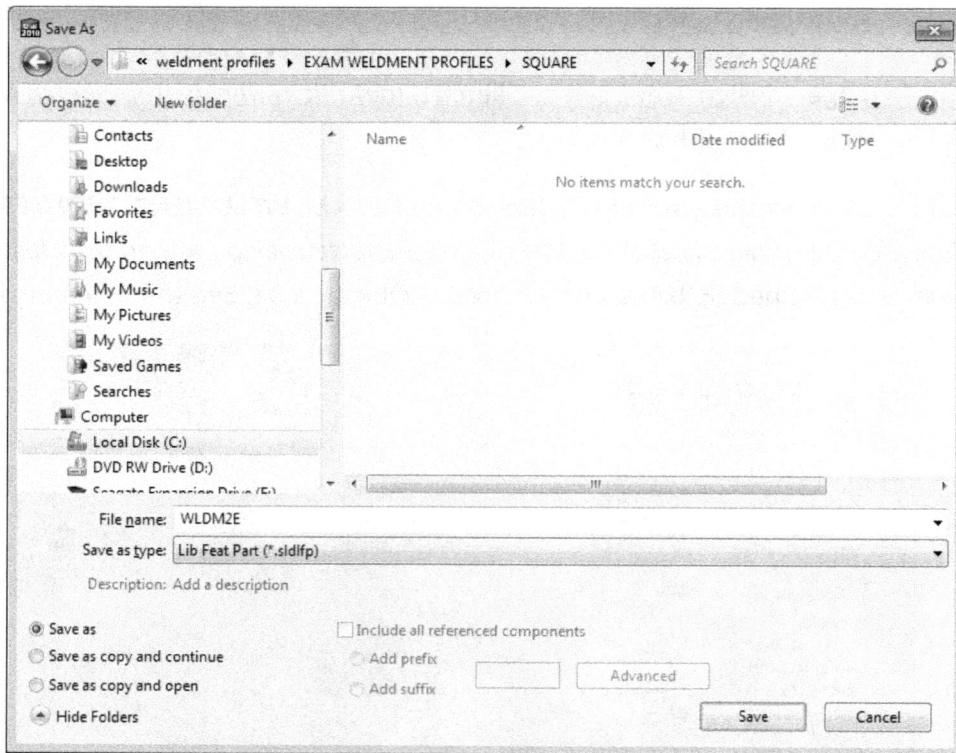

Figure 39 - Question 3 of 26 - Custom Weldment Profile Save In Location

Click Save.

In the FeatureManager design tree you will notice that the Sketch1 icon changes from ⌐ (-) Sketch1 to ⌐ Sketch1 . Close the part, you don't have to save it.

Open Question 2 and Use **Save as a copy and open** - > Change File name to Question 03 then click Save. Close Question 02.sldprt.

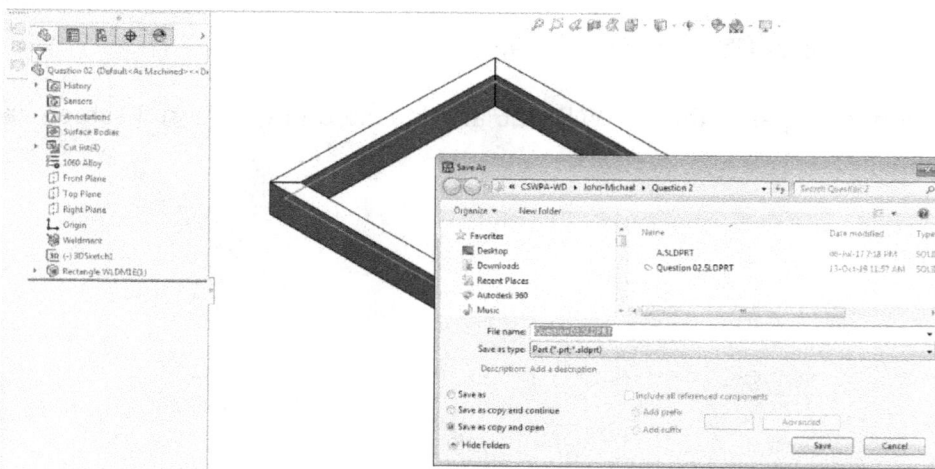

Figure 40 - Question 3 of 26 - Using Save as copy and open

EDITING A WELDMENT STRUCTURAL MEMBER

In the FeatureManager design tree, right click on the structural member feature and select Edit Feature - see the following image.

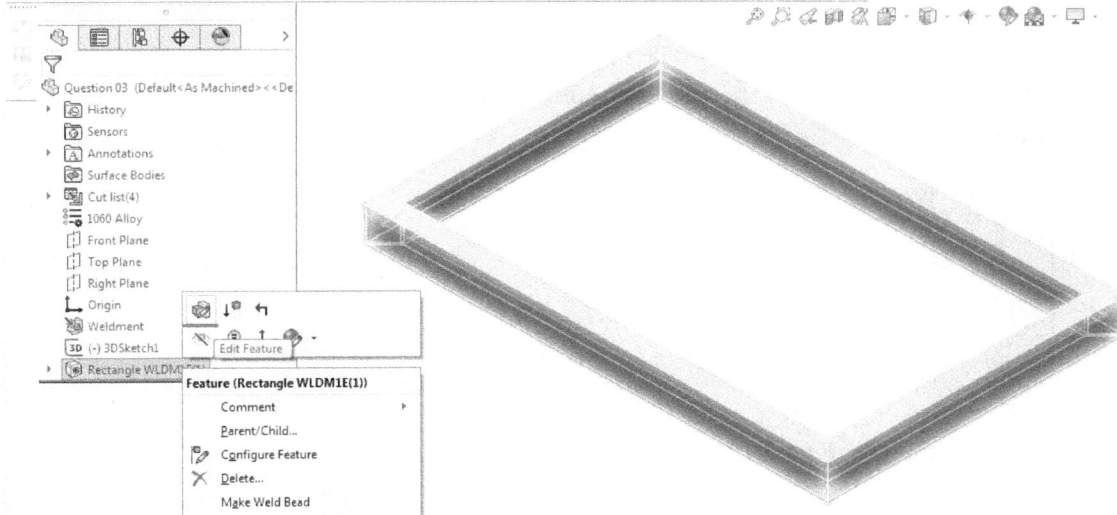

Figure 41 - Question 3 of 26 - Editing a weldment structural member

The Structural Member PropertyManager opens. In the Structural Member PropertyManager, change the Profile Type from RECTANGLE to SQUARE and under Size - > Select WLDM2E. Make sure the End miter Corner Treatment is still applied. Click Ok.

Click Rebuild or Ctrl + B. Click Save.

MEASURING THE MASS OF THE PART

Under the Evaluate Tab, click the Mass Properties button and the mass properties menu appears as shown below. Type in the mass of 3032.64 grams in the Question 3 textbox and click Next Question to go to Question 4.

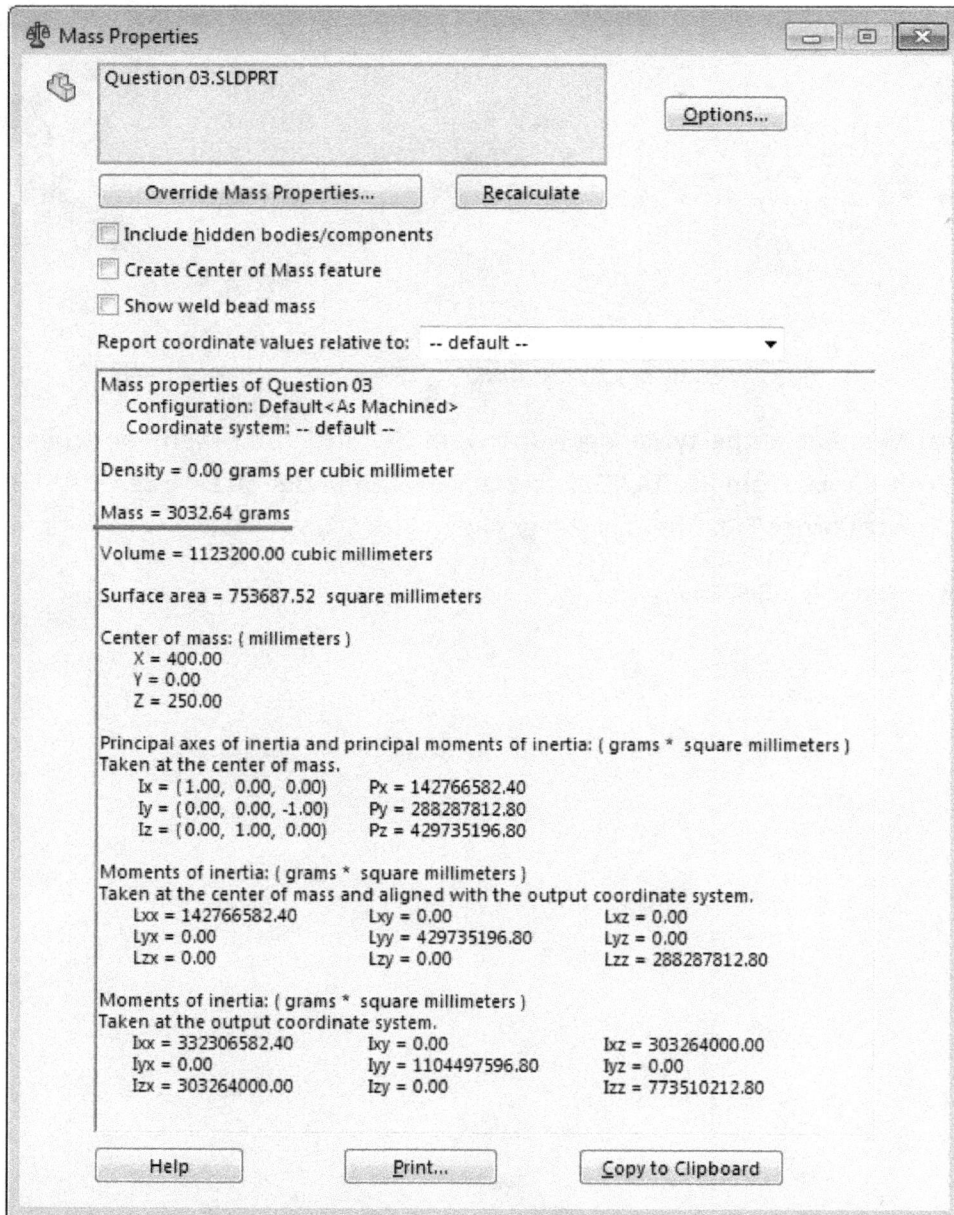

Figure 42 - Question 3 of 26 - Mass Properties

CHAPTER 4

QUESTION 4 EXAM SCREEN CAPTURES

QUESTION 4 EXAM SCREEN CAPTURE 1

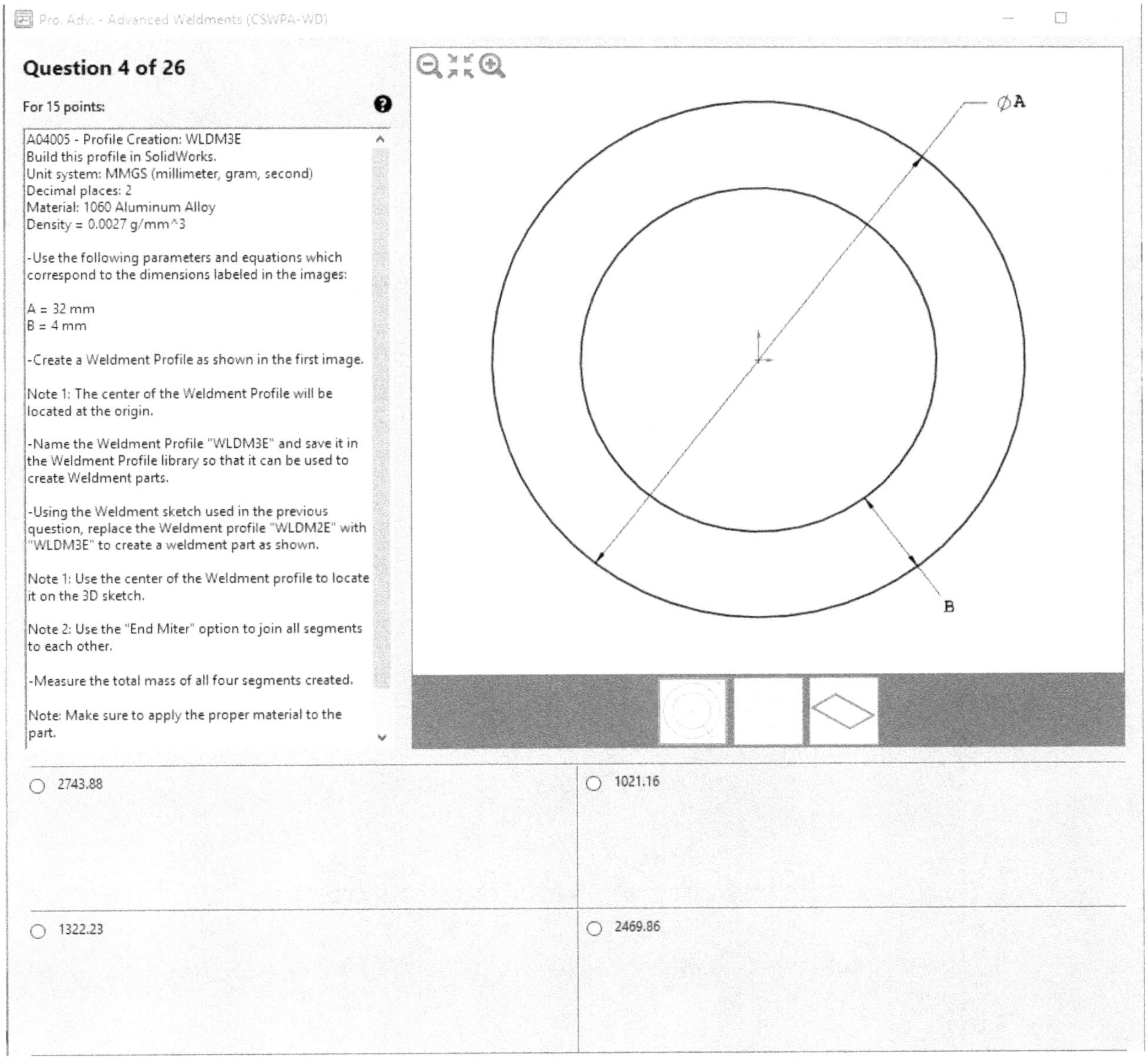

Pro. Adv. - Advanced Weldments (CSWPA-WD)

Question 4 of 26

For 15 points:

A04005 - Profile Creation: WLDM3E
Build this profile in SolidWorks.
Unit system: MMGS (millimeter, gram, second)
Decimal places: 2
Material: 1060 Aluminum Alloy
Density = 0.0027 g/mm^3

-Use the following parameters and equations which correspond to the dimensions labeled in the images:

A = 32 mm
B = 4 mm

-Create a Weldment Profile as shown in the first image.

Note 1: The center of the Weldment Profile will be located at the origin.

-Name the Weldment Profile "WLDM3E" and save it in the Weldment Profile library so that it can be used to create Weldment parts.

-Using the Weldment sketch used in the previous question, replace the Weldment profile "WLDM2E" with "WLDM3E" to create a weldment part as shown.

Note 1: Use the center of the Weldment profile to locate it on the 3D sketch.

Note 2: Use the "End Miter" option to join all segments to each other.

-Measure the total mass of all four segments created.

Note: Make sure to apply the proper material to the part.

- ⌀A
- B

- ◯ 2743.88
- ◯ 1021.16
- ◯ 1322.23
- ◯ 2469.86

Figure 43 - Question 4 of 26 - Exam Screen Capture 1

Unit system: MMGS (millimeter, gram, second)
Decimal places: 2
Material: 1060 Aluminum Alloy
Density = 0.0027 g/mm^3

-Use the following parameters and equations which correspond to the dimensions labeled in the images:

A = 32 mm
B = 4 mm

-Create a Weldment Profile as shown in the first image.

Note 1: The center of the Weldment Profile will be located at the origin.

-Name the Weldment Profile "WLDM3E" and save it in the Weldment Profile library so that it can be used to create Weldment parts.

-Using the Weldment sketch used in the previous question, replace the Weldment profile "WLDM2E" with "WLDM3E" to create a weldment part as shown.

Note 1: Use the center of the Weldment profile to locate it on the 3D sketch.

Note 2: Use the "End Miter" option to join all segments to each other.

-Measure the total mass of all four segments created.

Note: Make sure to apply the proper material to the part.

What is the total mass of all four weldment segments(grams)?

Figure 44 - Question 4 of 26 - Exam Screen Capture 2

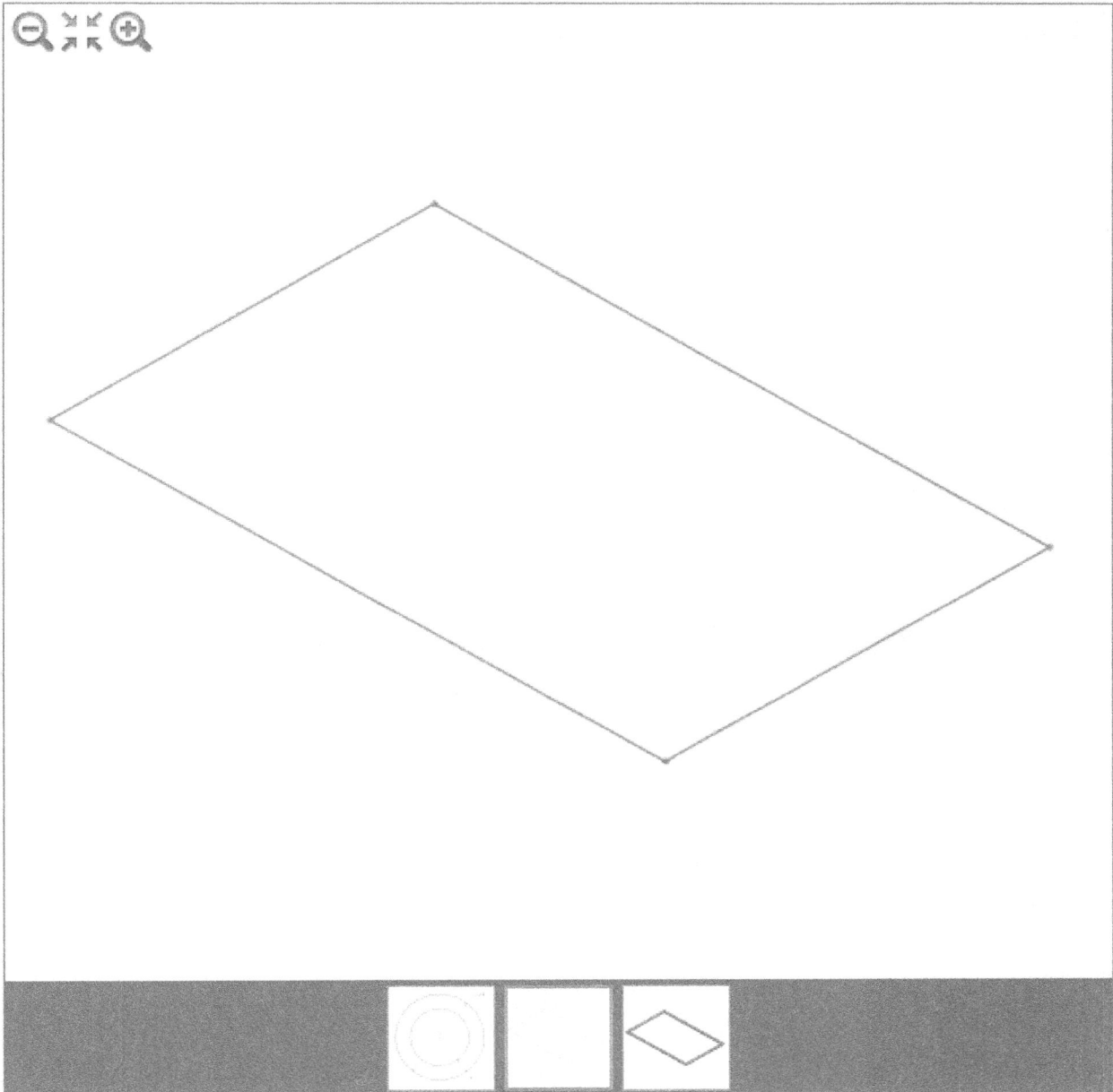

Figure 45 - Question 4 of 26 - Exam Screen Capture 3

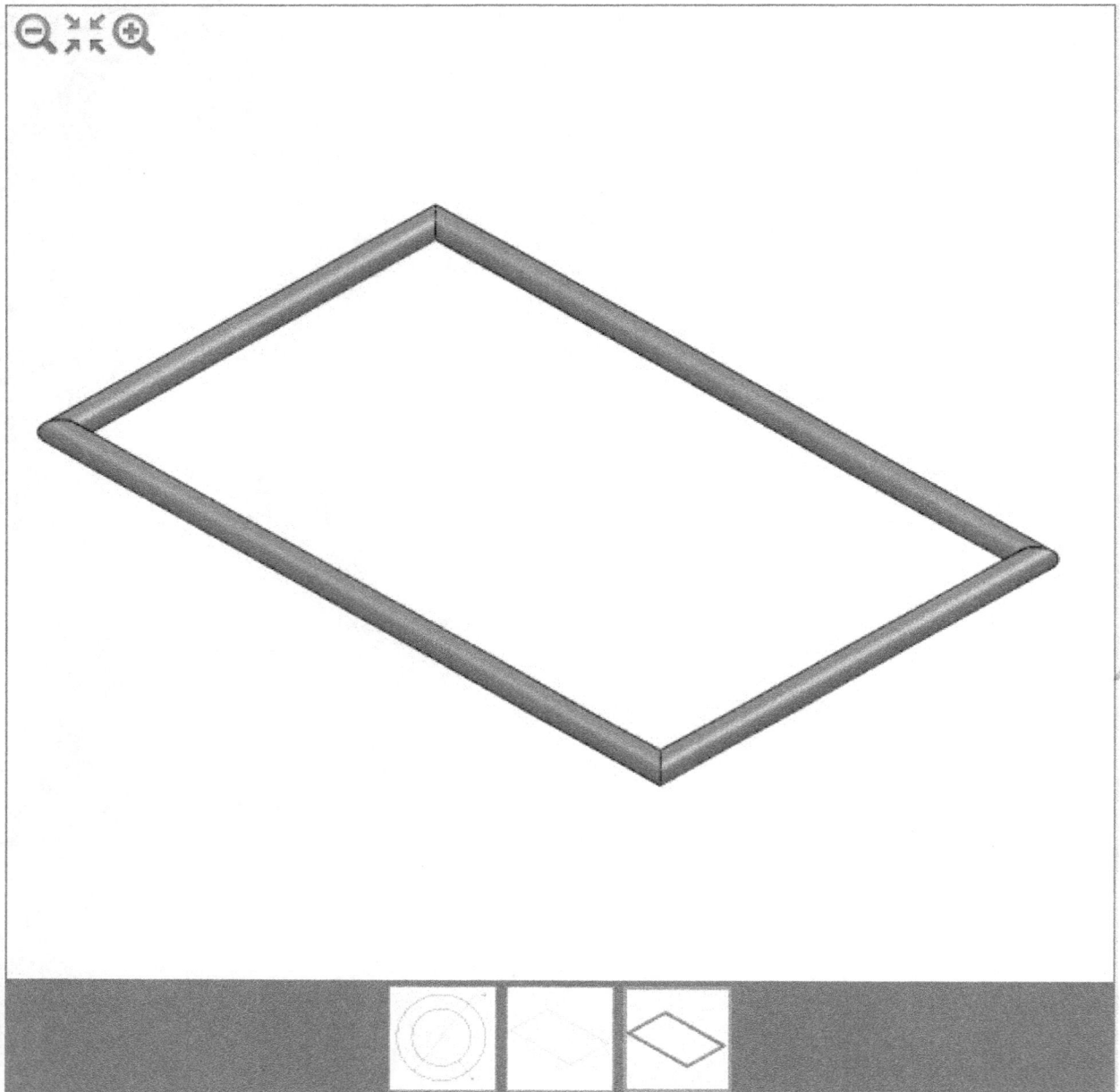

Figure 46 - Question 4 of 26 - Exam Screen Capture 4

Start a new part in Solidworks.

SET A UNIT SYSTEM

Go to Tools > Options > Document Properties > Drafting Standard. From the drop down menu, Select the ISO Standard since our dimensions are given in millimeters.

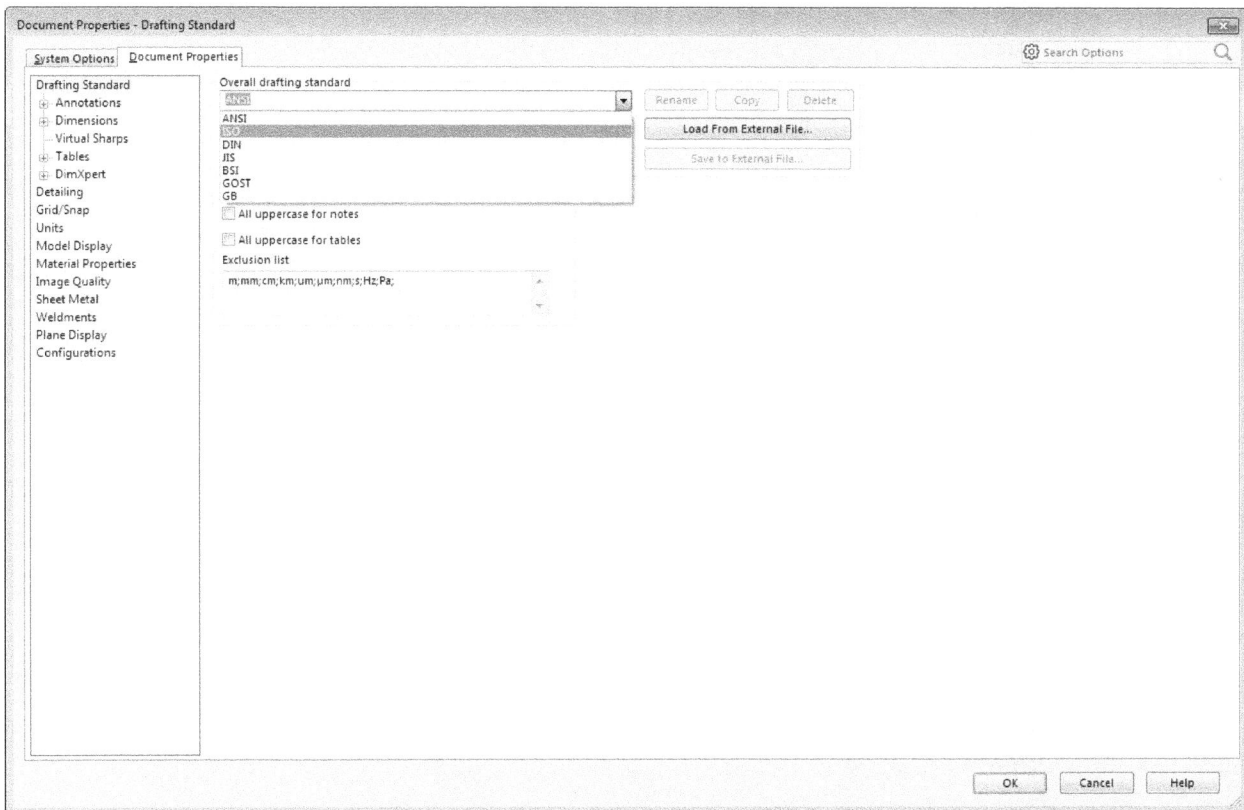

Figure 47 - Question 4 of 26 - Document Properties - Drafting Standard

Go to Tools > Options > Document Properties > Units to change the Unit System to MMGS (millimeter, gram, second) and to also set the number of decimal places to two decimal places.

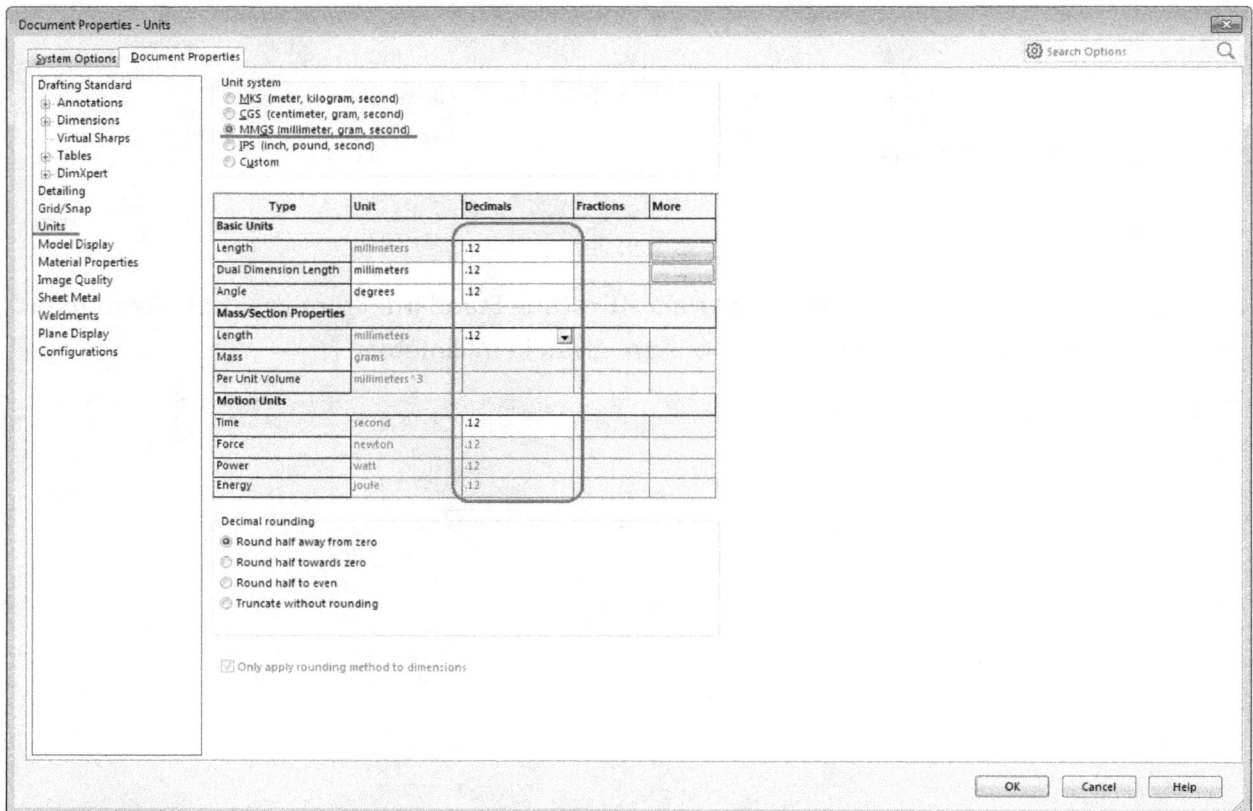

Figure 48 - Question 4 of 26 - Document Properties - Units

Click OK.

APPLYING A MATERIAL

If the material is not already applied as 1060 Alloy - Right-click Material in the FeatureManager design tree. Click Edit Material, select 1060 Alloy in the material tree under Aluminium Alloys, and click Apply, then Close.

CREATING A WELDMENT PROFILE

Select any plane in the FeatureManager design tree and click a sketch entity tool or click Sketch on the Sketch toolbar.

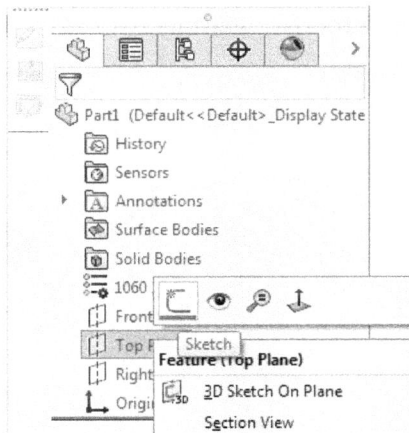

Figure 49 - Question 4 of 26 - Starting A Sketch on a Plane

Under the Sketch Tab, Click Circle and select the Center-Based Circle.

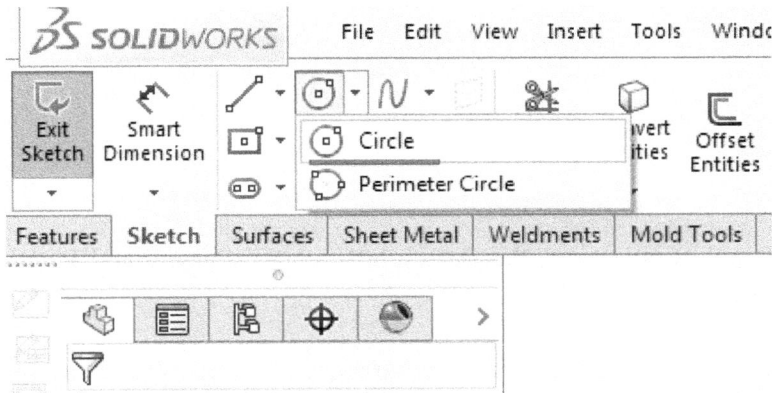

Figure 50 - Question 4 of 26 - Selecting a Center-Based Circle

Click on the origin to place the center of the circle. Drag and click to set the radius. Click OK. Your sketch should look as shown in the following image.

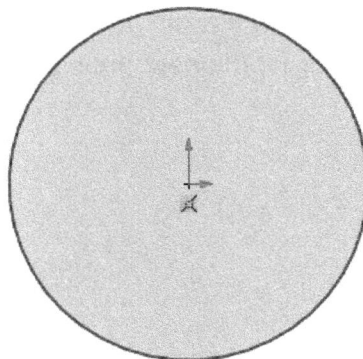

Figure 51 - Question 4 of 26 - Under Defined Center-Based Circle

41

Dimension the sketch as shown in the following image.

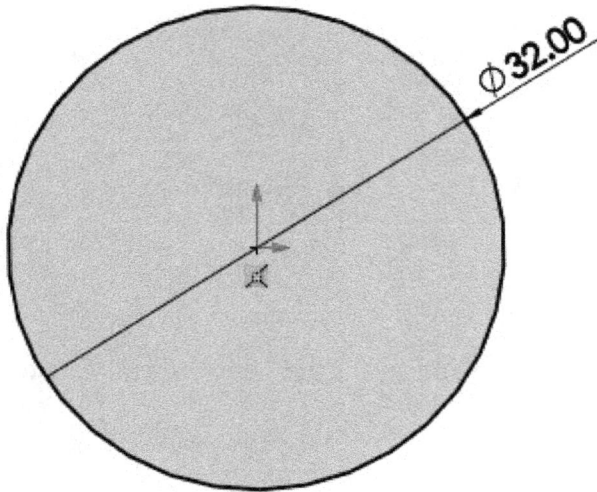

Figure 52 - Question 4 of 26 - Fully Defined Center-Based Circle Sketch

While the sketch is still open, select the sketch entity then Click Offset Entities (Sketch toolbar) or Tools > Sketch Tools > Offset Entities.

Set the properties in the Offset Entities PropertyManager as shown in the following image. Select the Reverse Checkbox if necessary to change the offset direction to the inside of the Fully Defined Center-Based Circle. When you click in the graphics area, the offset entity is complete. NB: Set the properties before you click in the graphics area.

Click Ok or click in the graphics area.

An Offset Entities relation is created between the new circle and our first Fully Defined Center-Based Circle. If the original circle changes, the new circle updates to maintain the offset.

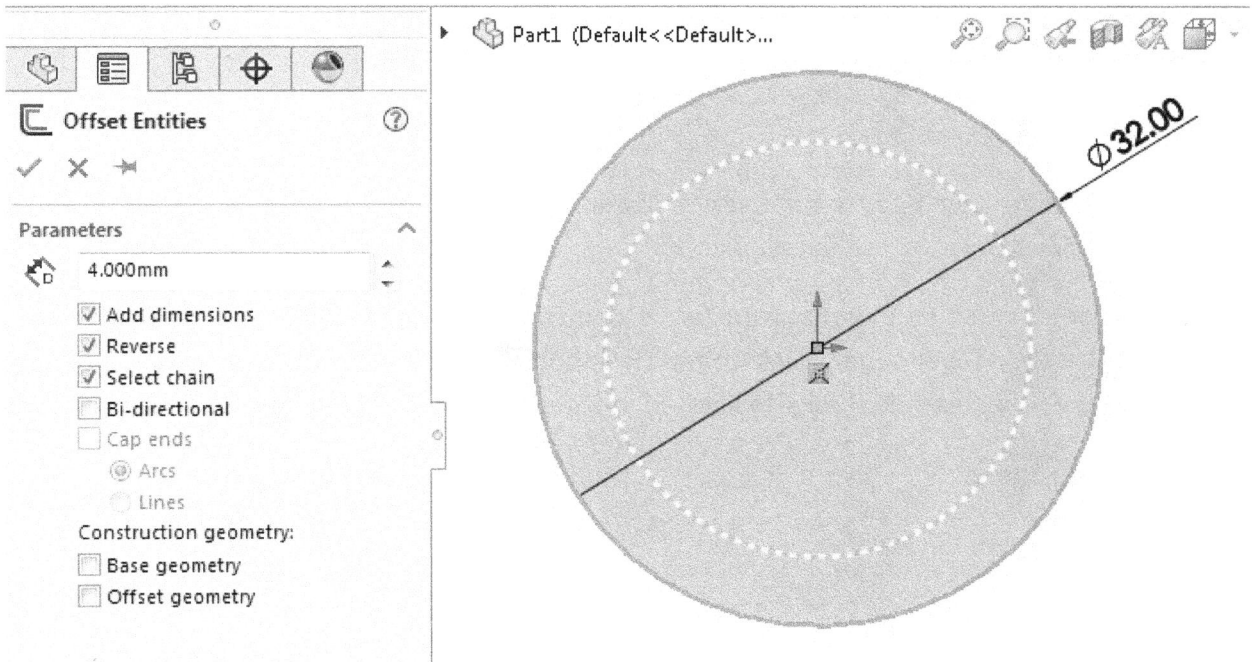

Figure 53 - Question 4 of 26 - Offsetting Sketch Entities

Your sketch should now look as shown in the following image.

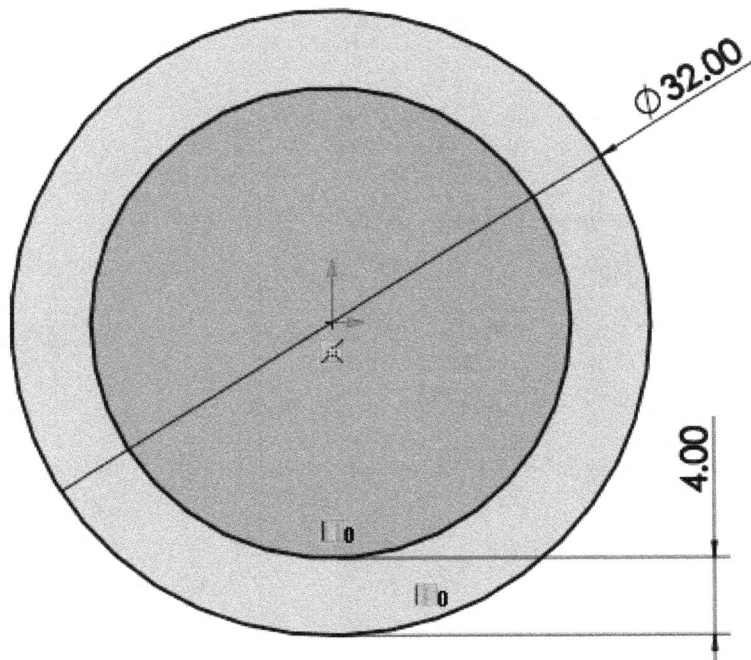

Figure 54 - Question 4 of 26 - Fully Defined Sketch for Weldment Profile Creation

43

Click ✎ in the Confirmation Corner or Insert > Exit Sketch to exit the sketch.

CREATING A WELDMENT PROFILE

In the FeatureManager design tree, select Sketch1 *(default name of the sketch we created above)*. Click File > Save As. In the dialog box:

Under Save in, browse to the **standard folder** named **EXAM WELDMENT PROFILES** which we created in Question 02 or simply click on the Desktop Shortcut also created in Question 02 and select the **type folder** named **PIPE** also created in Question 02. See the following image.

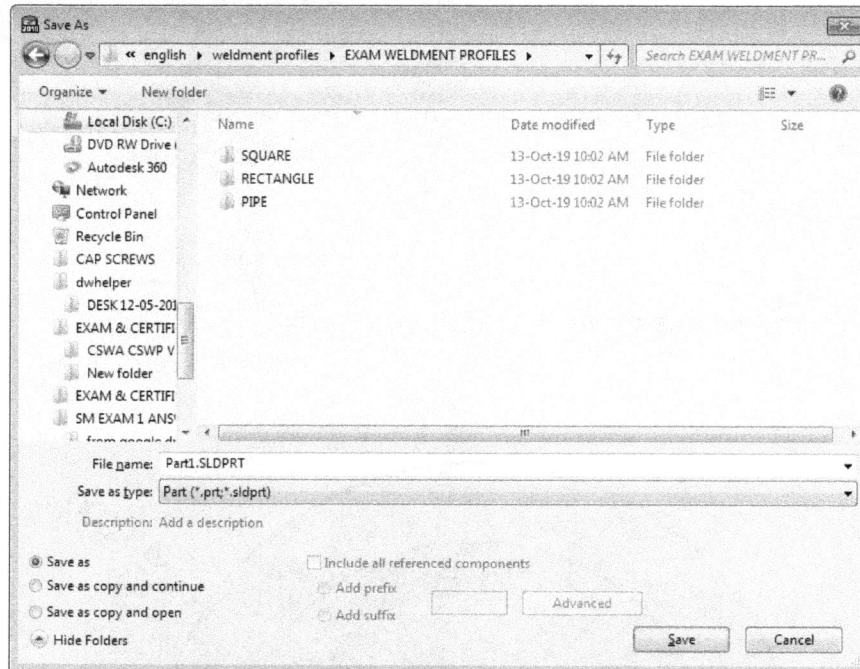

Figure 55 - Question 4 of 26 - Custom Weldment Profile created Type Folders

We will save this weldment profile into the PIPE type folder.

In Save as type, select Lib Feat Part (*.sldlfp).

Type a name for Filename that is WLDM3E as requested in question 4 screen capture - see the following image.

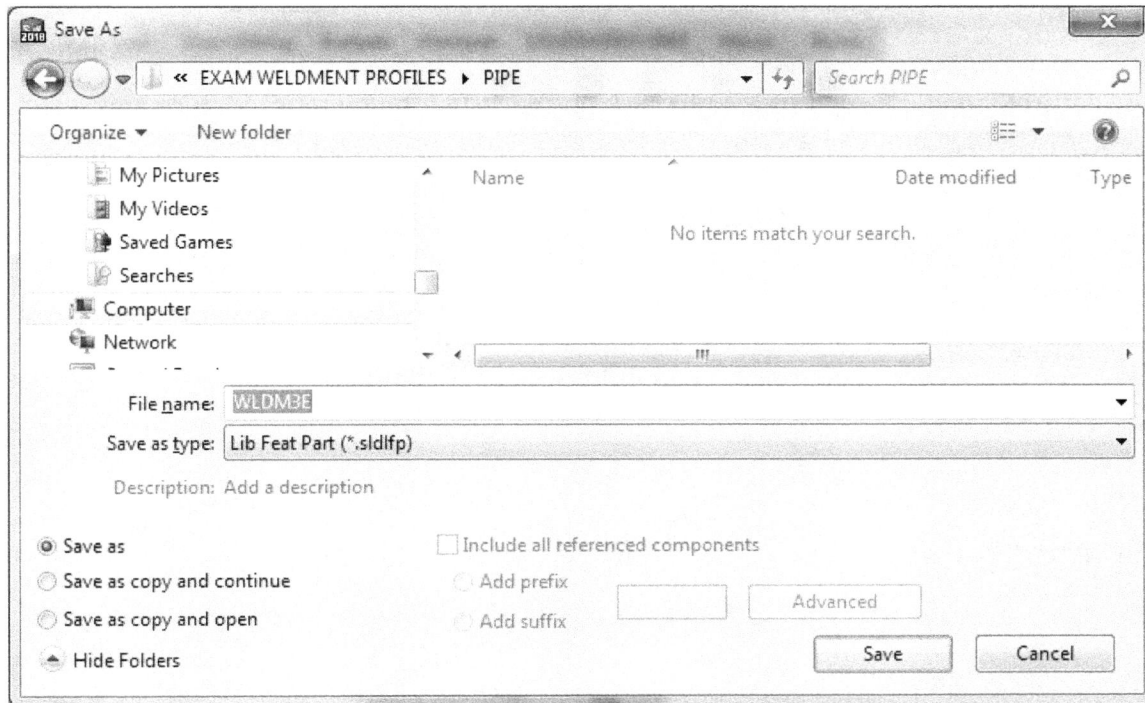

Figure 56 - Question 4 of 26 - Custom Weldment Profile Save In Location

Click Save.

In the FeatureManager design tree you will notice that the Sketch1 icon changes from ⌐ (-) Sketch1 to ⌐⌐ Sketch1 . Close the part, you don't have to save it.

Open Question 3 and Use **Save as a copy and open** - > Change File name to Question 04 then click Save. Close Question 03.sldprt.

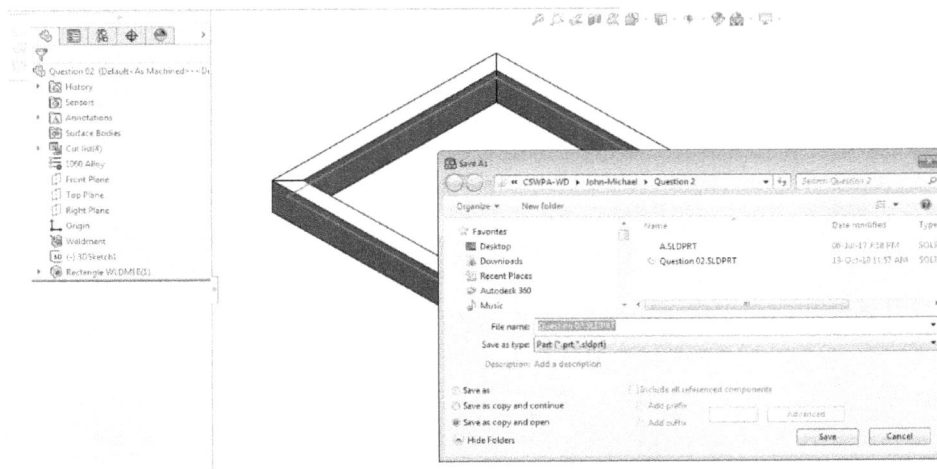

Figure 57 - Question 4 of 26 - Using Save as copy and open

EDITING A WELDMENT STRUCTURAL MEMBER

In the FeatureManager design tree, right click on the structural member feature and select Edit Feature.

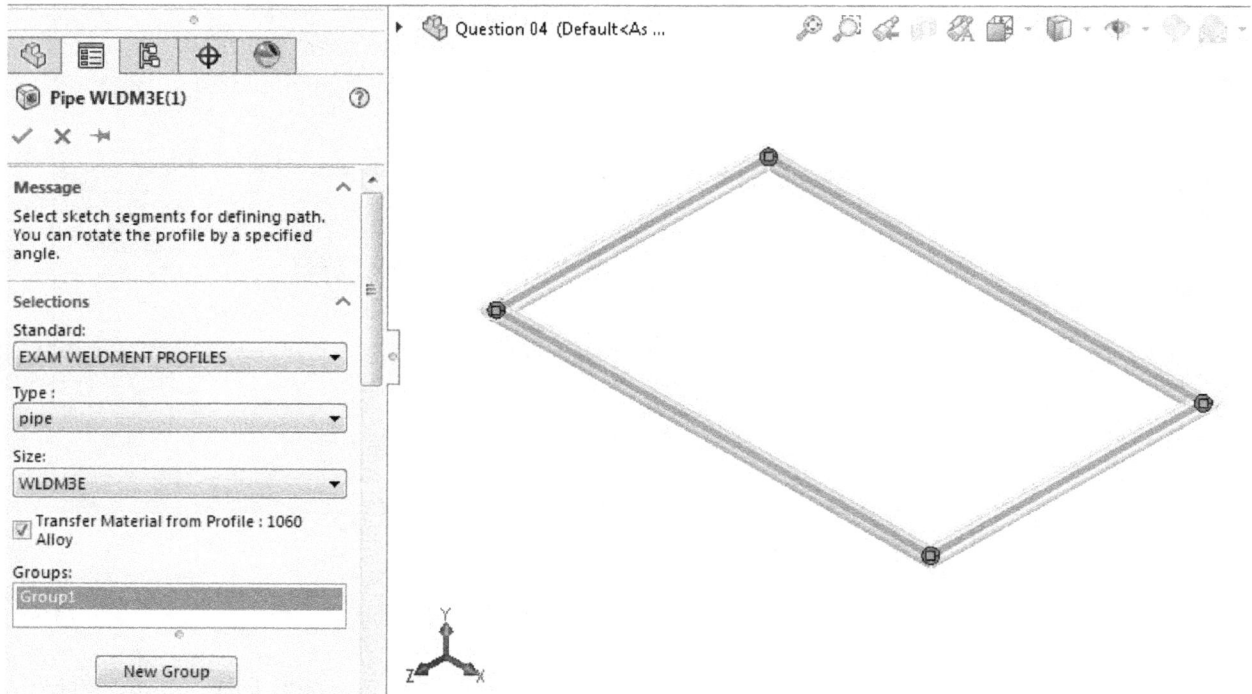

Figure 58 - Question 4 of 26 - Editing a weldment structural member

The Structural Member PropertyManager opens. In the Structural Member PropertyManager, change the Profile Type from SQUARE to PIPE and under Size - > Select WLDM3E. Make sure the End miter Corner Treatment is still applied. Click Ok.

Click Rebuild or Ctrl + B. Click Save.

QUESTION 4 ANSWER

MEASURING THE MASS OF THE PART

Under the Evaluate Tab, click the Mass Properties button and the mass properties menu appears as shown below. Type in the mass of 2469.86 grams in the Question 4 textbox and click Next Question to go to Question 5.

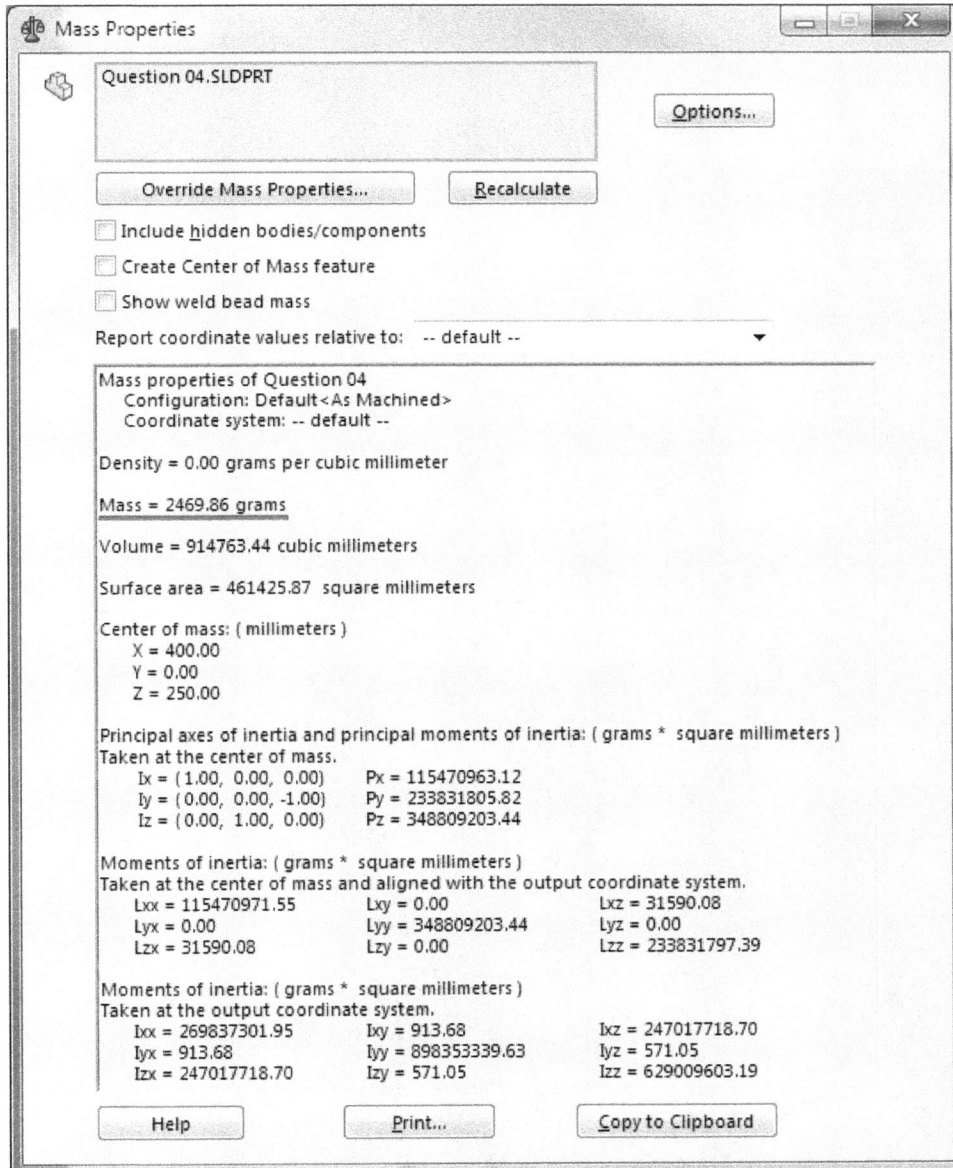

```
Mass Properties                                    □ ▣ ✕

  Question 04.SLDPRT

                                                Options...

      Override Mass Properties...      Recalculate

   ☐ Include hidden bodies/components

   ☐ Create Center of Mass feature

   ☐ Show weld bead mass

   Report coordinate values relative to:   -- default --        ▼

   Mass properties of Question 04
       Configuration: Default<As Machined>
       Coordinate system: -- default --

   Density = 0.00 grams per cubic millimeter

   Mass = 2469.86 grams

   Volume = 914763.44 cubic millimeters

   Surface area = 461425.87  square millimeters

   Center of mass: ( millimeters )
       X = 400.00
       Y = 0.00
       Z = 250.00

   Principal axes of inertia and principal moments of inertia: ( grams * square millimeters )
   Taken at the center of mass.
       Ix = ( 1.00,  0.00,  0.00)     Px = 115470963.12
       Iy = ( 0.00,  0.00, -1.00)     Py = 233831805.82
       Iz = ( 0.00,  1.00,  0.00)     Pz = 348809203.44

   Moments of inertia: ( grams * square millimeters )
   Taken at the center of mass and aligned with the output coordinate system.
       Lxx = 115470971.55      Lxy = 0.00            Lxz = 31590.08
       Lyx = 0.00              Lyy = 348809203.44    Lyz = 0.00
       Lzx = 31590.08          Lzy = 0.00            Lzz = 233831797.39

   Moments of inertia: ( grams * square millimeters )
   Taken at the output coordinate system.
       Ixx = 269837301.95      Ixy = 913.68          Ixz = 247017718.70
       Iyx = 913.68            Iyy = 898353339.63    Iyz = 571.05
       Izx = 247017718.70      Izy = 571.05          Izz = 629009603.19

        Help              Print...         Copy to Clipboard
```

Figure 59 - Question 4 of 26 - Question 4 of Answer

47

CHAPTER 5

QUESTION 5 EXAM SCREEN CAPTURES

QUESTION 5 EXAM SCREEN CAPTURE 1

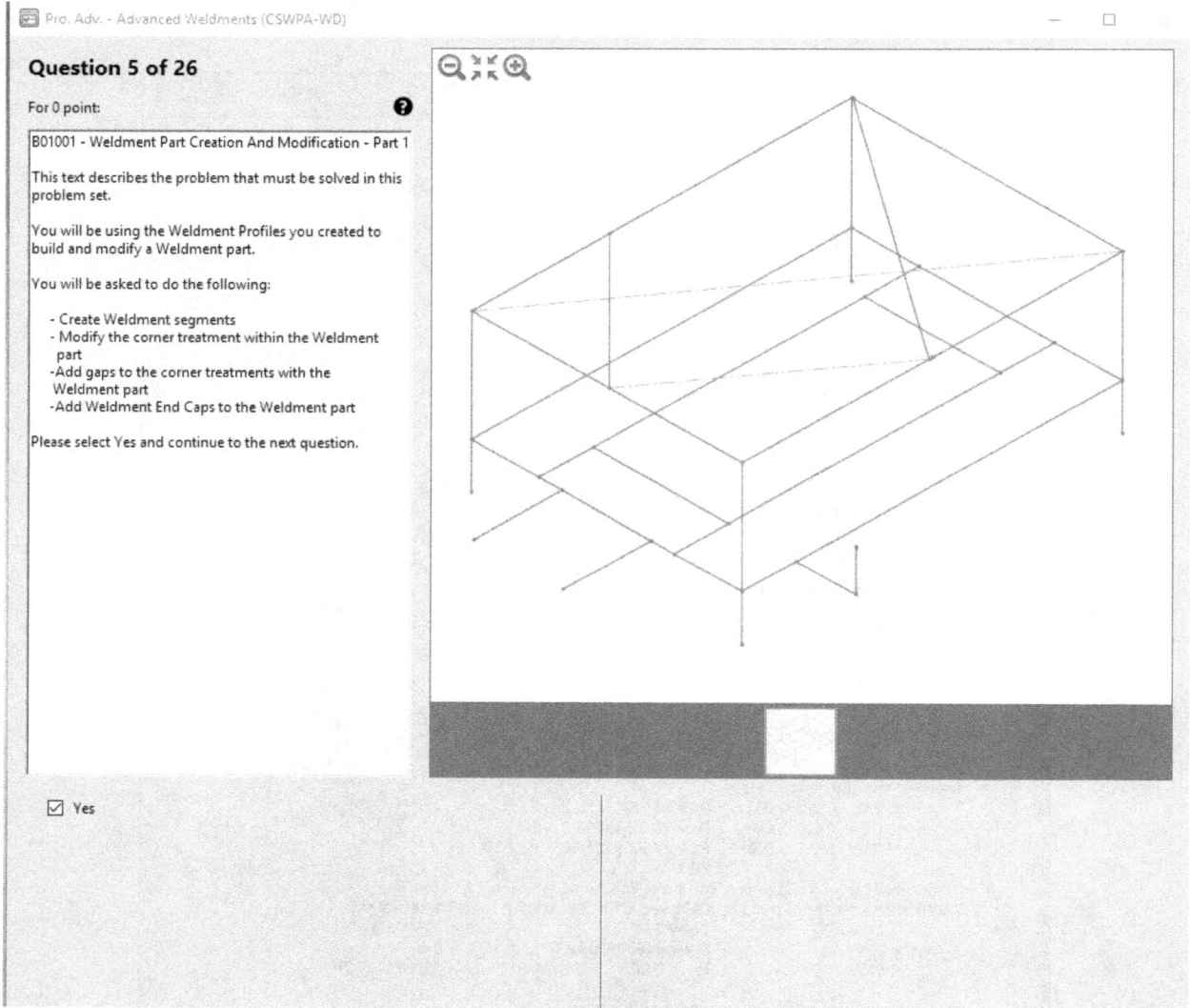

Figure 60 - Question 5 of 26 - Exam Screen Capture 1

In this question, you just read the instructions then select Yes and continue to Question 6.

CHAPTER 6

QUESTION 6 EXAM SCREEN CAPTURES

QUESTION 6 EXAM SCREEN CAPTURE 1

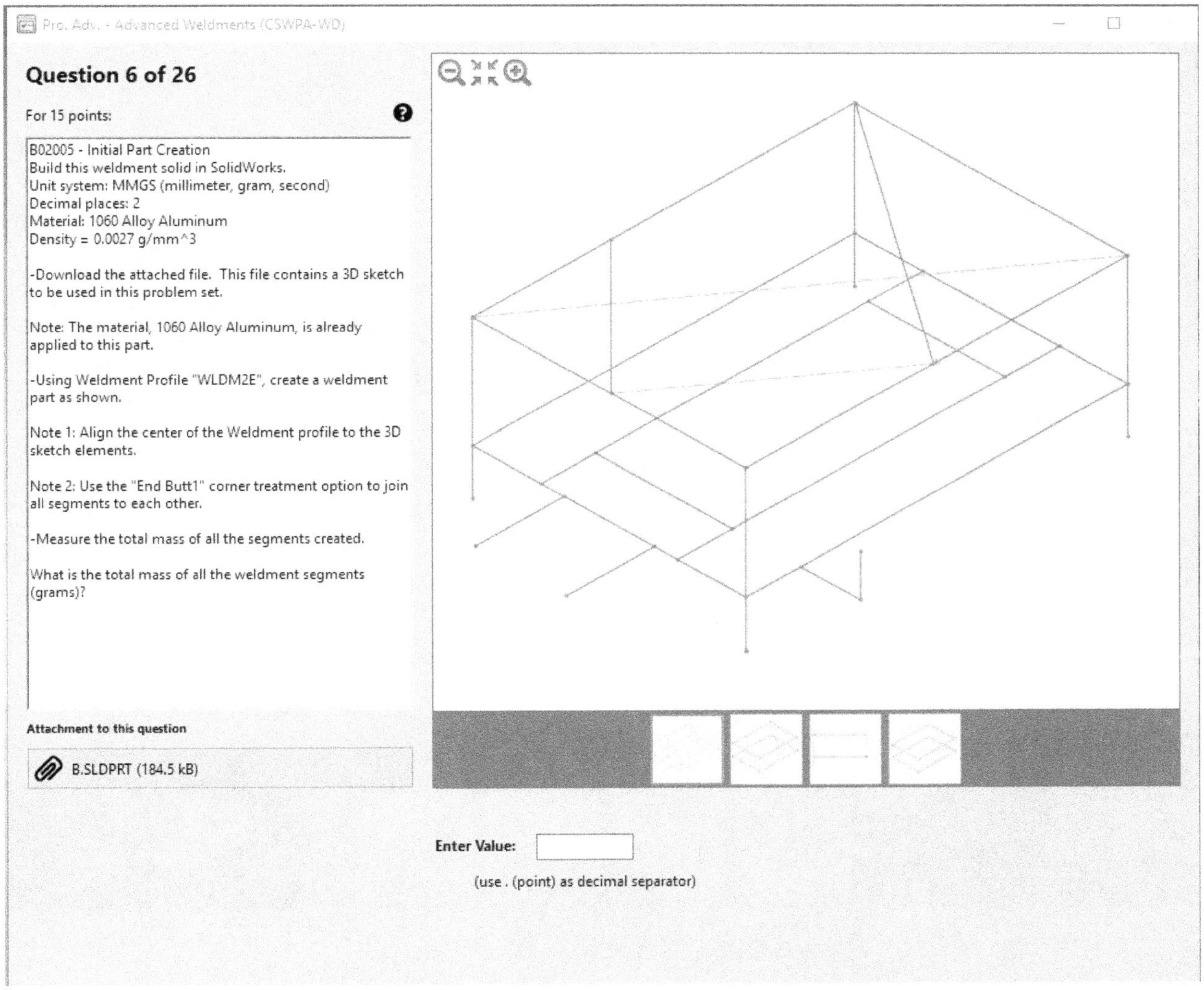

Pro. Adv. - Advanced Weldments (CSWPA-WD)

Question 6 of 26

For 15 points: ❓

B02005 - Initial Part Creation
Build this weldment solid in SolidWorks.
Unit system: MMGS (millimeter, gram, second)
Decimal places: 2
Material: 1060 Alloy Aluminum
Density = 0.0027 g/mm^3

-Download the attached file. This file contains a 3D sketch to be used in this problem set.

Note: The material, 1060 Alloy Aluminum, is already applied to this part.

-Using Weldment Profile "WLDM2E", create a weldment part as shown.

Note 1: Align the center of the Weldment profile to the 3D sketch elements.

Note 2: Use the "End Butt1" corner treatment option to join all segments to each other.

-Measure the total mass of all the segments created.

What is the total mass of all the weldment segments (grams)?

Attachment to this question

📎 B.SLDPRT (184.5 kB)

Enter Value: [   ]

(use . (point) as decimal separator)

Figure 61 - Question 6 of 26 - Exam Screen Capture 1

QUESTION 6 EXAM SCREEN CAPTURE 2

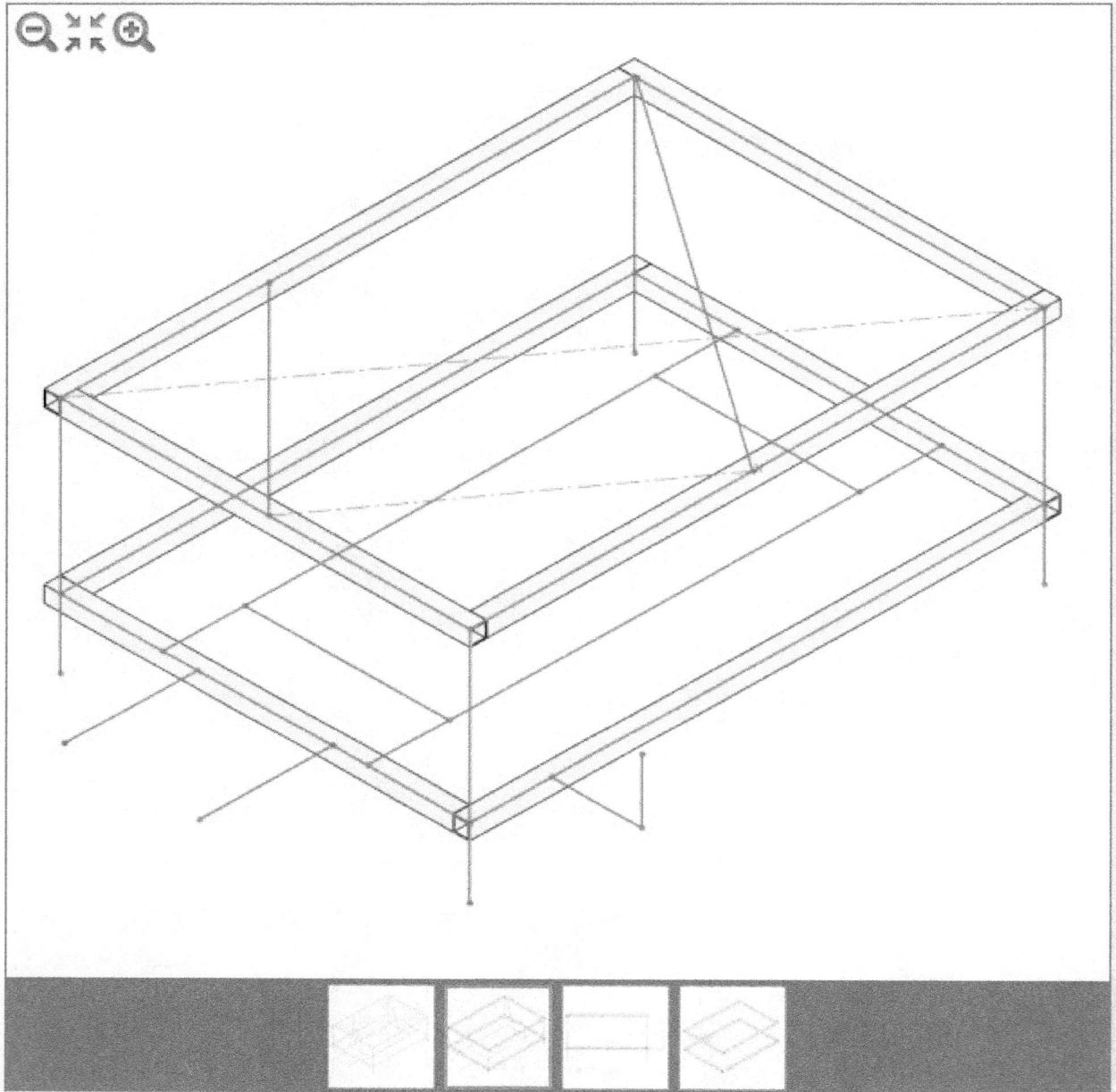

Figure 62 - Question 6 of 26 - Exam Screen Capture 2

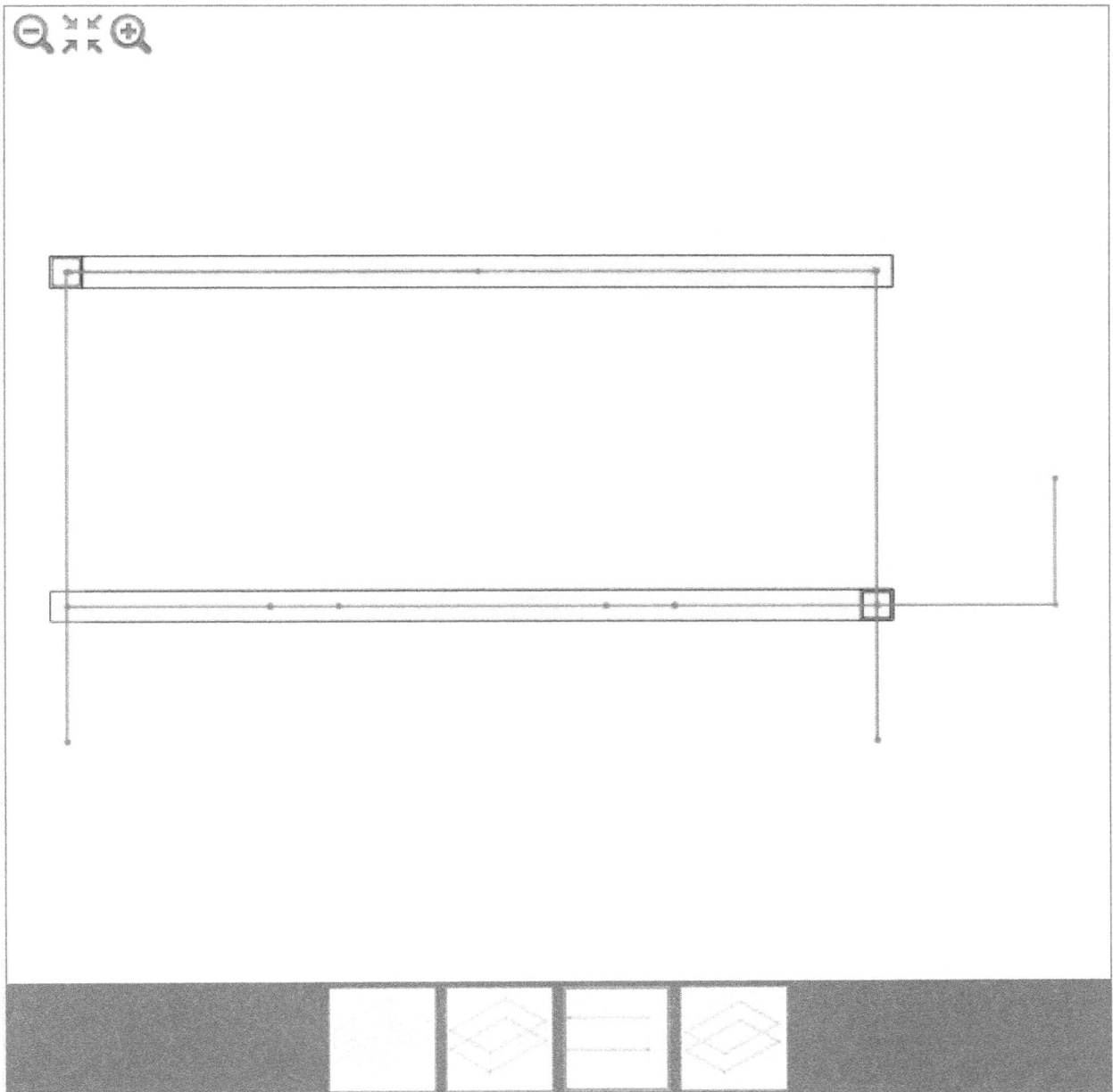

Figure 63 - Question 6 of 26 - Exam Screen Capture 3

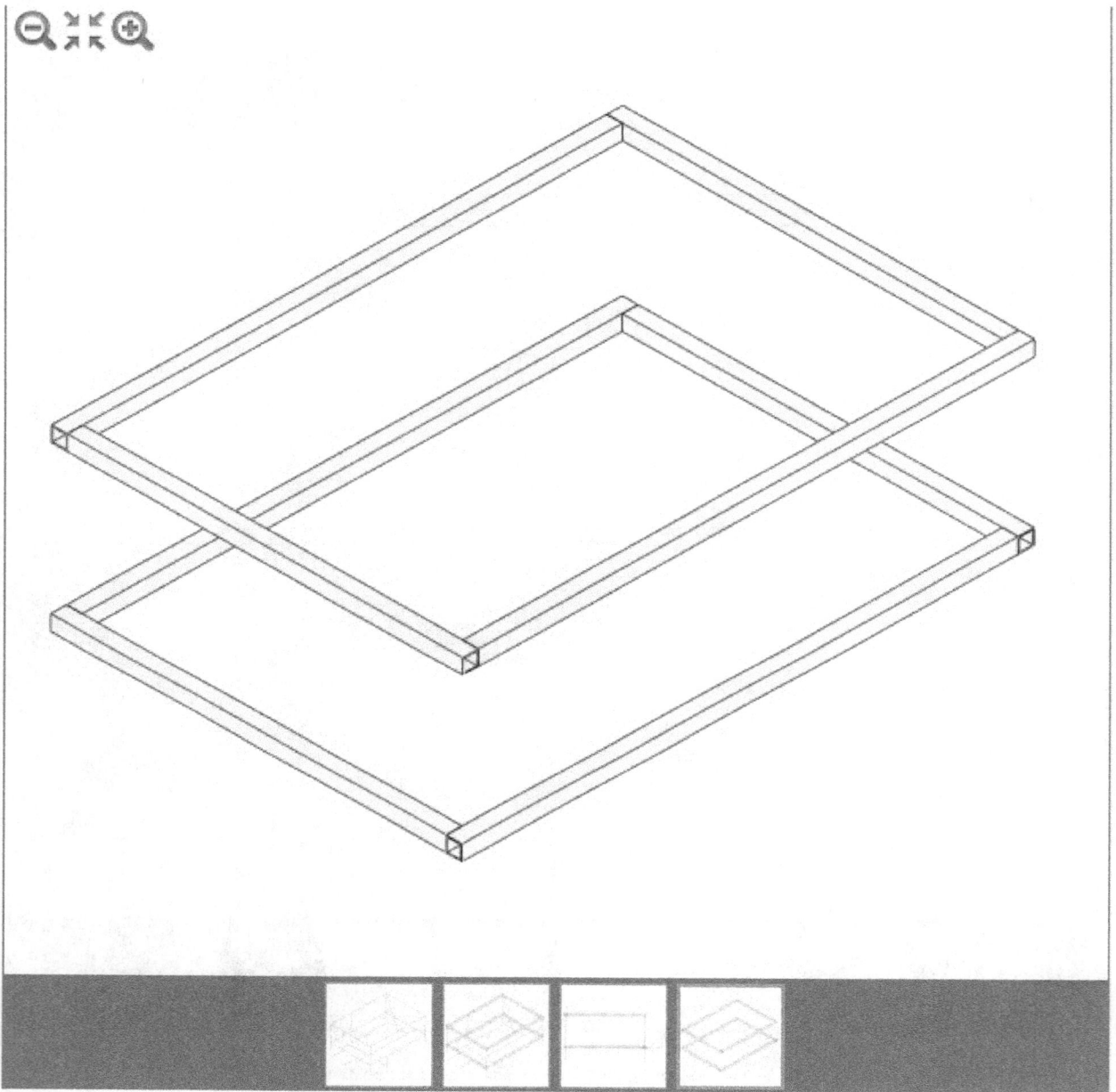

Figure 64 - Question 6 of 26 - Exam Screen Capture 4

Download the part B.SLDPRT which contains a 3D Sketch from this Google Drive location *(http://bit.ly/CSWPA-WD)* in Question 6 Folder. Open the downloaded part and save it on your computer.

CREATING A WELDMENT PART USING A 3D SKETCH

The part downloaded from aforementioned location contains a 3D Sketch as shown in the following image:

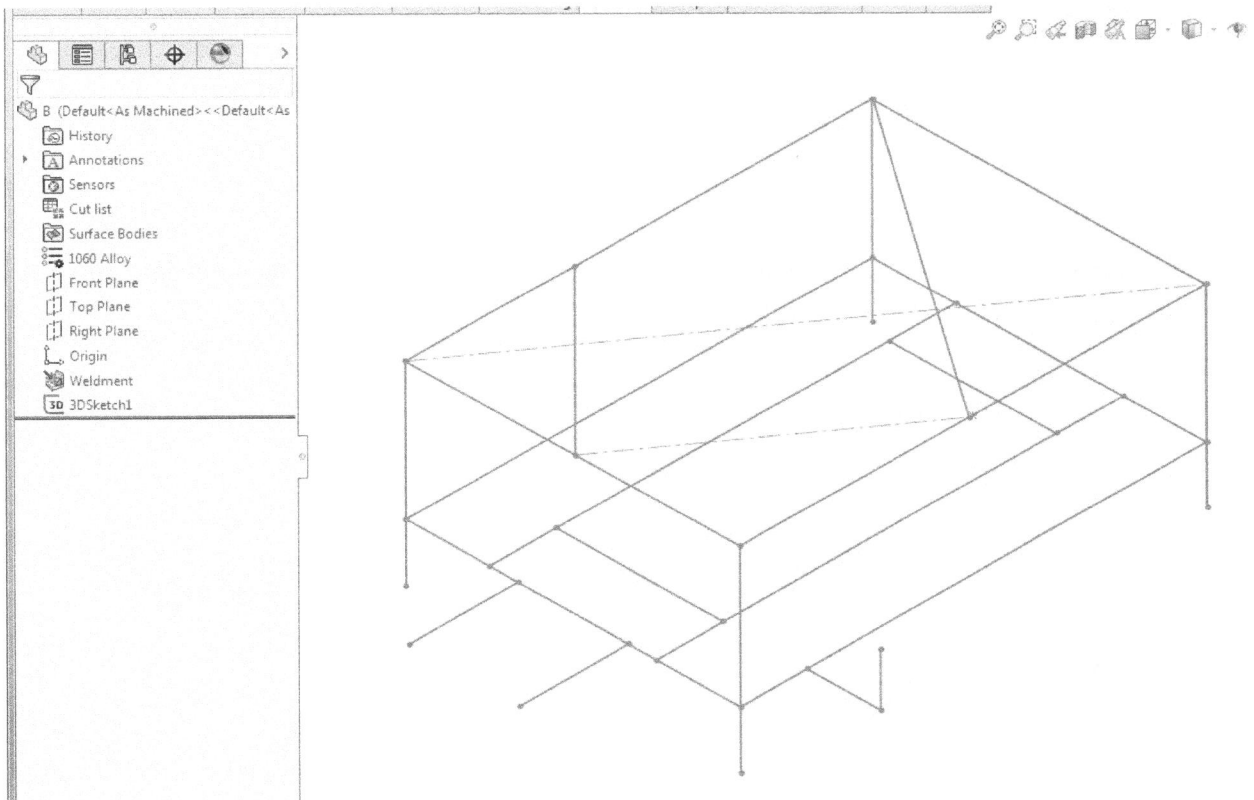

Figure 65 - Question 6 of 26 - Downloaded Part

ADDING A STRUCTURAL MEMBER

Click Structural Member under the Weldments toolbar or Insert > Weldments > Structural Member. Make selections in the PropertyManager to define the profile for the structural member as shown in the following image. Select Transfer Material from Profile to transfer the material from the profile to the weldment you are creating.

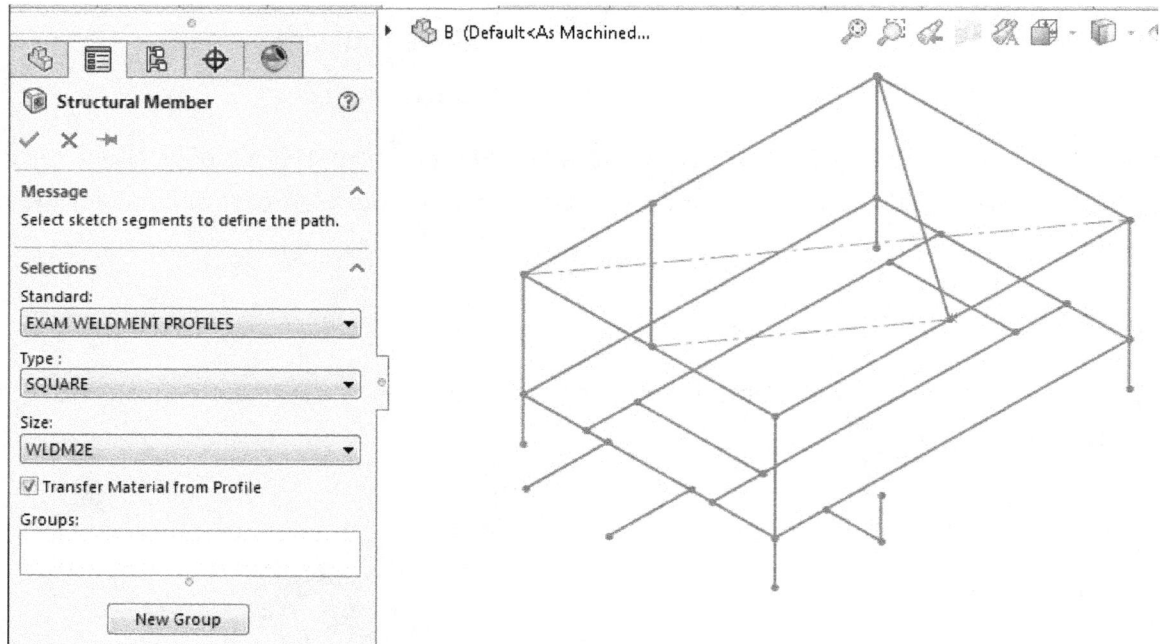

Figure 66 - Question 6 of 26 - Structural Member Property Manager

In the graphics area - select sketch segments shown in the following image to define the path for the structural member.

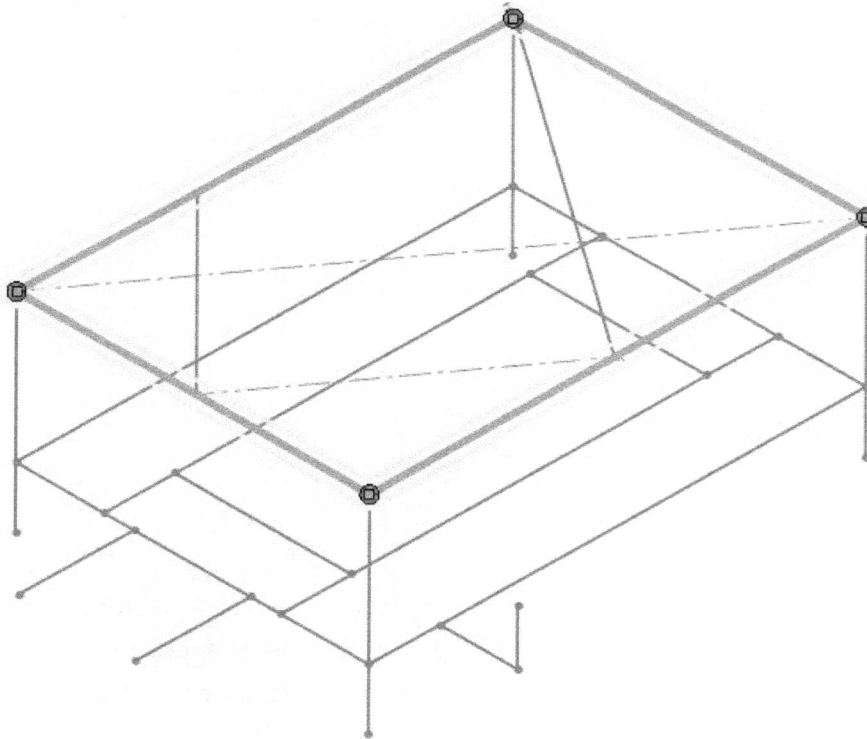

Figure 67 - Question 6 of 26 - Adding a Structural Member

Select Apply Corner Treatment and click on the End Butt1 option as shown in the following image.

Figure 68 - Question 6 of 26 - Corner Treatment

Click the New Group Command Button on the Structural Member Property Manager and select sketch segments shown in the following image to define the path for the structural member.

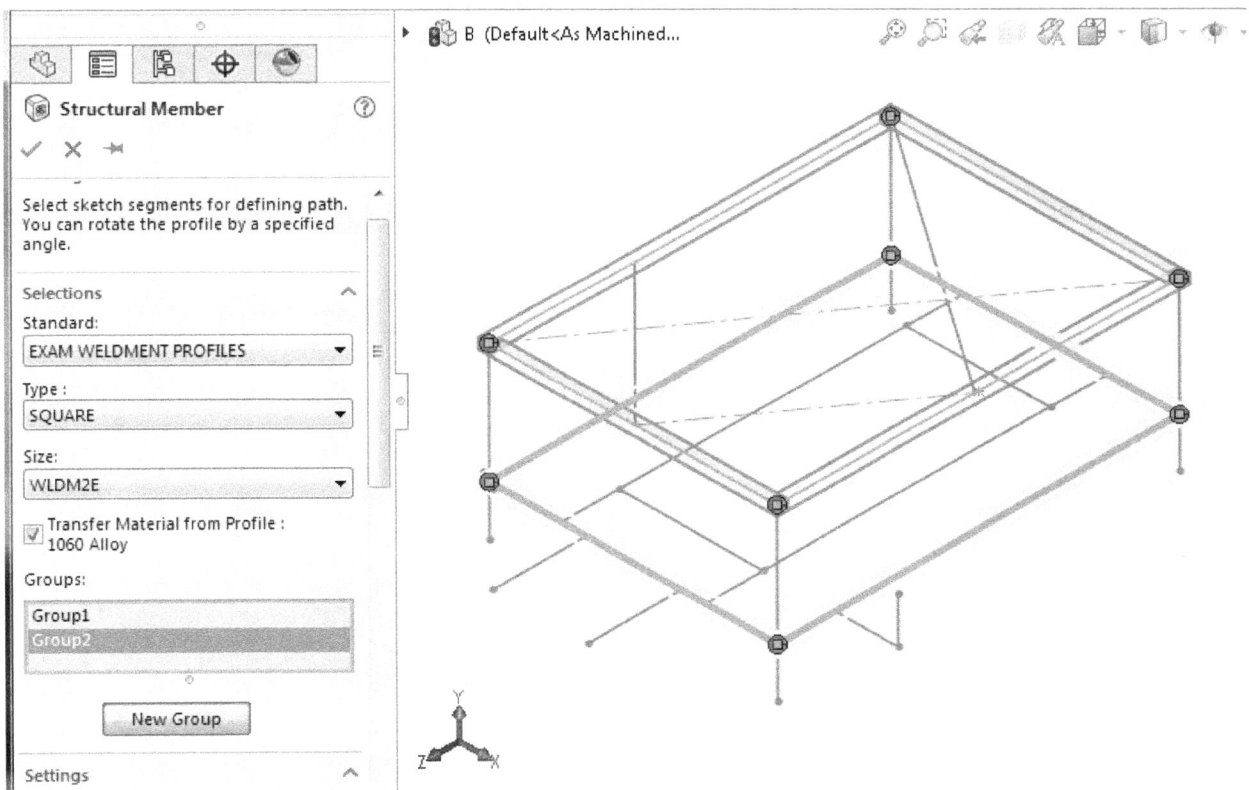

Figure 69 - Question 6 of 26 - Adding a Structural Member

Zoom in to each corner using the Zoom to Area tool on the View toolbar to see the mitered corners and make sure each is trimmed as per the image in this question. If not, you may click to select the point (purple or pink circle). The Corner Treatment dialog box appears, which lets you override the corner treatment for members of each group that meet at this corner - see the following image. Click Ok to close the Corner Treatment Dialog Box.

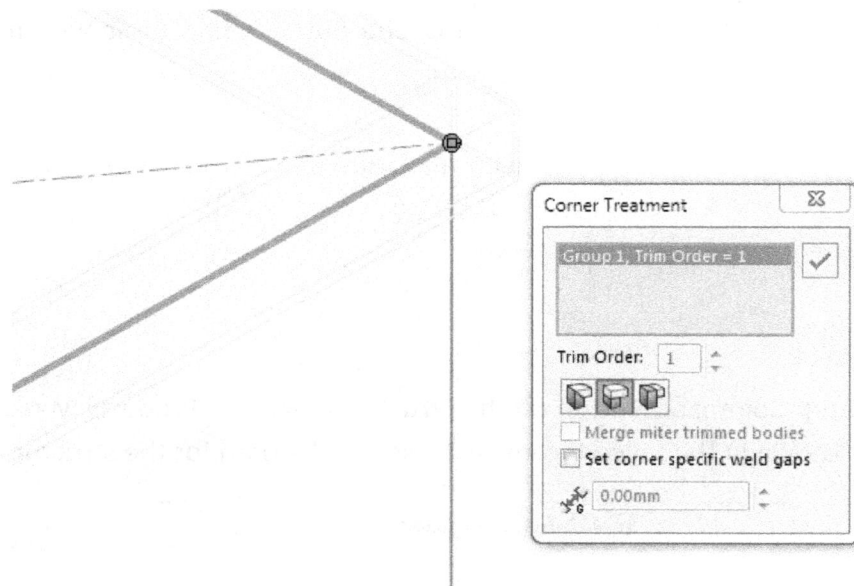

Figure 70 - Question 6 of 26 - Corner Treatment Override

In the PropertyManager, click ✓ (OK). Your Part should now appear as shown in the following image.

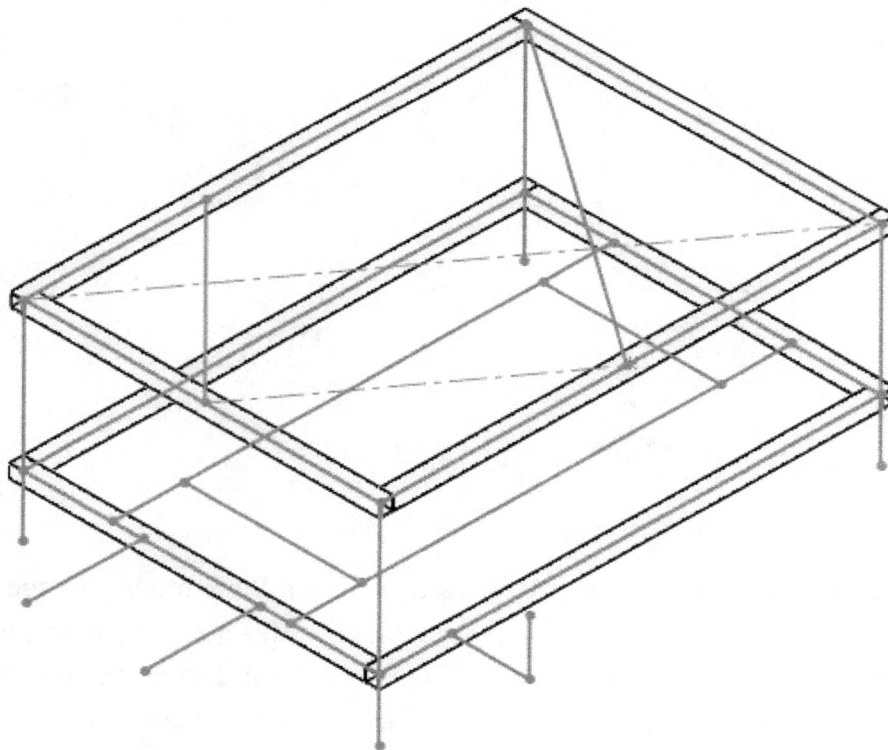

Figure 71 - Question 6 of 26 - Weldment Current Status

MASS PROPERTIES

Under the Evaluate Tab, click on the Mass Properties feature and the Mass Properties Menu appears showing that the mass of the parts is 11197.44 grams. **NB:** You can use different views to make sure your part is as required in this Question.

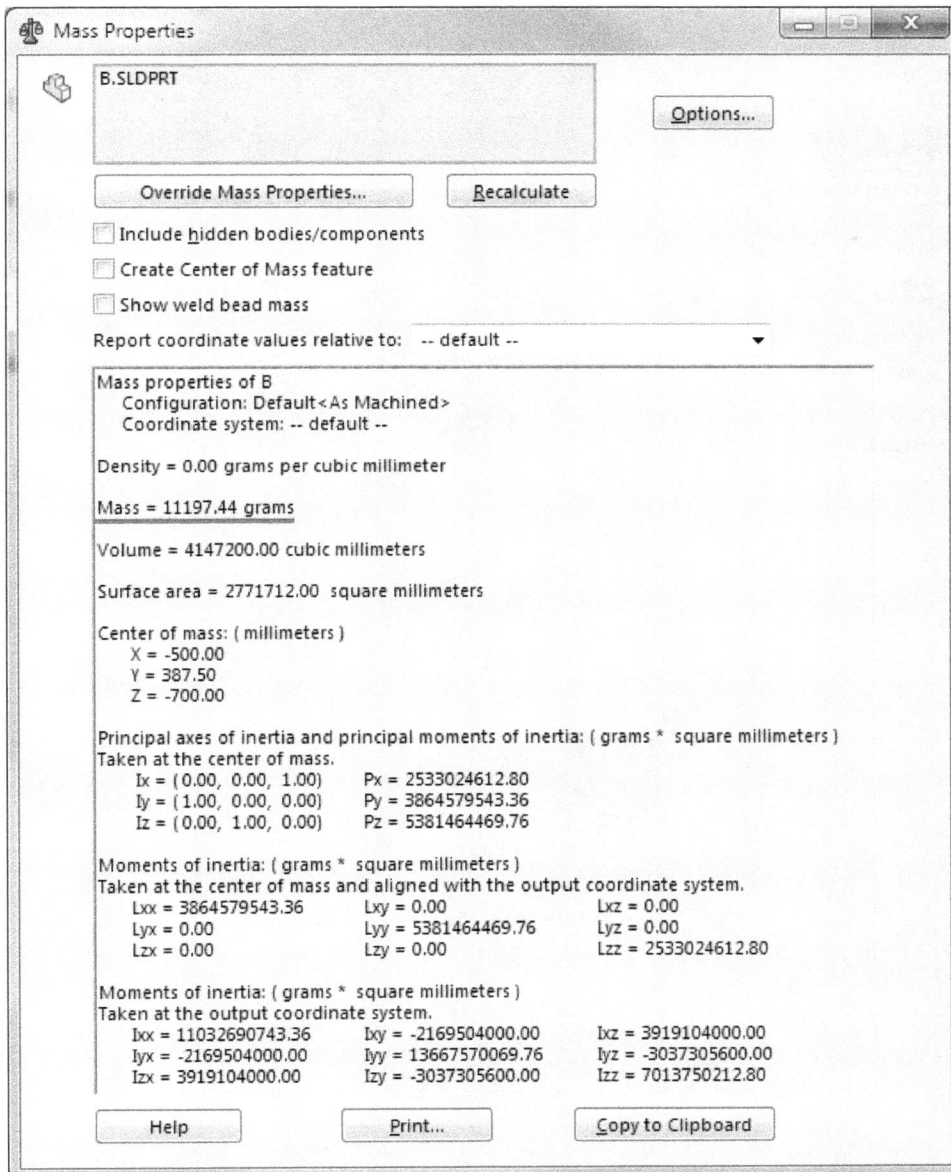

Figure 72 - Question 6 of 26 - Mass Properties

CHAPTER 7

QUESTION 7 EXAM SCREEN CAPTURES

QUESTION 7 EXAM SCREEN CAPTURE 1

Pro. Adv. - Advanced Weldments (CSWPA-WD)

Question 7 of 26

For 15 points:

B03005 - Weldment Part Modification
Build this weldment solid in SolidWorks.
Unit system: MMGS (millimeter, gram, second)
Decimal places: 2
Material: 1060 Alloy Aluminum
Density = 0.0027 g/mm^3

-Modify the top four segments so that the corner treatment for all four segments are changed to "End Miter"

-Add a 3mm gap between these same four segments.

-Select the piece indicated in the image.

Note: This is one of the long top segments.

-Measure the mass of this single segment.

What is the mass of the selected segment (grams)?

AA

Enter Value:

(use . (point) as decimal separator)

Figure 73 - Question 7 of 26 - Exam Screen Capture 1

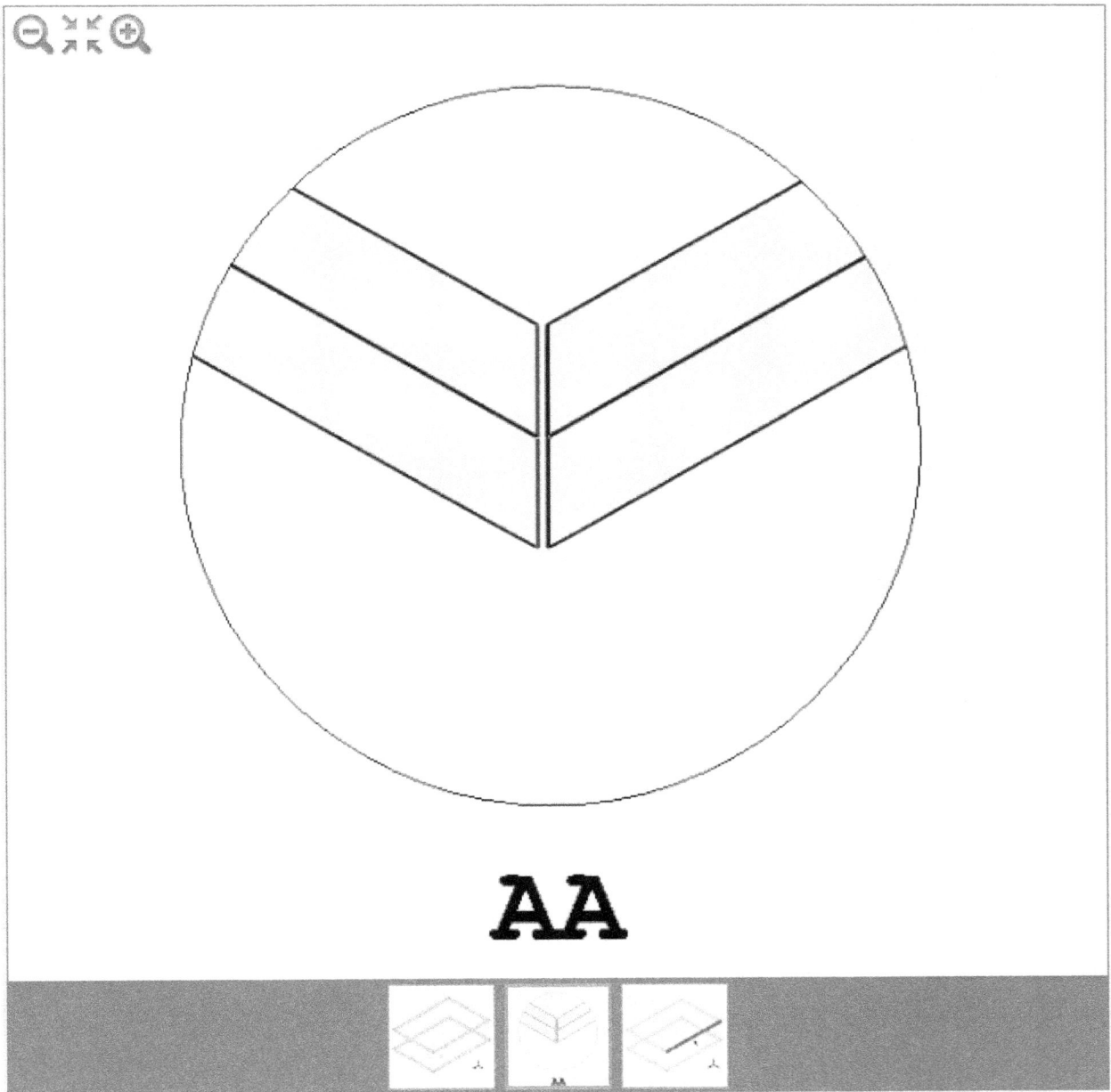

AA

Figure 74 - Question 7 of 26 - Exam Screen Capture 2

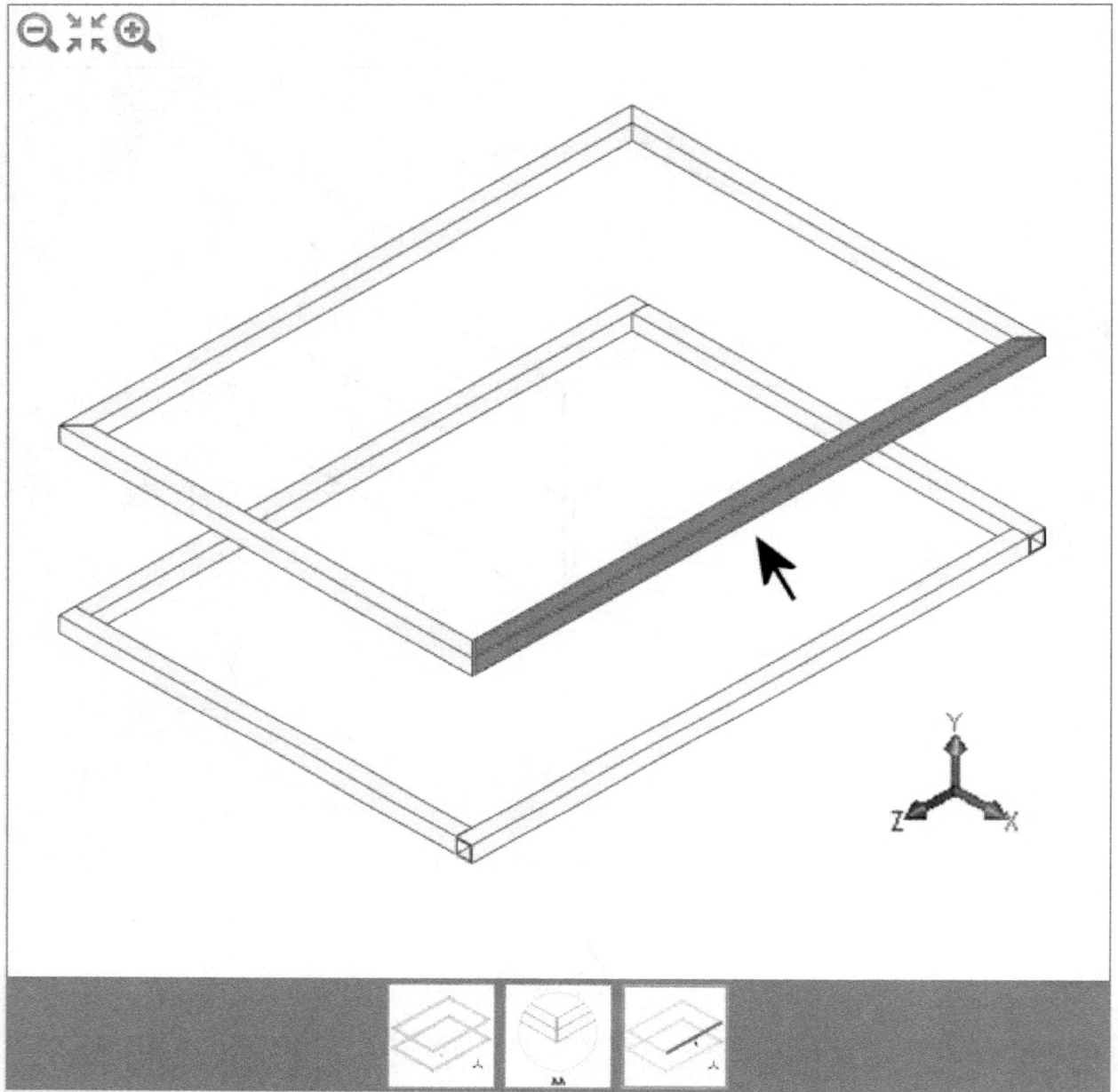

Figure 75 - Question 7 of 26 - Exam Screen Capture 3

QUESTION 7 SOLUTION

EDITING A STRUCTURAL MEMBER

Question 7 is a continuation from Question 6, hence we continue to work on the same part - B.SLDPRT.

In the Feature Manager Design Tree, right click on the Structural Member (*Square WLDM2E(1)*) ▸ 🗖 Square WLDM2E(1)   and select Edit Feature as shown in the following image.

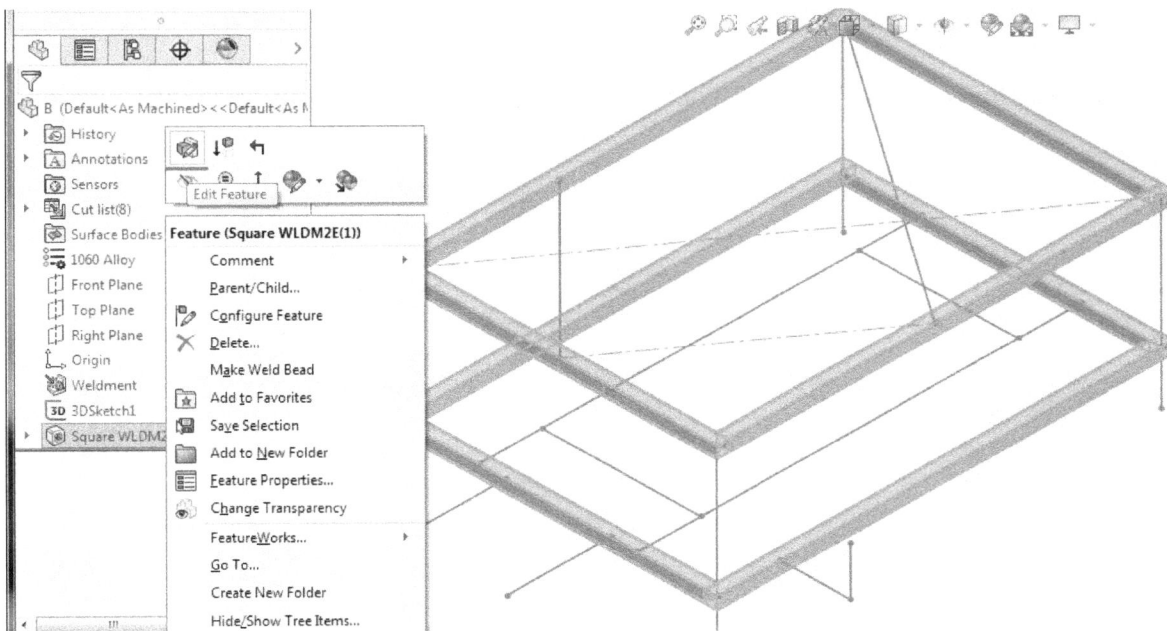

Figure 76 - Question 7 of 26 - Editing a Structural Member

Select Group1 under Groups in the Property Manager and under Apply corner treatment, select the End Miter option. Under the Gap between connected segments in the same group text area, type 3.00mm as shown in the following image.

61

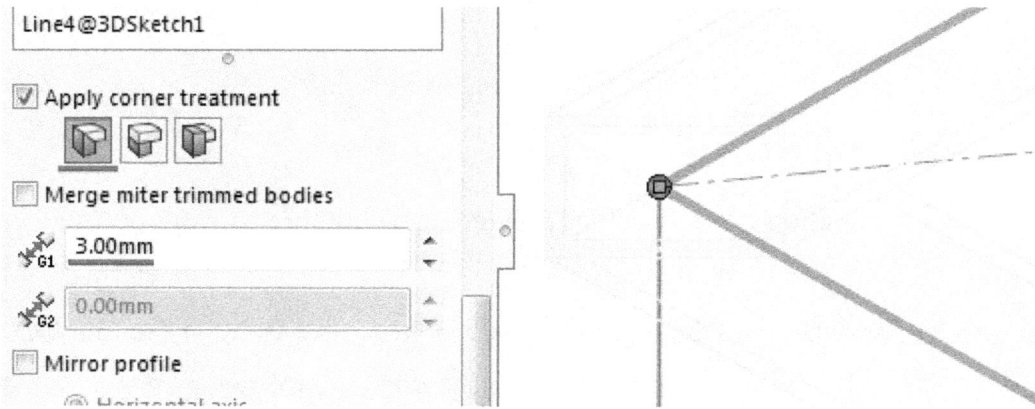

Figure 77 - Question 7 of 26 - Changing a group's corner treatment and adding a gap between connected segments

In the PropertyManager, click ✓ (OK). Your Part should now appear as shown in the following image.

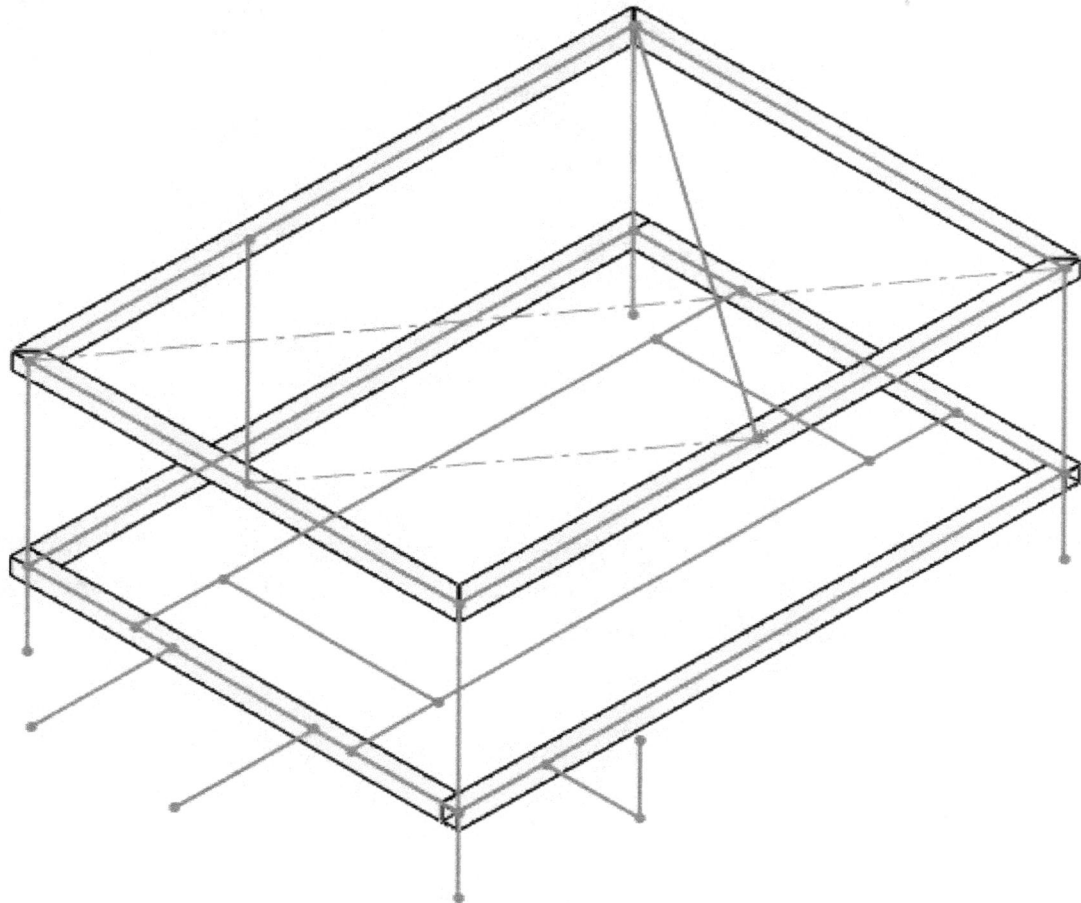

Figure 78 - Question 7 of 26 - Weldment Current Status

Click Rebuild and save the part.

MEASURING THE MASS OF A SINGLE SEGMENT IN A WELDMENT

In the Feature Manager Design Tree, right click on one of the long top segments and select Isolate as shown in the following image.

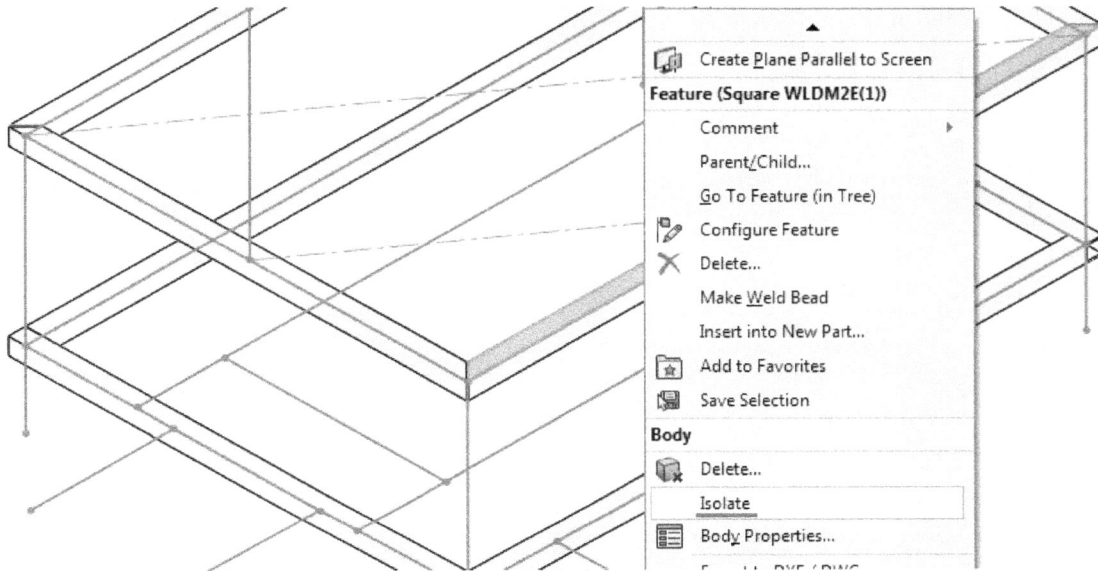

Figure 79 - Question 7 of 26 - Isolating a single segment

Your part should now look as shown in the following image - all the other segments are hidden.

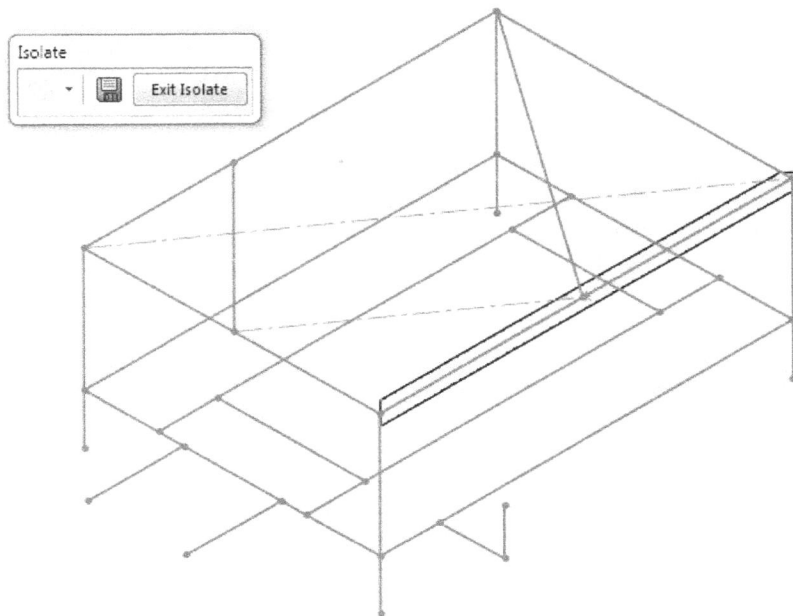

Figure 80 - Question 7 of 26 - Isolating a single segment

Under the Evaluate Tab, click on the Mass Properties feature and the Mass Properties Menu appears showing that the mass of the segment is 1628.01 grams. **NB:** Make sure the Include hidden bodies/components checkbox *(UNDERLINED IN BLUE IN THE FOLLOWING IMAGE)* is not selected - selecting it would give the mass of the whole weldment including hidden segments.

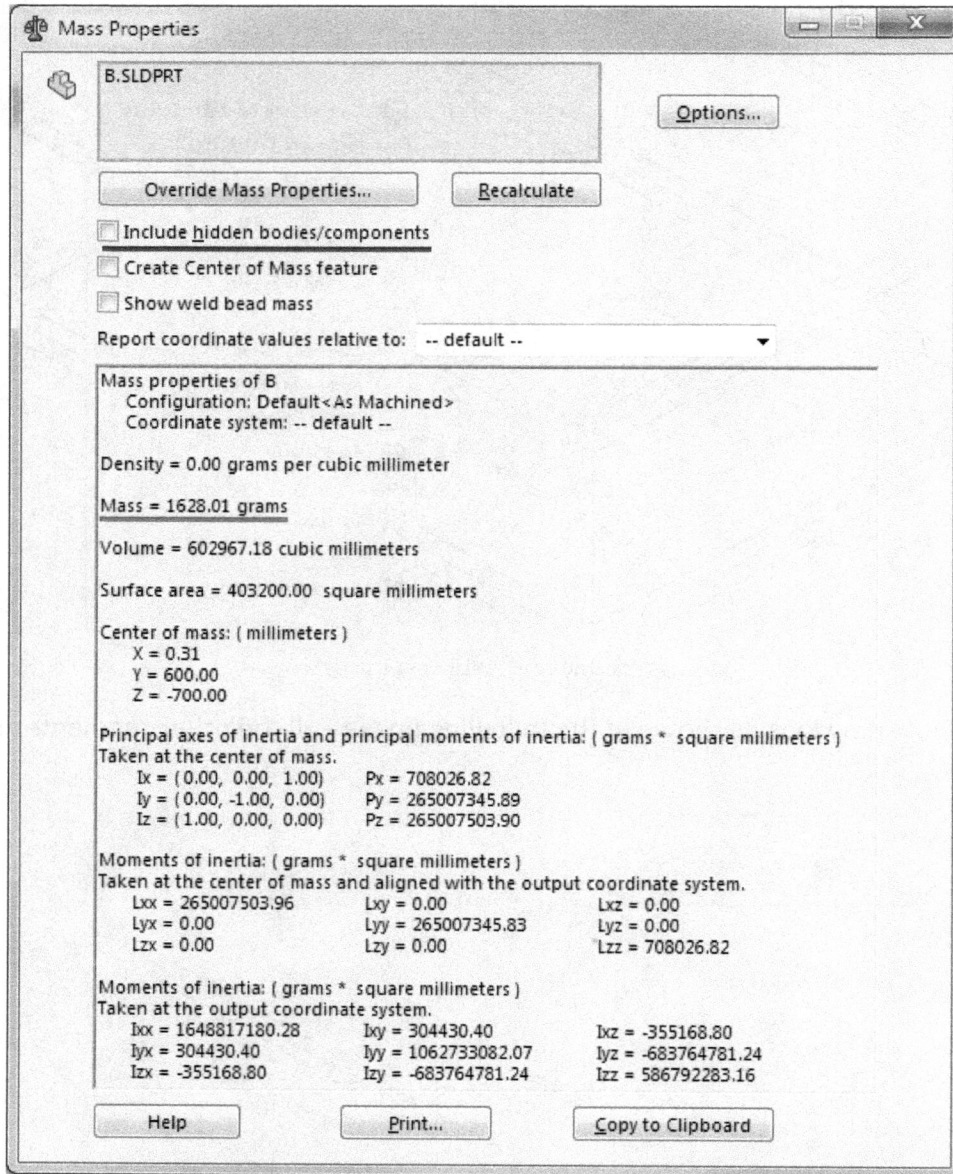

Figure 81 - Question 7 of 26 - Mass of a single segment

Close the Mass Properties Dialog Box and Click Exit Isolate. Save your part.

CHAPTER 8

QUESTION 8 EXAM SCREEN CAPTURES

QUESTION 8 EXAM SCREEN CAPTURE 1

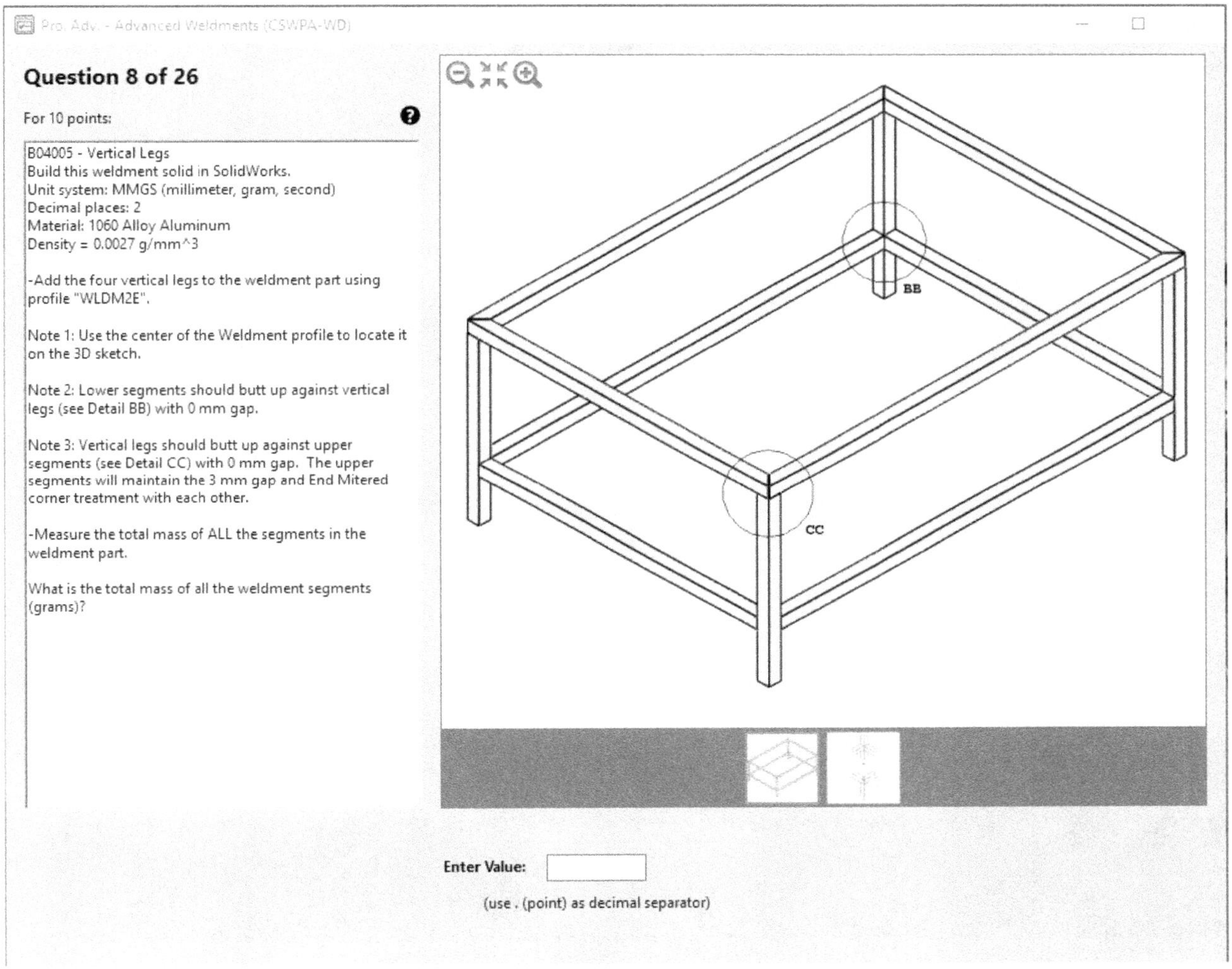

Figure 82 - Question 8 of 26- Exam Screen Capture 1

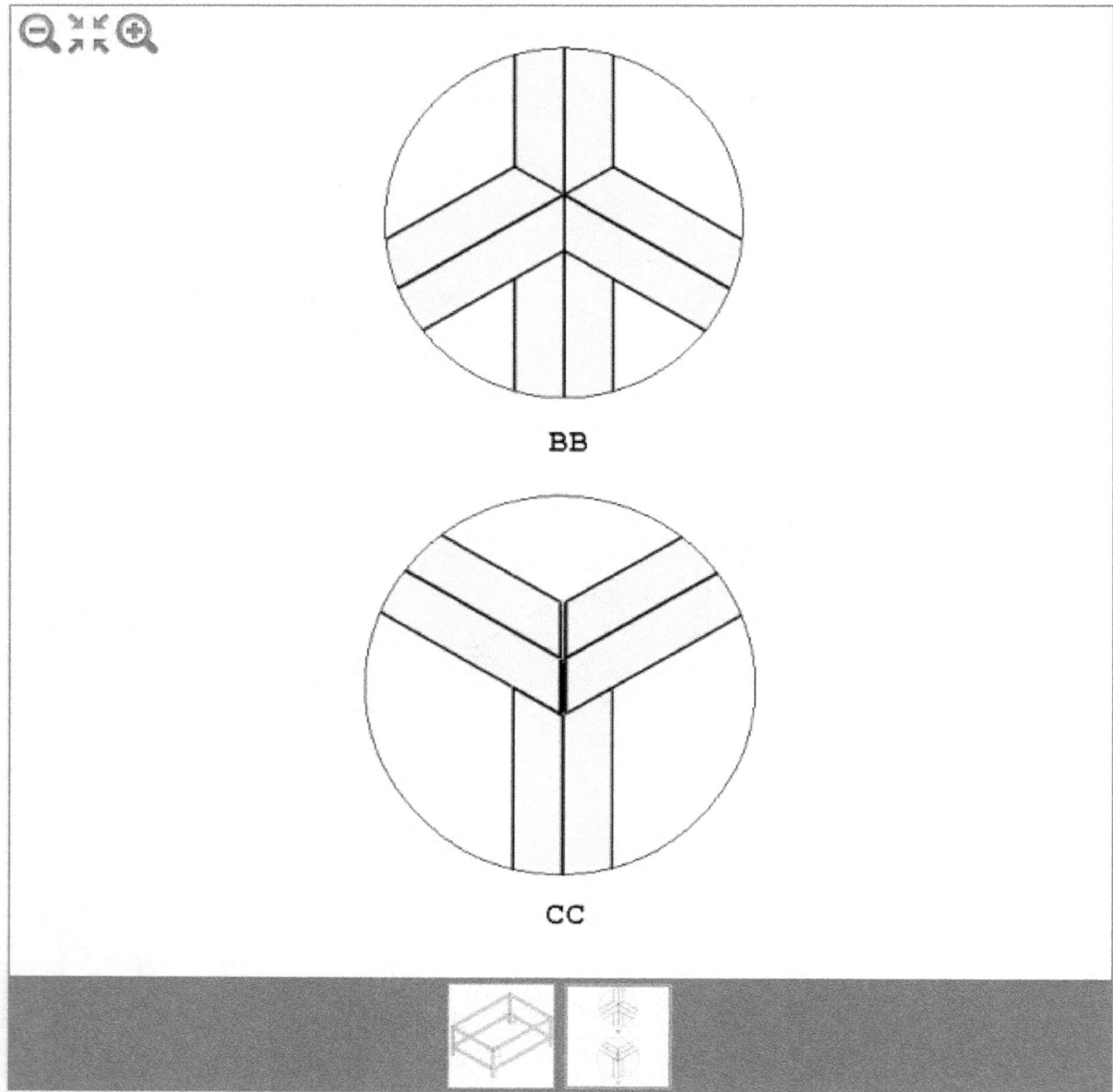

BB

CC

Figure 83 - Question 8 of 26 - Exam Screen Capture 2

ADDING STRUCTURAL MEMBERS

Question 8 is a continuation from Question 7.

In the Feature Manager Design Tree, right click on the Structural Member (*Square WLDM2E(1)*)

▶ 🔲 Square WLDM2E(1) and select Edit Feature as shown in the following image.

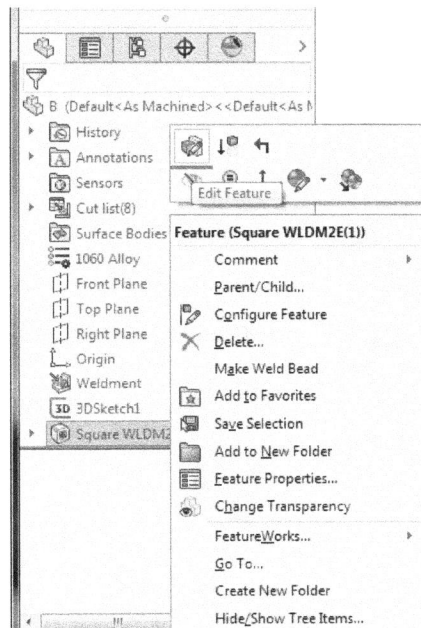

Figure 84 - Question 8 of 26 - Editing a structural member

Click on the New Group command button in the Property Manager and then select sketch entities shown in the following image.

67

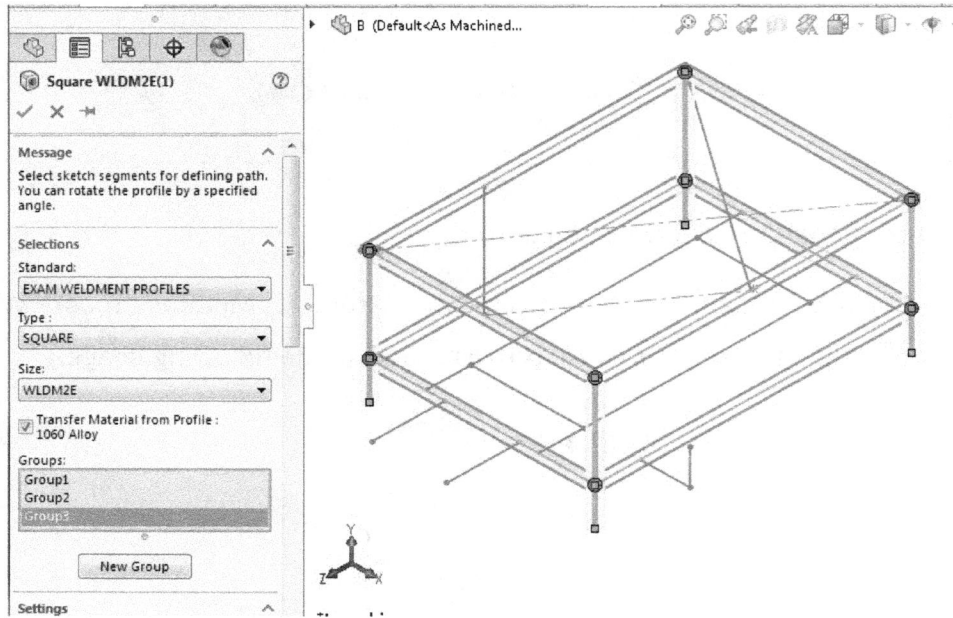

Figure 85 - Question 8 of 26 - Adding a New Group

In the PropertyManager, click ✔ (OK). Your Part should now appear as shown in the following image.

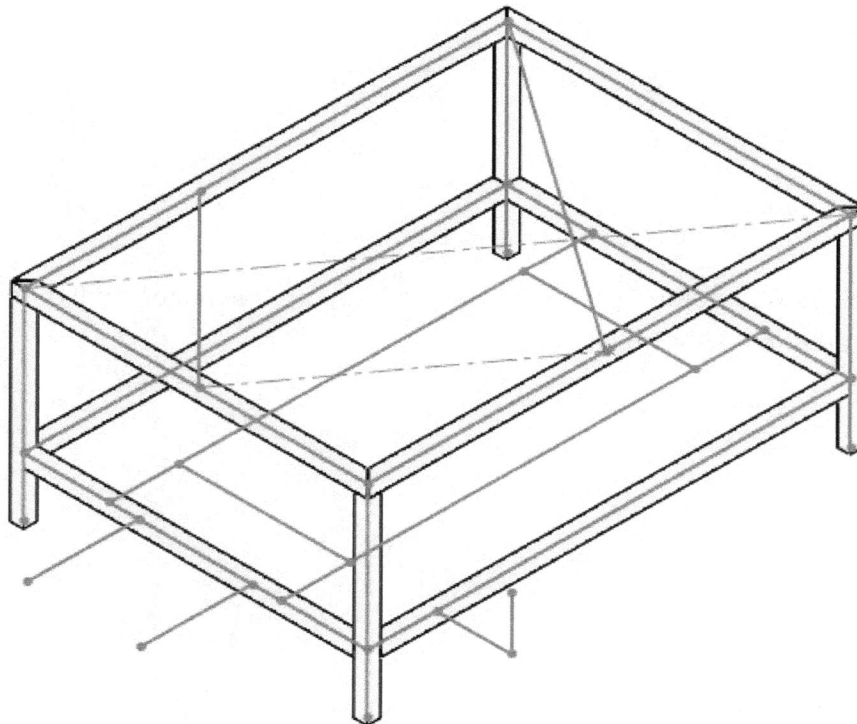

Figure 86 - Question 8 of 26 - Weldment Current Status

MASS PROPERTIES

Under the Evaluate Tab, click on the Mass Properties feature or Tools > Evaluate > Mass Properties. The total mass of all the weldment segments is 13704.07 grams as shown in image below.

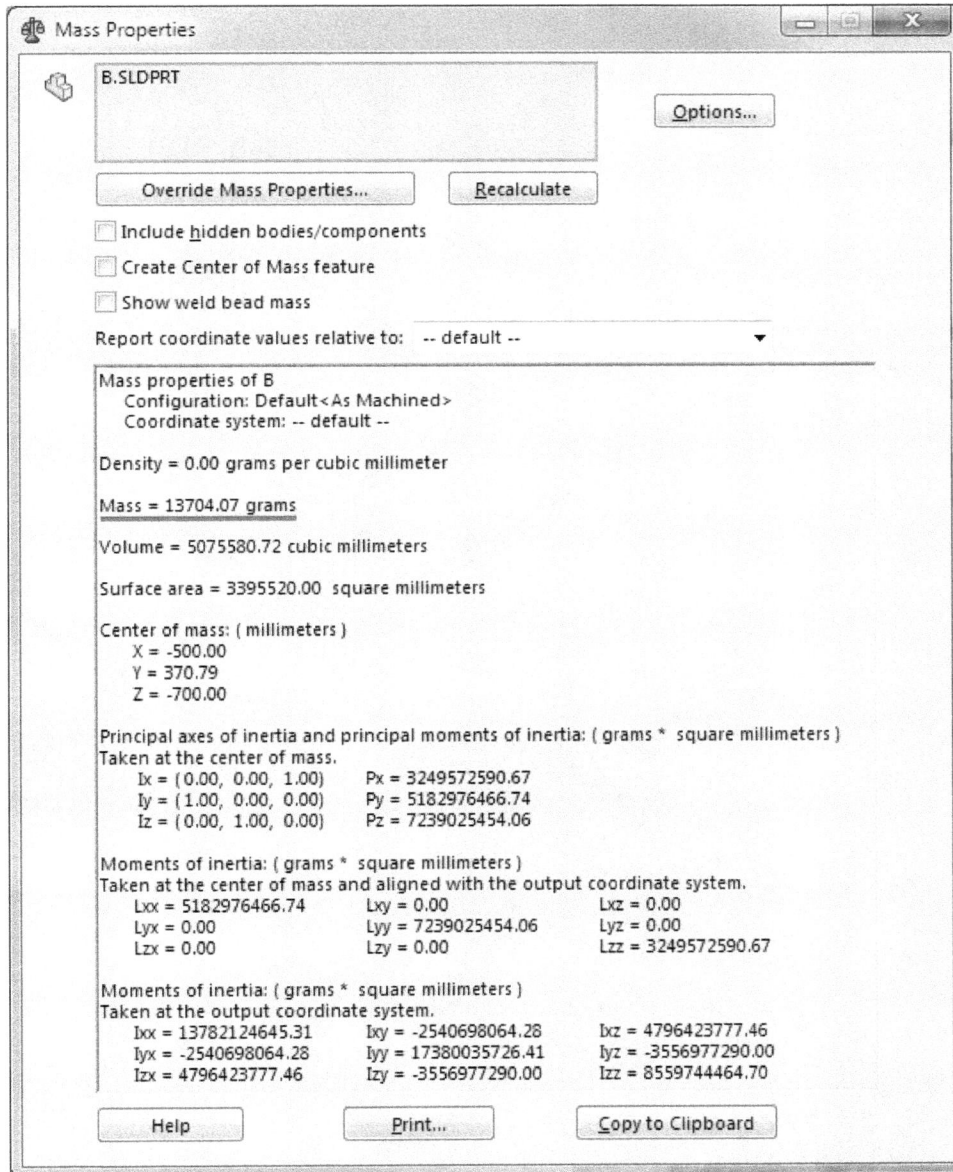

Figure 87 - Question 8 of 26 - Mass Properties

CHAPTER 9

QUESTION 9 EXAM SCREEN CAPTURES

QUESTION 9 EXAM SCREEN CAPTURE 1

Pro. Adv. - Advanced Weldments (CSWPA-WD)

Question 9 of 26

For 10 points:

B05005 - 3 Segment Corner Miter
Build this weldment solid in SolidWorks.
Unit system: MMGS (millimeter, gram, second)
Decimal places: 2
Material: 1060 Alloy Aluminum
Density = 0.0027 g/mm^3

-Modify the corner indicated by Detail DD so that all three segments are mitered equally.

-Add a 3mm gap between all three segments at this corner.

-Select the one vertical leg at this corner and measure its mass.

What is the mass of the selected vertical leg (grams)?

DD

Enter Value:

(use . (point) as decimal separator)

Figure 88 - Question 9 of 26 - Exam Screen Capture 1

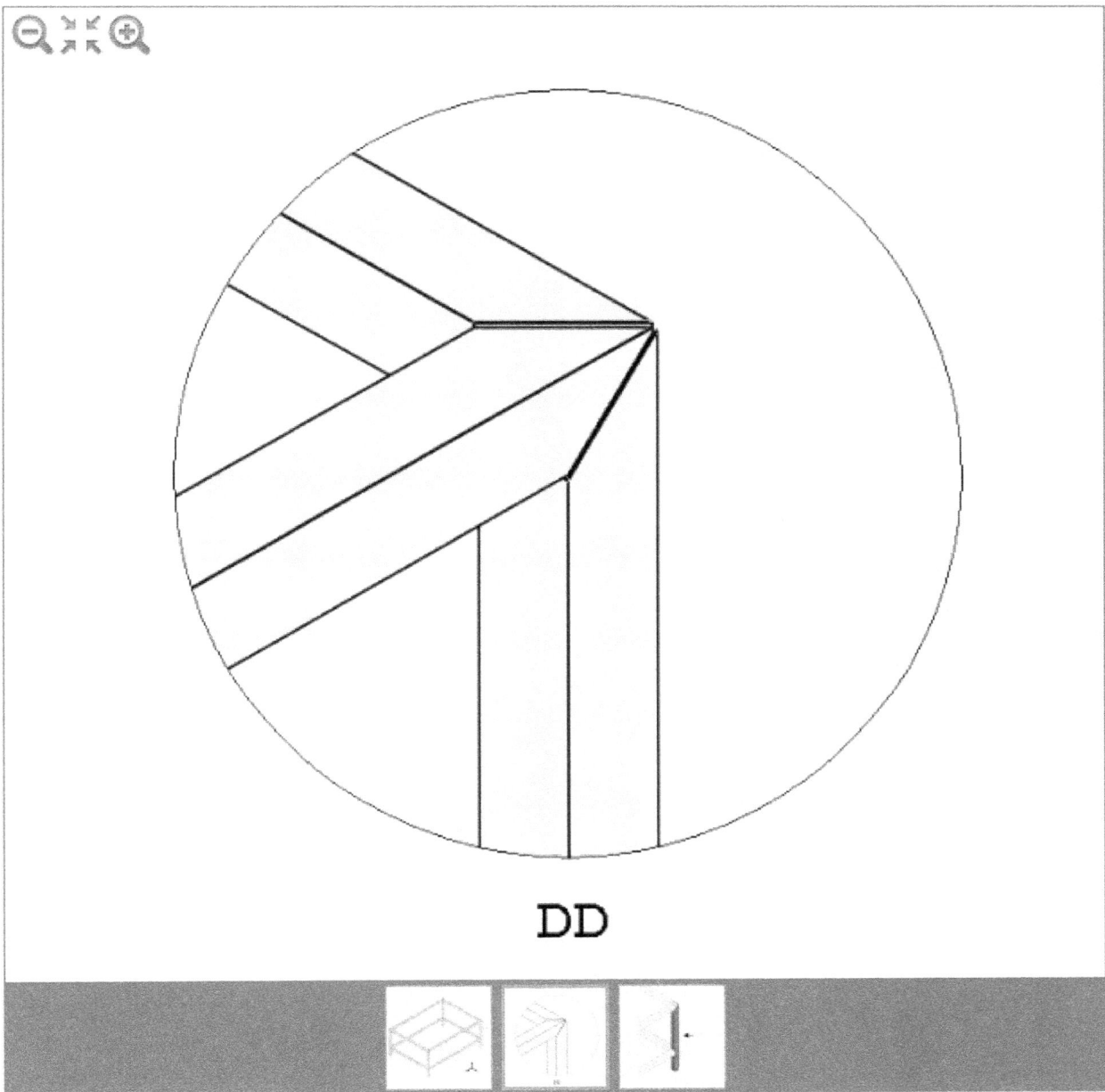

DD

Figure 89 - Question 9 of 26 - Exam Screen Capture 2

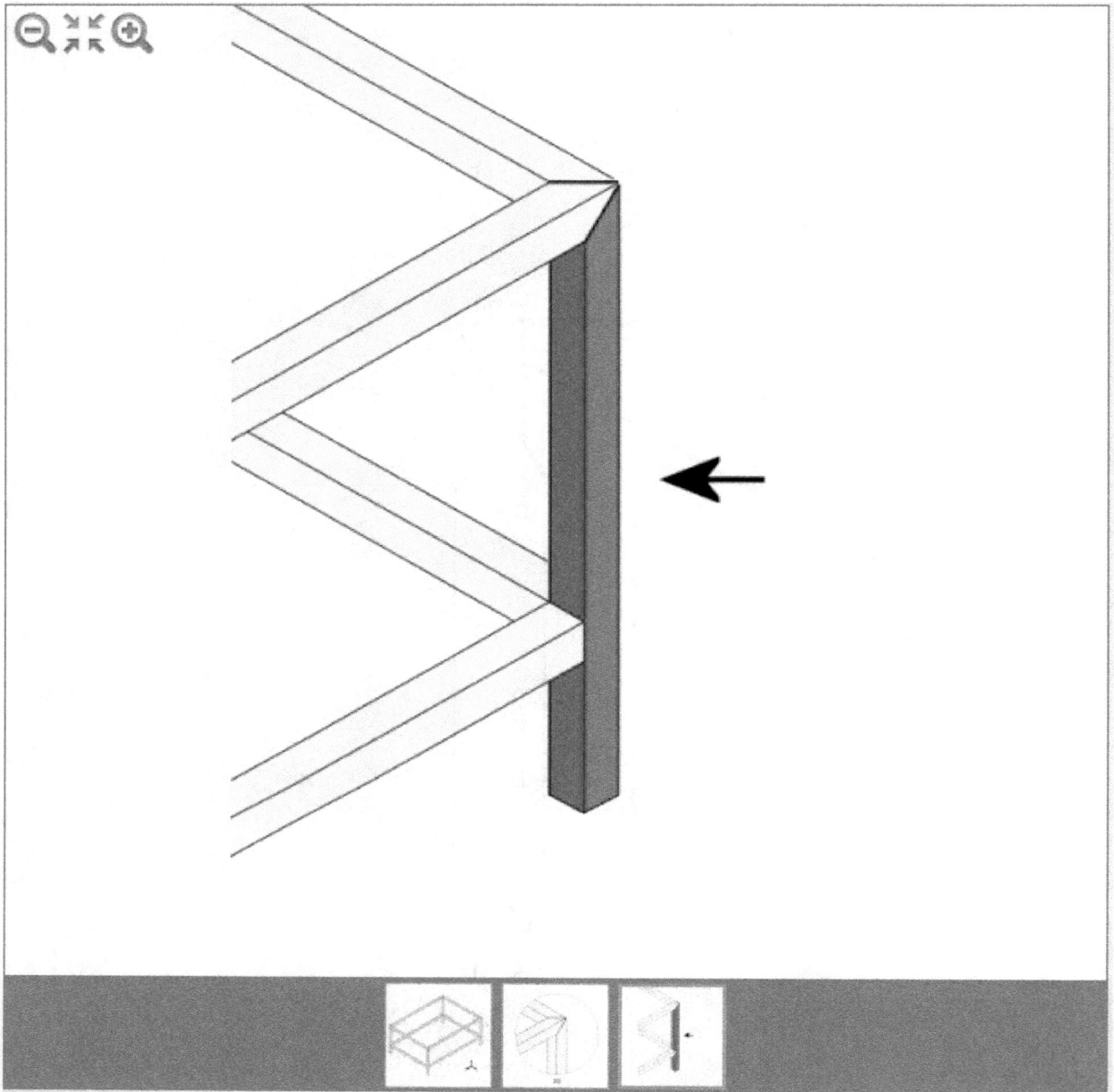

Figure 90 - Question 9 of 26 - Exam Screen Capture 3

72

CHANGING A CORNER TREATMENT

Question 9 is a continuation from Question 8. In the Feature Manager Design Tree, right click on the Structural Member (*Square WLDM2E(1)*) ▸ 🔲 Square WLDM2E(1) and select Edit Feature as shown in the following image.

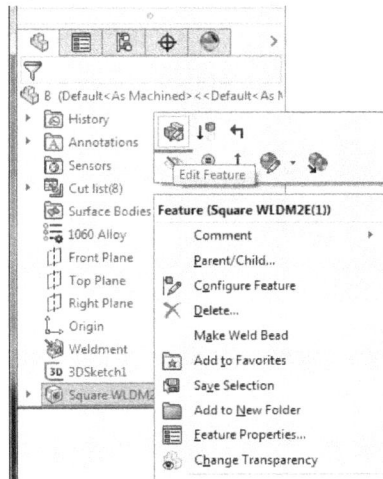

Figure 91 - Question 9 of 26 - Editing a structural member

Click on the top right corner indicated by Detail DD in Exam Screen Capture 1 which is the corner we want to modify - see the following image.

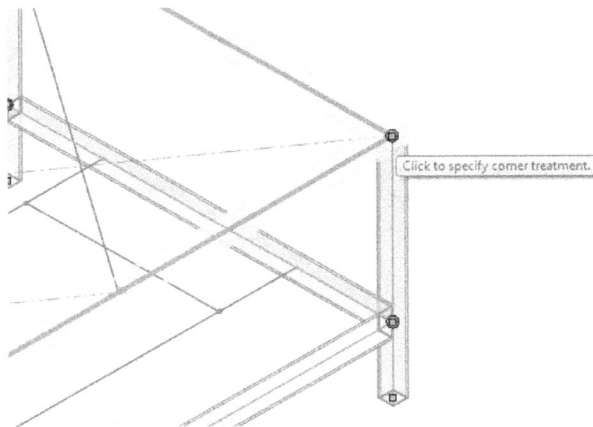

Figure 92 - Question 9 of 26 - Changing a Corner Treatment

In the Corner Treatment dialog box, work with the highlighted group, Group 1. Under Trim Order, change the value from 1 to 2 and select the Set corner specific weld gaps checkbox then enter a value of 3mm as shown in the following image.

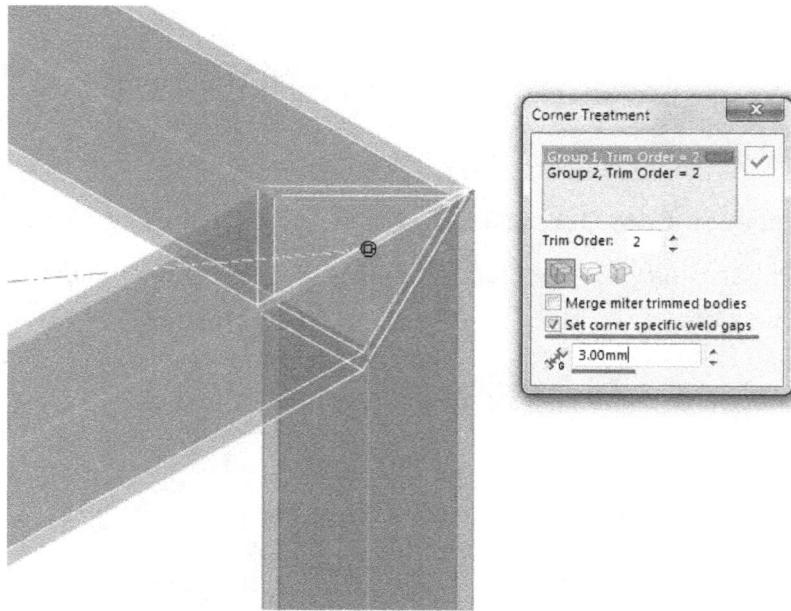

Figure 93 - Question 9 of 26 - Changing a Corner Treatment

Click on Group 2 and select the Set corner specific weld gaps checkbox then enter a value of 3mm as shown in the following image.

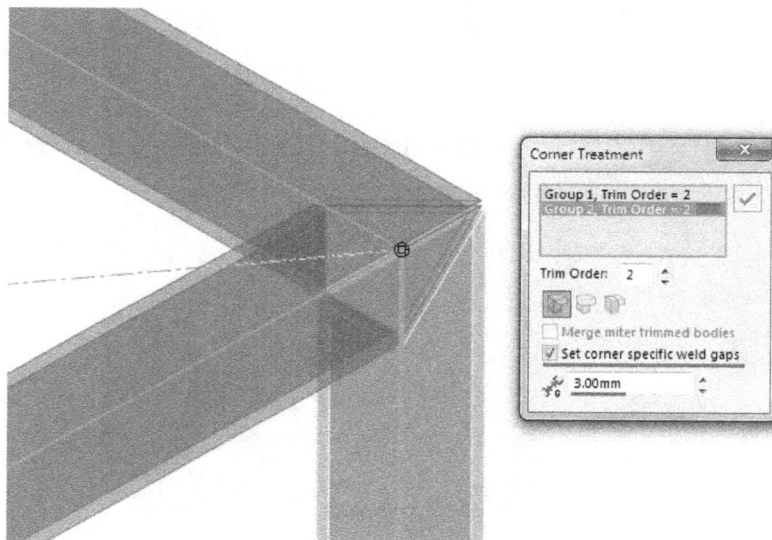

Figure 94 - Question 9 of 26 - Changing a Corner Treatment

Click Ok on the Corner Treatment Dialog Box then Click Ok on the Property Manager. Your part should now appear as shown in the following image.

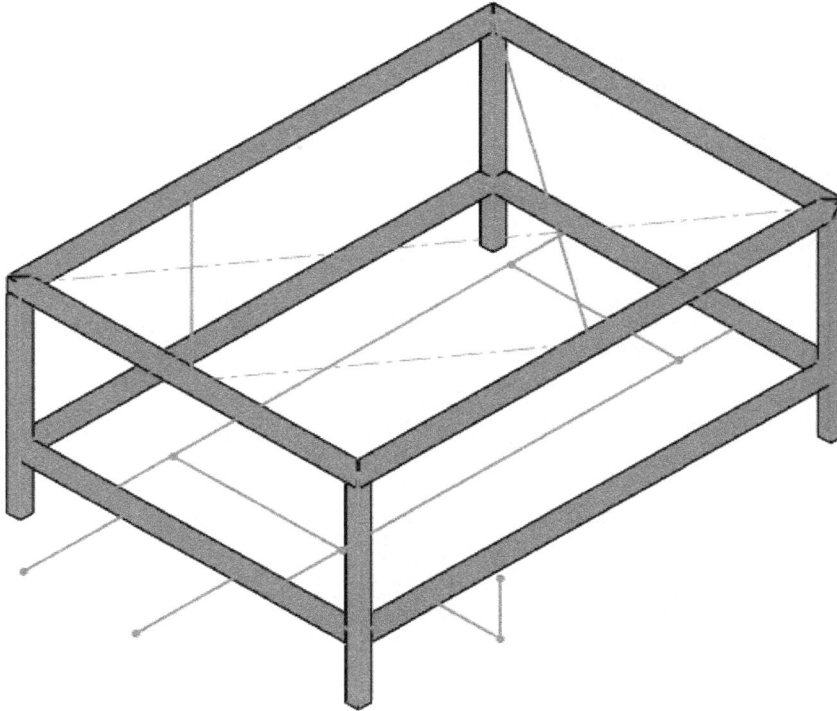

Figure 95 - Question 9 of 26 - Weldment current status

NB: Use the measure tool to double check the gaps between all three segments at this corner and make sure they are 3mm all-round - see the following image.

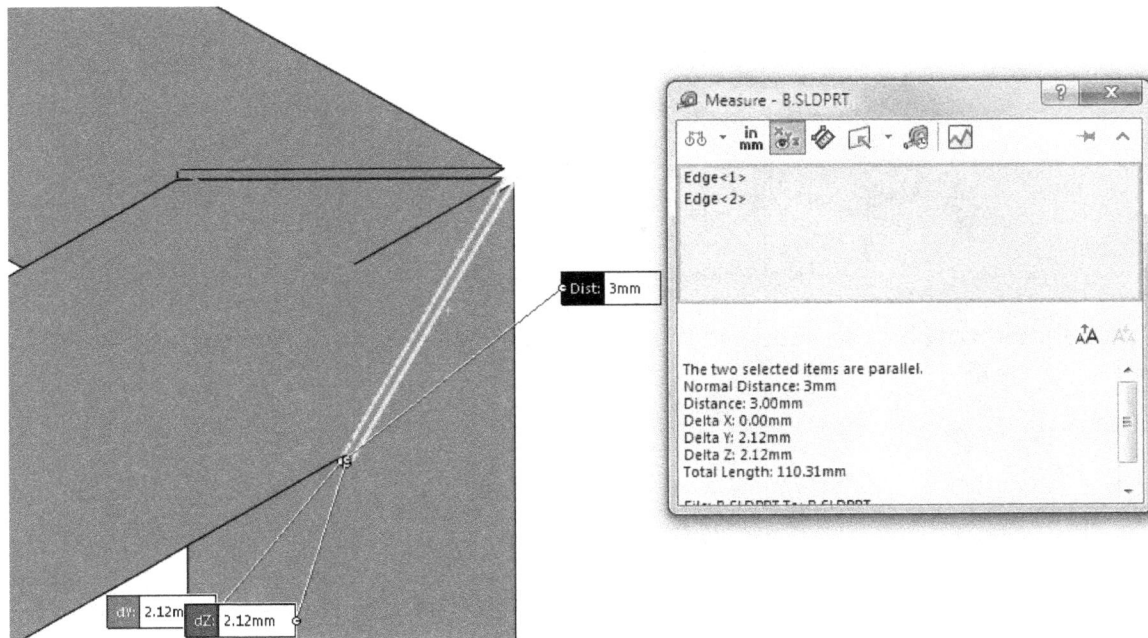

Figure 96 - Question 9 of 26 - Checking gaps between all 3 segments using the Measure Tool

MEASURING THE MASS OF A SINGLE SEGMENT IN A WELDMENT

In the Feature Manager Design Tree, right click on the vertical leg at the corner shown in Exam Screen Capture 3 and select Isolate as shown in the following image.

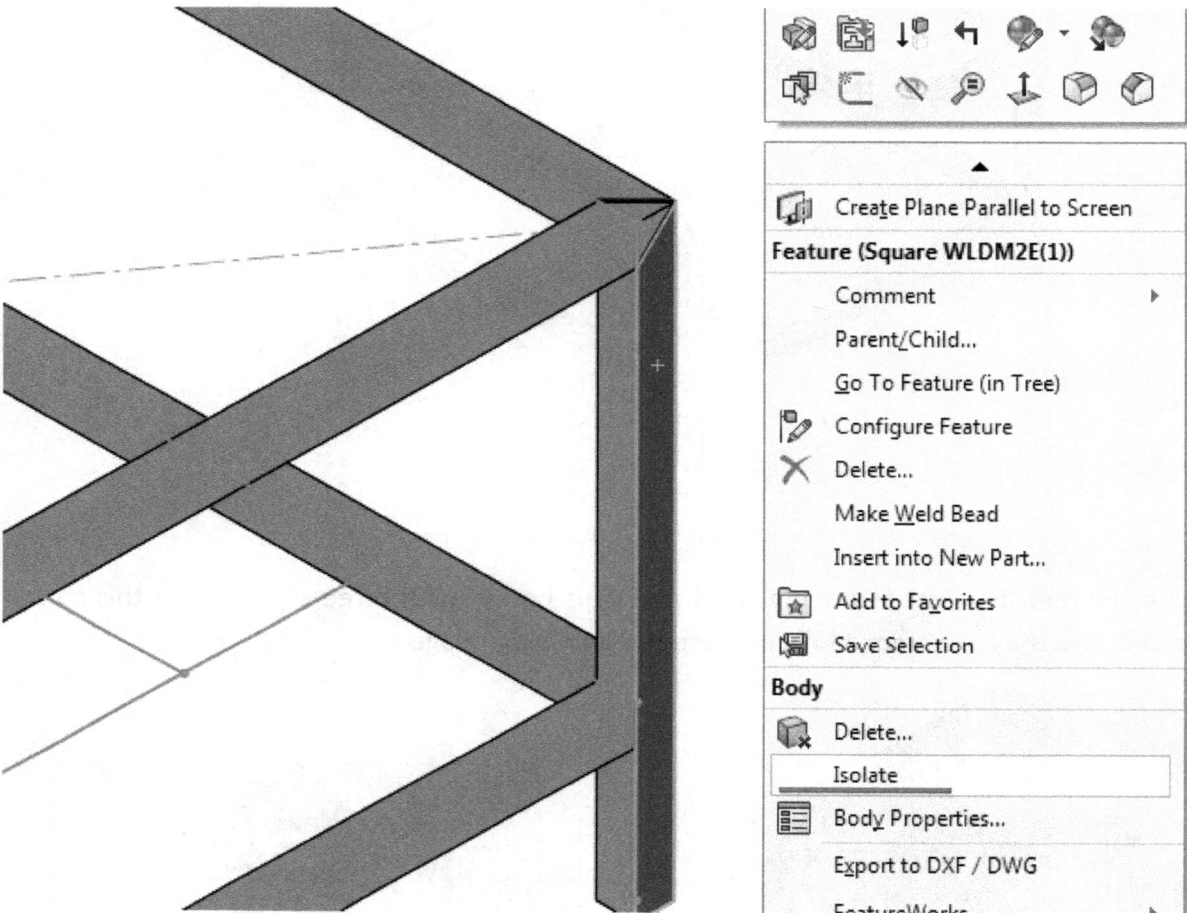

Figure 97 - Question 9 of 26 - Isolating a single segment

Your part should now look as shown in the following image - all the other segments are hidden.

76

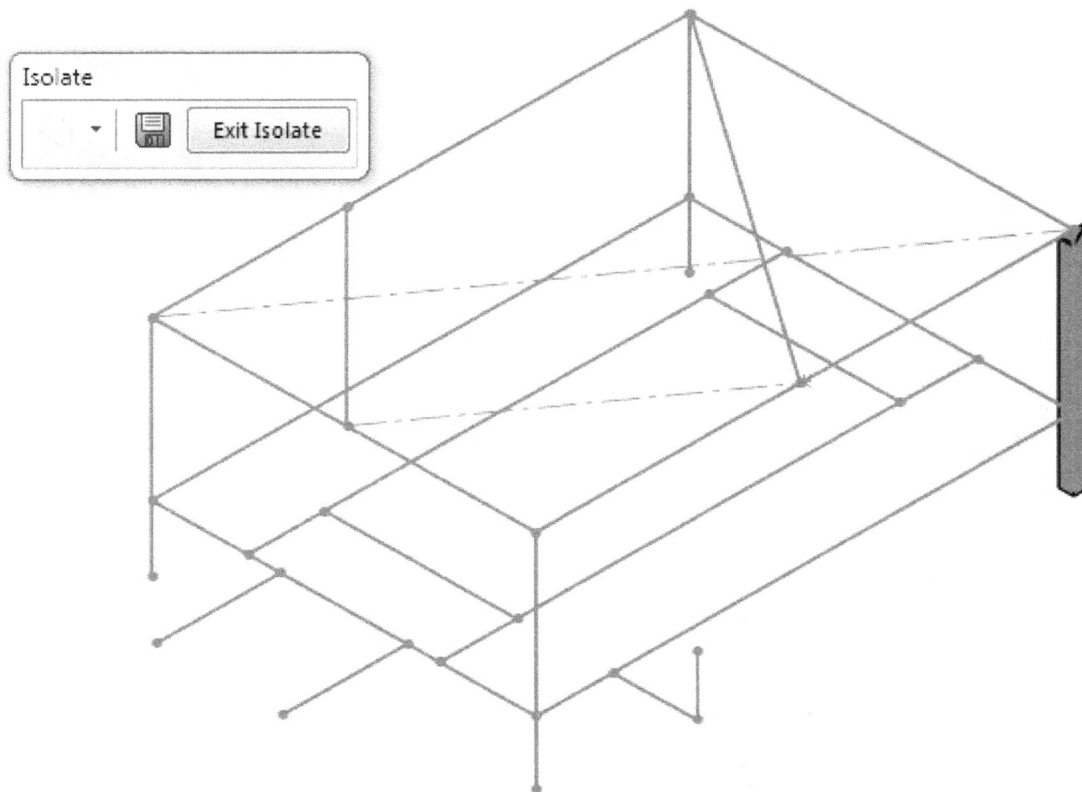

Figure 98 - Question 9 of 26 - Isolating a single segment

Under the Evaluate Tab, click on the Mass Properties feature and the Mass Properties Menu appears showing that the mass of the segment is 686.84 grams. **NB:** Make sure the Include hidden bodies/components checkbox is not selected - selecting it would give the mass of the whole weldment including hidden segments.

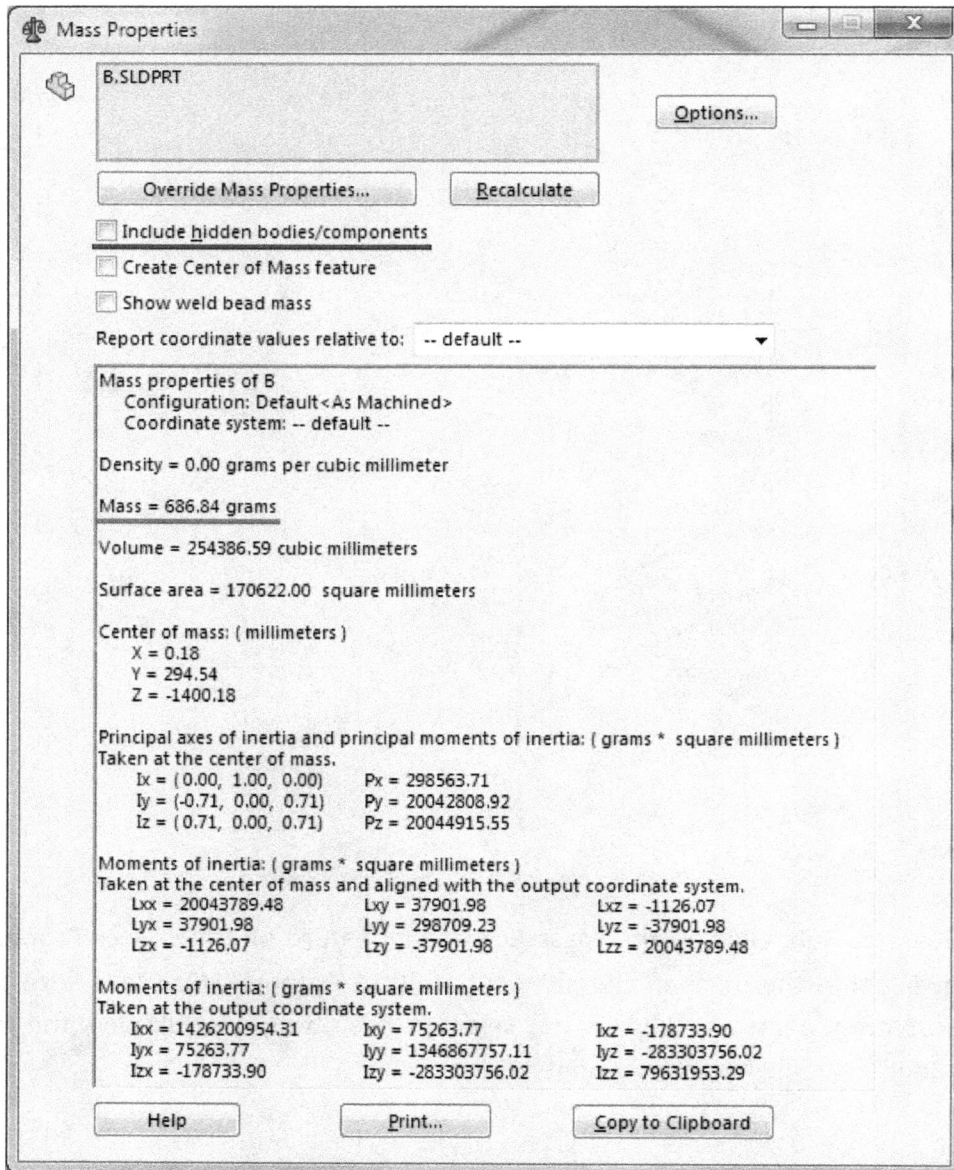

Figure 99 - Question 9 of 26 - Mass of a single segment

Close the Mass Properties Dialog Box and Click Exit Isolate. Save your part.

CHAPTER 10

QUESTION 10 EXAM SCREEN CAPTURES

QUESTION 10 EXAM SCREEN CAPTURE 1

Question 10 of 26

For 10 points:

B06005 - Change Corner From Miter To Butted
Build this weldment solid in SolidWorks.
Unit system: MMGS (millimeter, gram, second)
Decimal places: 2
Material: 1060 Alloy Aluminum
Density = 0.0027 g/mm^3

-Modify the corner indicated by Detail EE so that the top of the vertical segment is coplanar with the top of the other two segments.

Note: The other two segments should butt up to the vertical segment.

-Add a 3mm gap between these three segments.

-Select the one vertical segment at this corner and measure its mass.

What is the mass of the selected vertical leg (grams)?

Enter Value:

(use . (point) as decimal separator)

Figure 100 - Question 10 of 26 - Exam Screen Capture 1

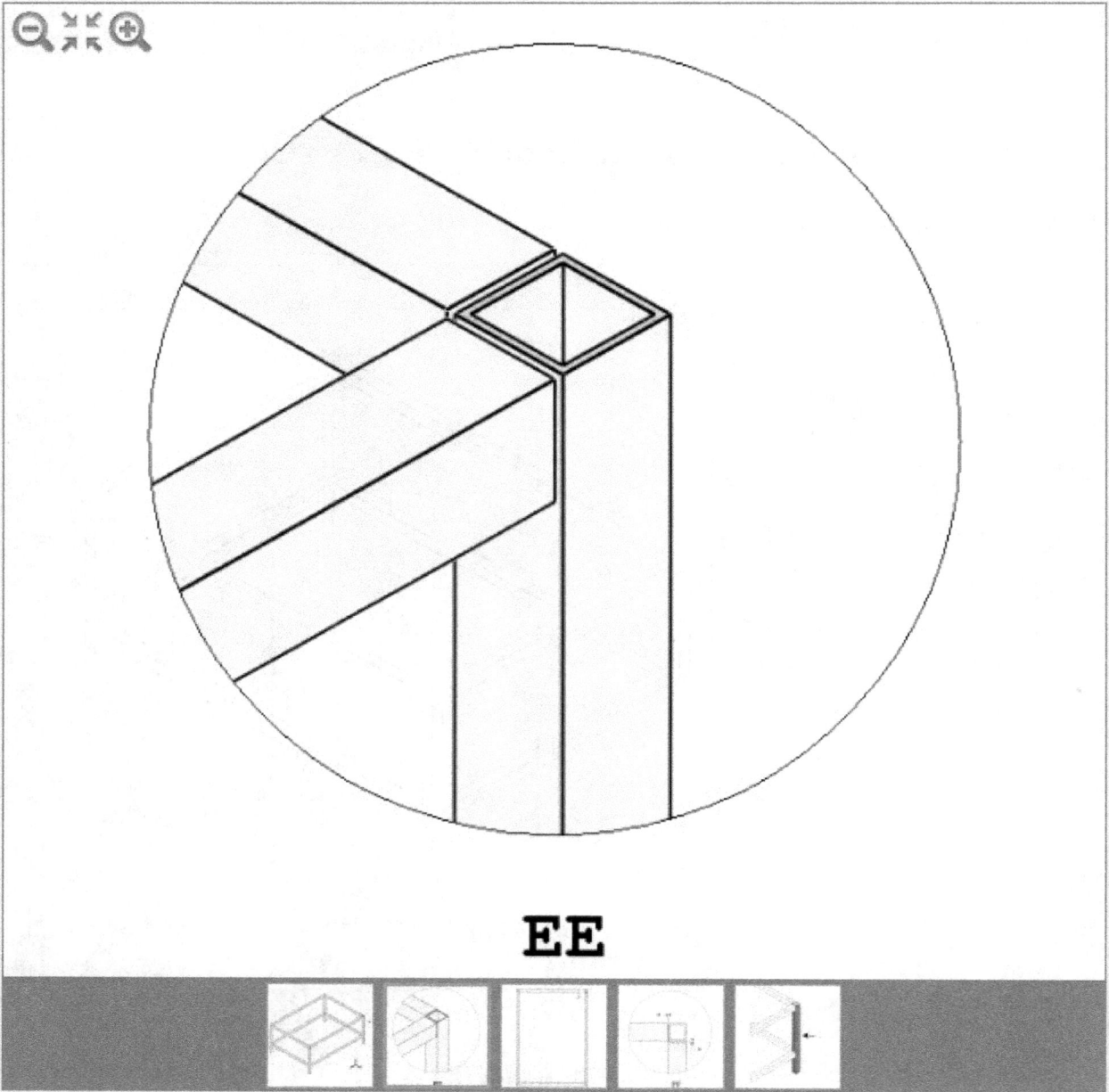

EE

Figure 101 - Question 10 of 26 - Exam Screen Capture 2

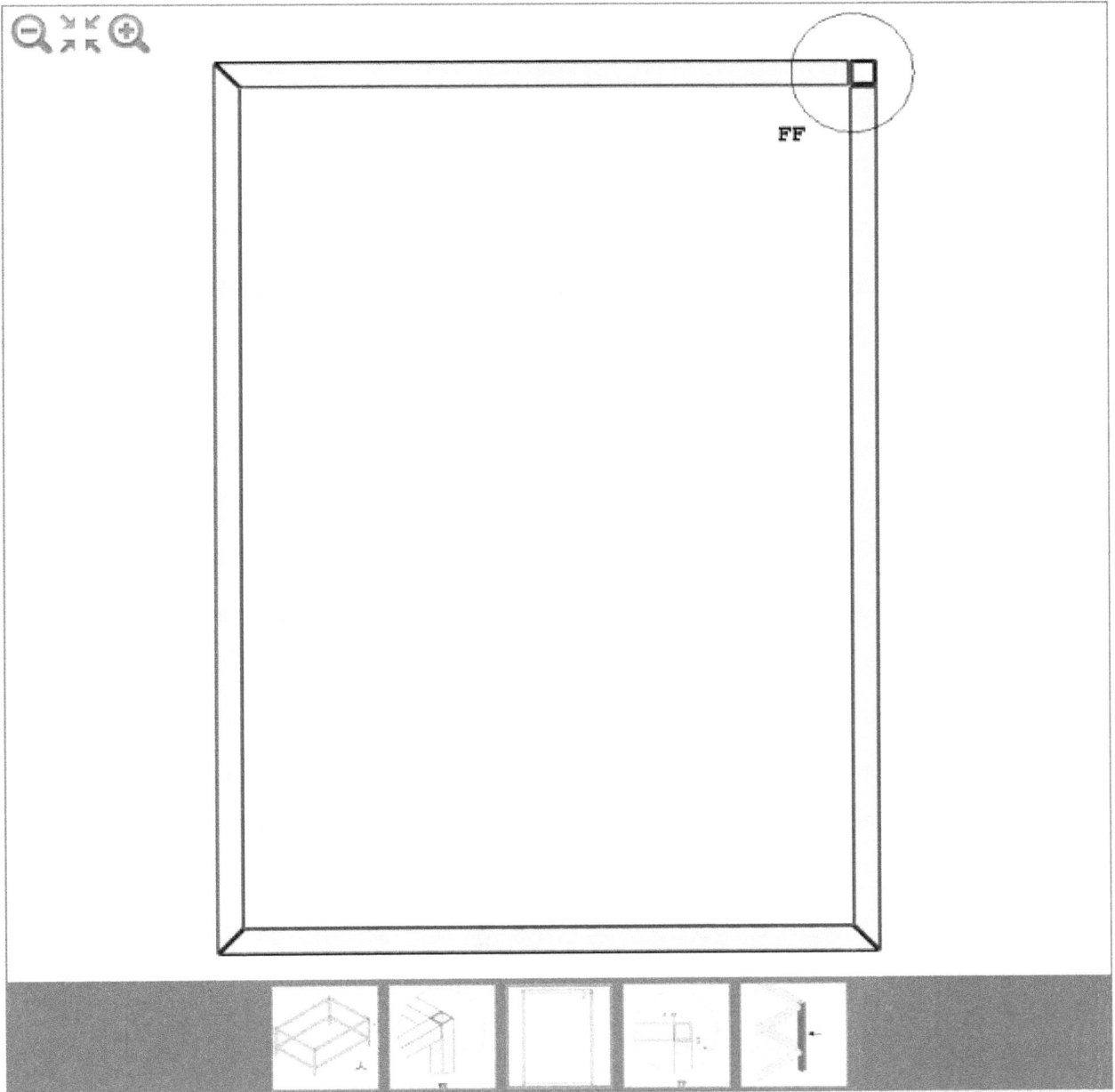

FF

Figure 102 - Question 10 of 26 - Exam Screen Capture 3

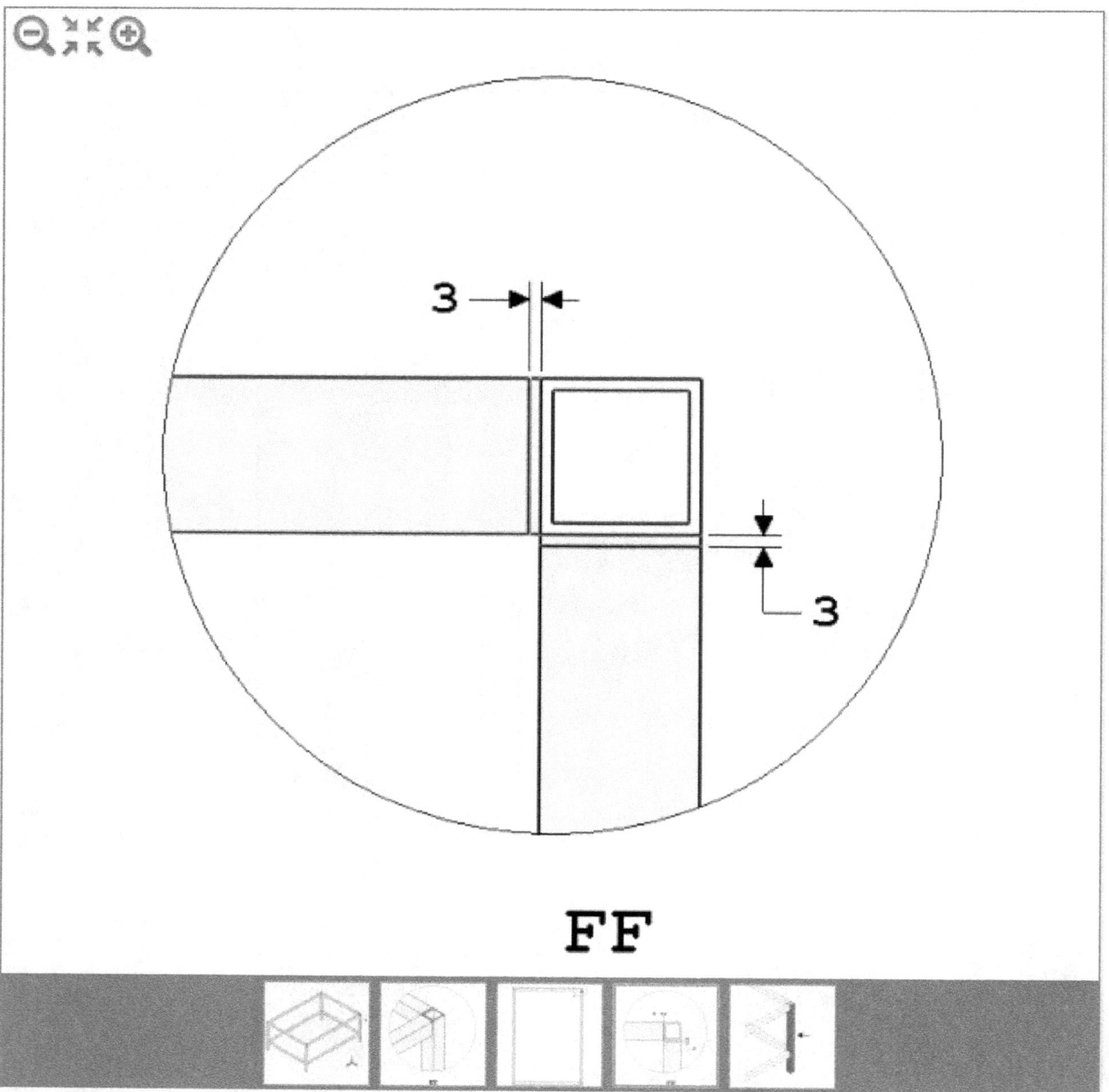

3

3

FF

Figure 103 - Question 10 of 26 - Exam Screen Capture 4

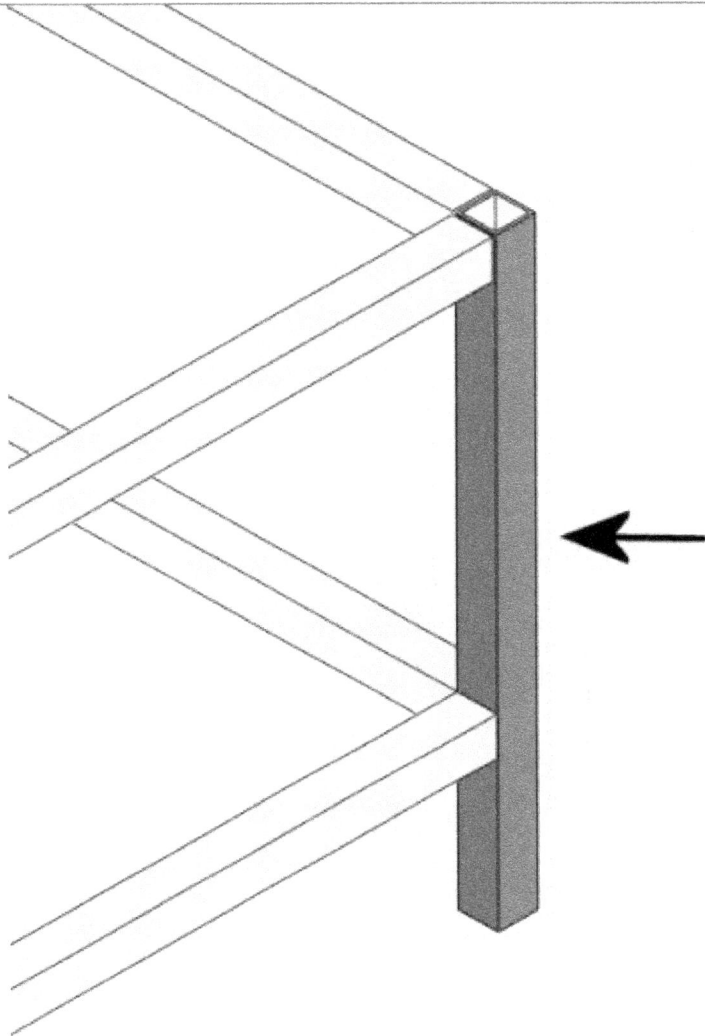

Figure 104 - Question 10 of 26 - Exam Screen Capture 5

CHANGING A CORNER TREATMENT

Question 10 is a continuation from Question 9. In the Feature Manager Design Tree, right click on the Structural Member (*Square WLDM2E(1)*) ▸ 🔲 Square WLDM2E(1) and select Edit Feature as shown in the following image.

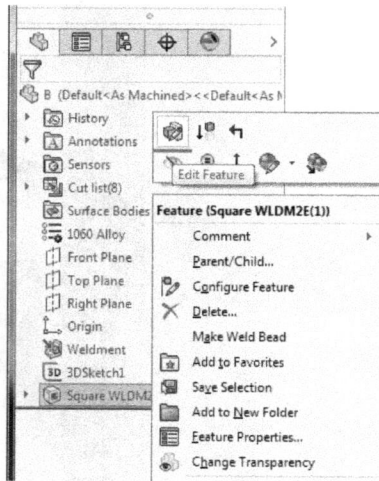

Figure 105 - Question 10 of 26 - Editing a structural member

Click on the top right corner indicated by Detail EE in Exam Screen Capture 1 which is the corner we want to modify - see the following image.

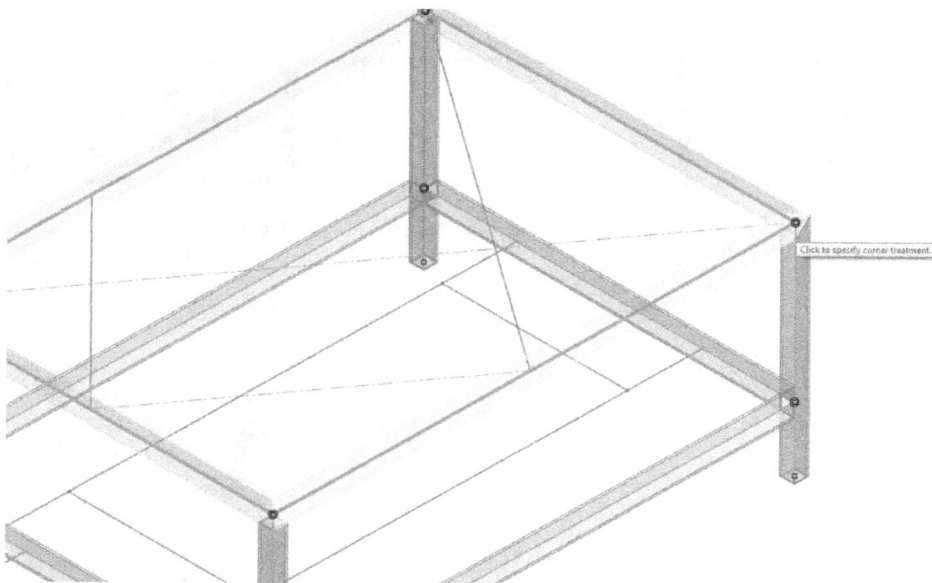

Figure 106 - Question 10 of 26 - Changing a Corner Treatment

NOTE - HOW TRIM ORDER NUMBERS WORK

A group with a lower trim order number trims a group with a higher number. If two groups have the same trim order number, they miter each other. Trim groups with a higher number are trimmed by groups with a lower number.

CHANGING CORNER TREATMENTS

In the Corner Treatment dialog box, under Trim Order, change the value to 2 under Group 1 and to 1 Under Group 2 and select the Set corner specific weld gaps checkbox then enter a value of 3mm as shown in the following image. Thus Group 1 will be trimmed by Group 2 which has a lower trim order number - see the following image.

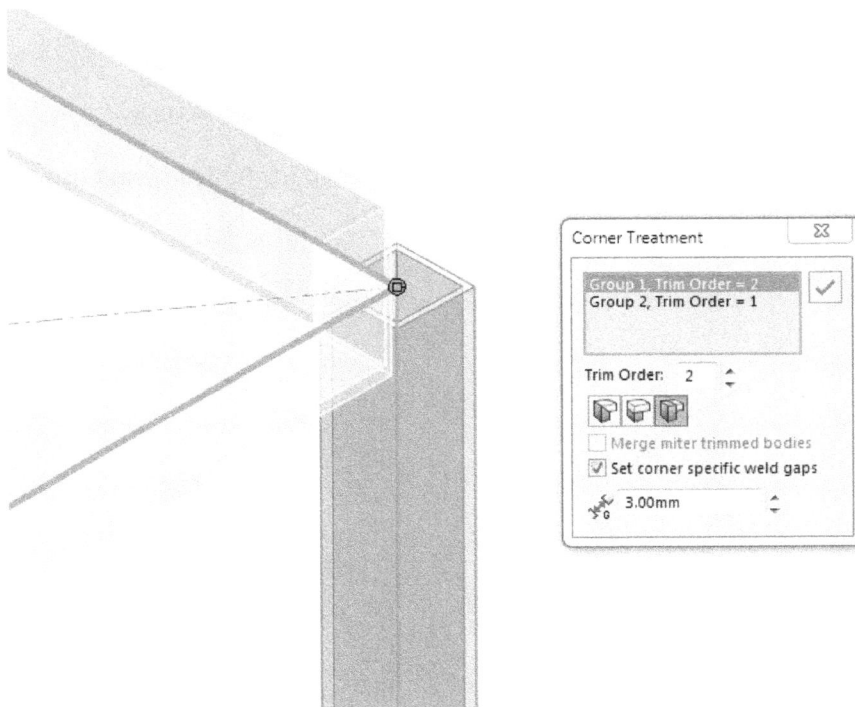

Figure 107 - Question 10 of 26 - Changing a Corner Treatment

Click Ok on the Corner Treatment Dialog Box then Click Ok on the Property Manager.

USING THE MEASURE TOOL

Use the measure tool to check the gap between the trimmed segments. Click Measure ☒ (Tools toolbar) or Tools > Evaluate > Measure. Check all the gaps and make sure they are 3mm as shown in the following image.

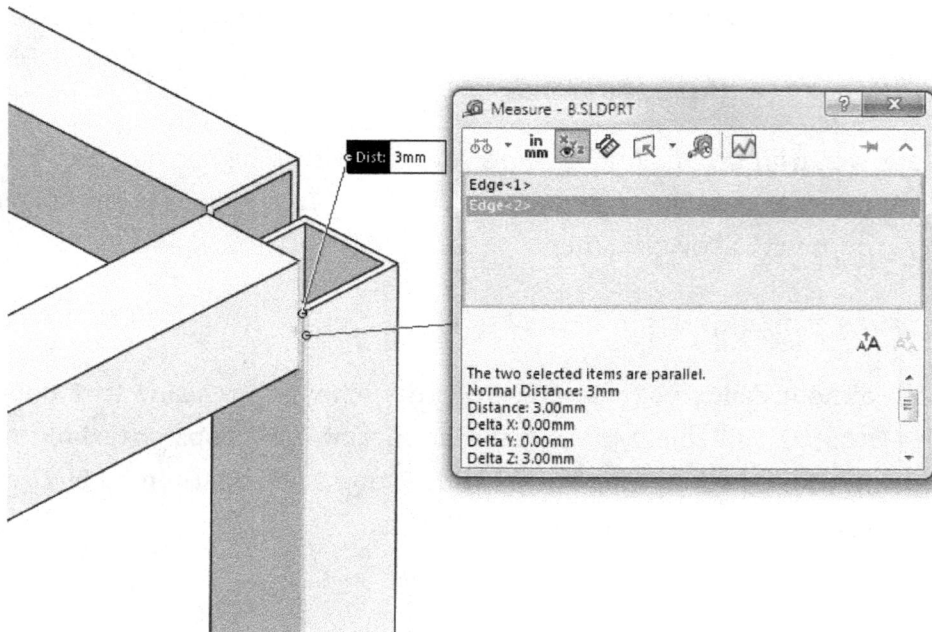

Figure 108 - Question 10 of 26 - Using the measure tool to double check weld gaps

Close the measure tool. The corner should now look as shown in the following image.

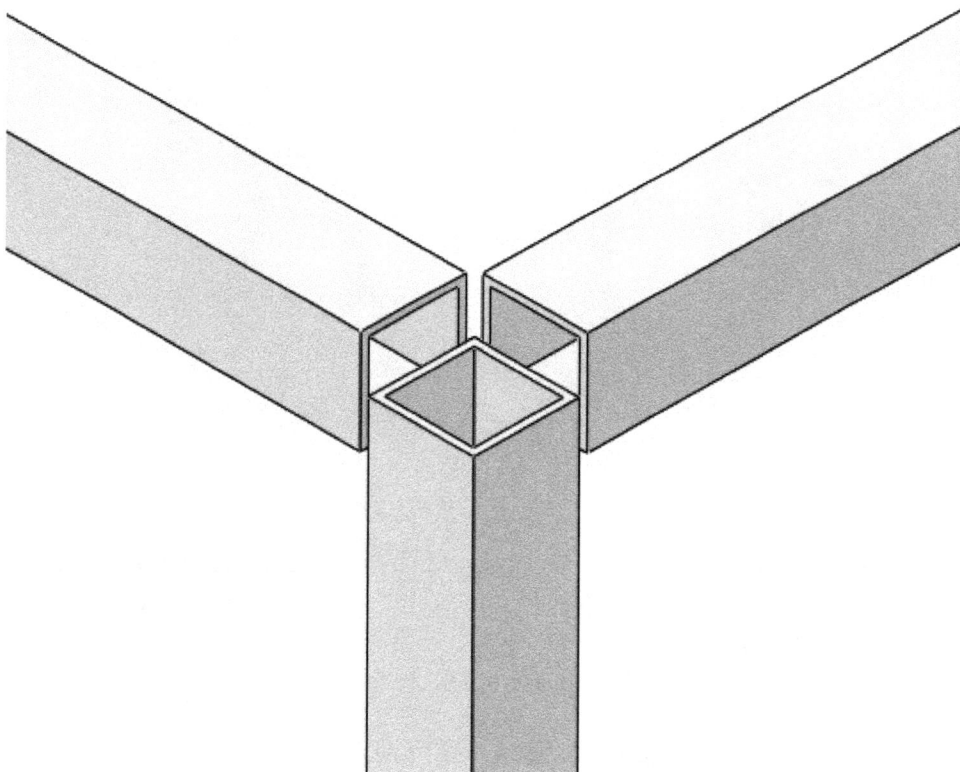

Figure 109 - Question 10 of 26 - Corner Treatment - Current Status

As you may have noticed, the vertical segment is not flush with the top faces of the horizontal segments, hence we need to extend the vertical segment. Click Trim/Extend (Weldments toolbar) or Insert > Weldments > Trim/Extend.

In the PropertyManager, set options as shown in the following image. Under Corner Type, select End Trim and under Bodies to be Trimmed select the vertical segment. Make sure the Allow extension checkbox is selected then under Trimming Boundary - select Face / plane then select the top face of one of the horizontal segments. Also select the preview checkbox to get a preview of the final result as you make the selections.

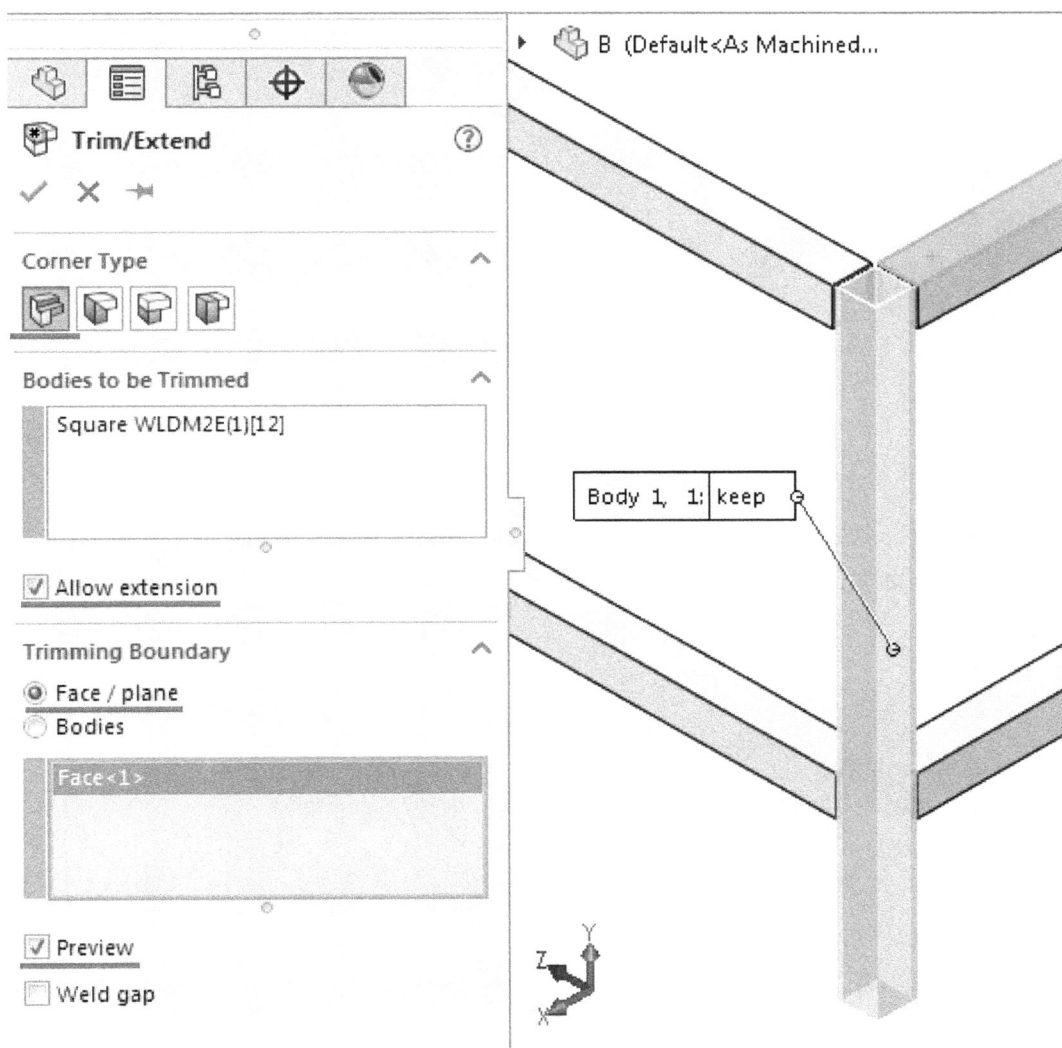

Figure 110 - Question 10 of 26 - Weldments Trim / Extend

Click Ok. Your part should now appear as shown in the following image with the 3D Sketch Hidden ofcourse.

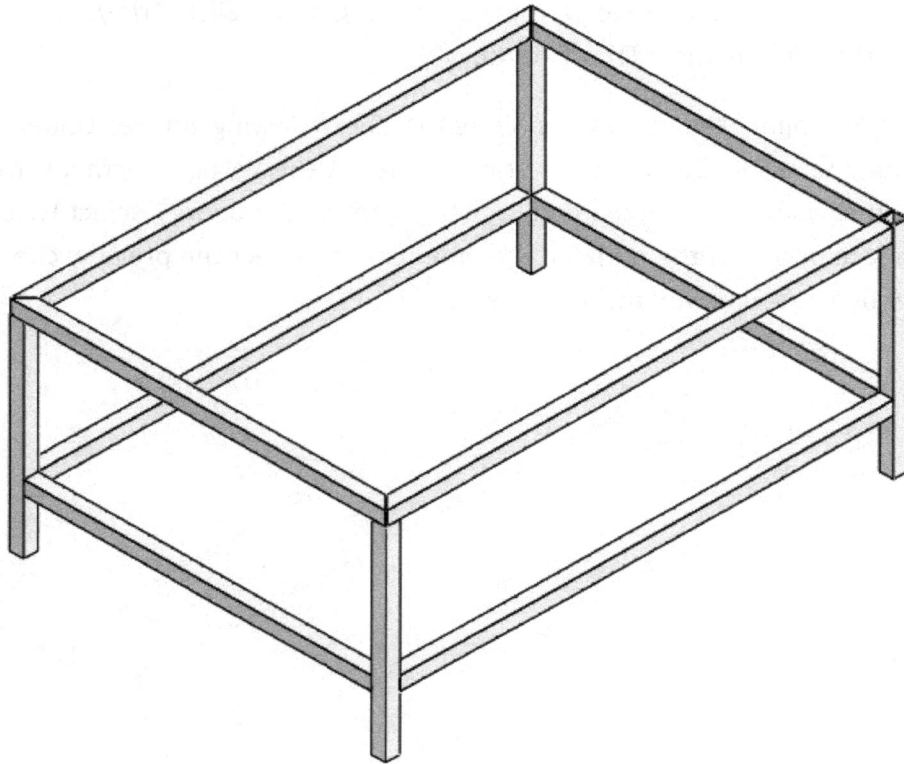

Figure 111 - Question 10 of 26 - Weldment current status

NB: Use the measure tool to double check the gaps between all three segments at this corner and make sure they are 3mm all-round.

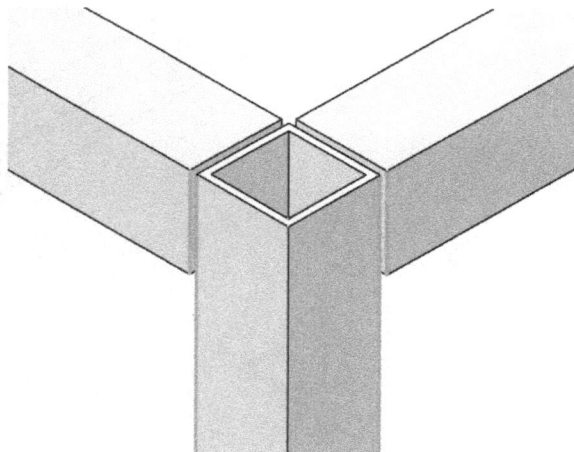

Figure 112 - Question 10 of 26 - Measure the Weld Gaps - Double Checking

MEASURING THE MASS OF A SINGLE SEGMENT IN A WELDMENT

In the Feature Manager Design Tree, right click on the vertical leg at the corner shown in Exam Screen Capture 5 and select Isolate as shown in the following image.

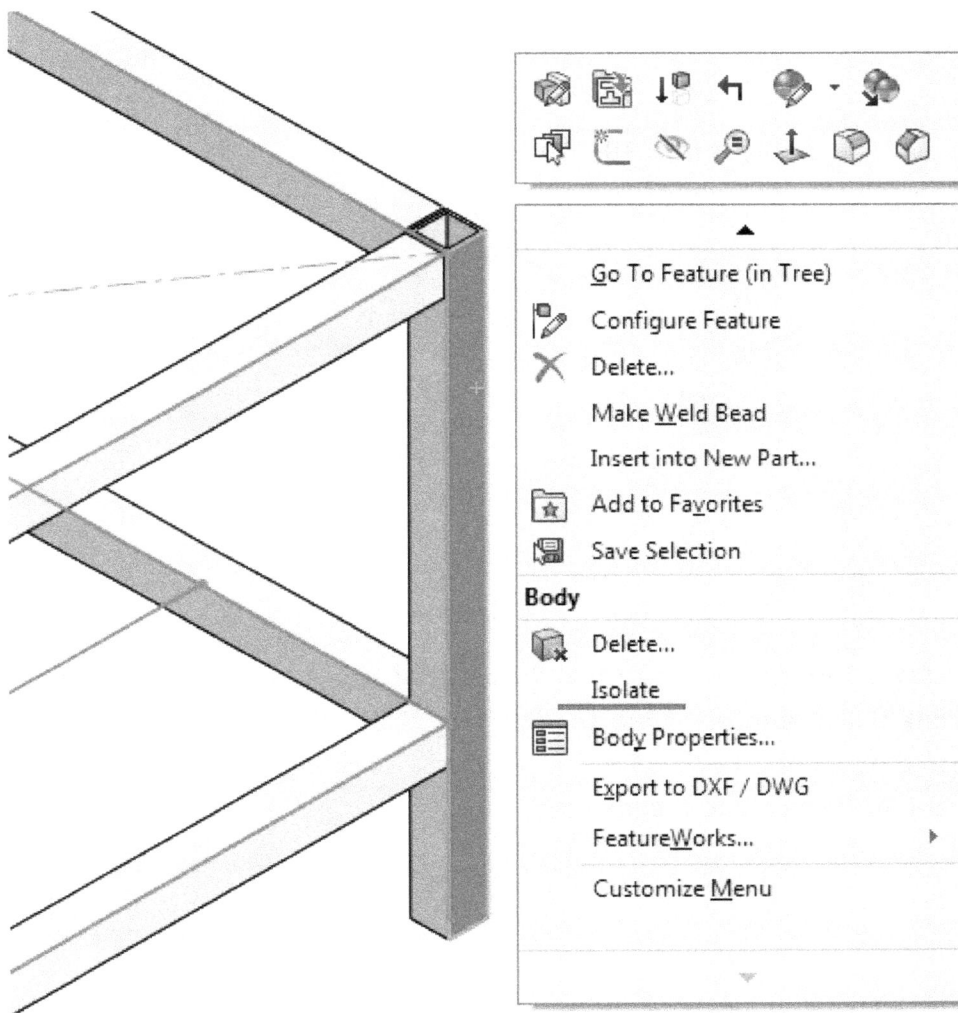

Figure 113 - Question 10 of 26 - Isolating a single segment

Your part should now look as shown in the following image - with all the other segments hidden.

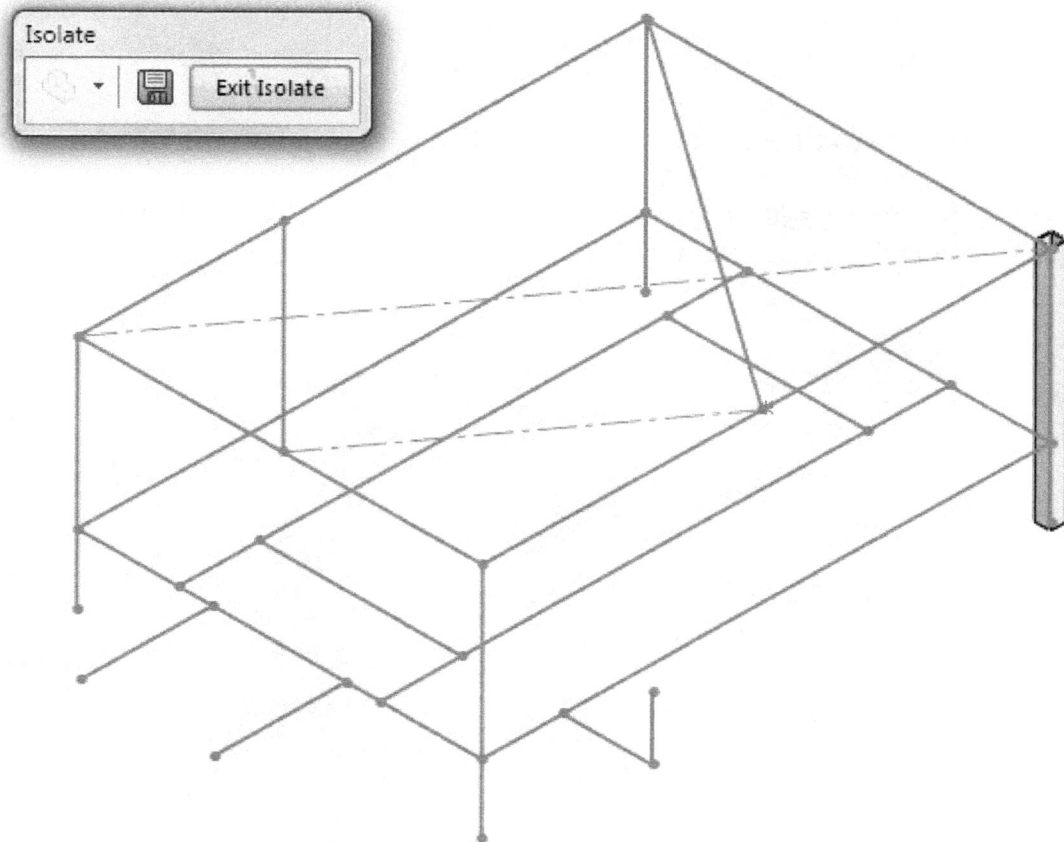

Figure 114 - Question 10 of 26 - Isolating a single segment

Under the Evaluate Tab, click on the Mass Properties feature and the Mass Properties Menu appears showing that the mass of the segment is 722.58 grams. **NB:** Make sure the Include hidden bodies/components checkbox is not selected - selecting it would give the mass of the whole weldment including hidden segments.

```
Mass Properties                                                    [ _ ][ □ ][ X ]

  ⚖  B.SLDPRT
                                                              [ Options... ]

        [ Override Mass Properties... ]    [ Recalculate ]
        ☐ Include hidden bodies/components
        ☐ Create Center of Mass feature
        ☐ Show weld bead mass
        Report coordinate values relative to:   -- default --            ▼

        Mass properties of B
            Configuration: Default<As Machined>
            Coordinate system: -- default --

        Density = 0.00 grams per cubic millimeter

        Mass = 722.58 grams

        Volume = 267624.00 cubic millimeters

        Surface area = 179280.00  square millimeters

        Center of mass: ( millimeters )
            X = 0.00
            Y = 309.75
            Z = -1400.00

        Principal axes of inertia and principal moments of inertia: ( grams * square millimeters )
        Taken at the center of mass.
            Ix = ( 0.00,  1.00,  0.00)      Px = 314324.39
            Iy = ( 0.00,  0.00,  1.00)      Py = 23266643.46
            Iz = ( 1.00,  0.00,  0.00)      Pz = 23266643.46

        Moments of inertia: ( grams * square millimeters )
        Taken at the center of mass and aligned with the output coordinate system.
            Lxx = 23266643.46       Lxy = 0.00              Lxz = 0.00
            Lyx = 0.00              Lyy = 314324.39         Lyz = 0.00
            Lzx = 0.00              Lzy = 0.00              Lzz = 23266643.46

        Moments of inertia: ( grams * square millimeters )
        Taken at the output coordinate system.
            Ixx = 1508861295.26     Ixy = 0.00              Ixz = 0.00
            Iyx = 0.00              Iyy = 1416580532.39     Iyz = -313348898.52
            Izx = 0.00              Izy = -313348898.52     Izz = 92595087.26

        [ Help ]            [ Print... ]          [ Copy to Clipboard ]
```

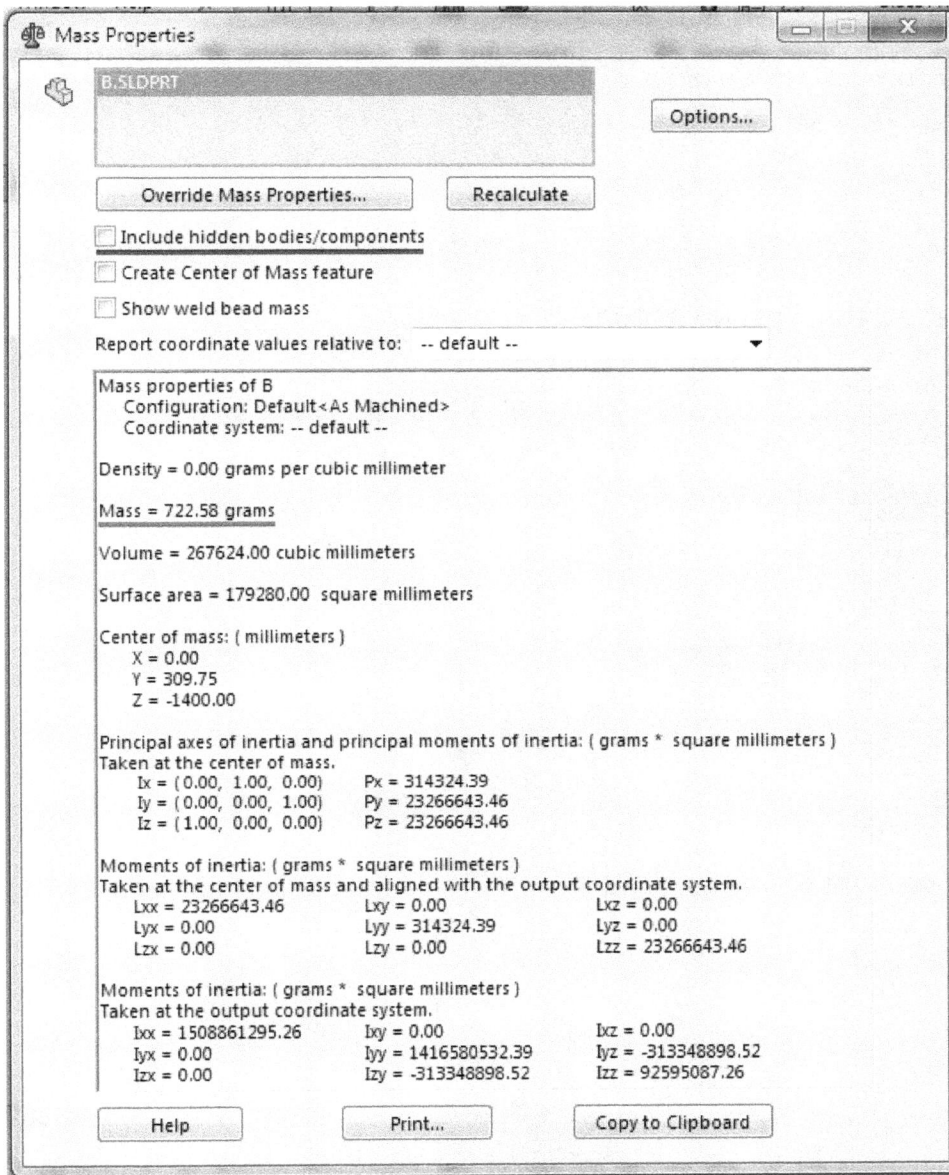

Figure 115 - Question 10 of 26 - Mass of a single segment

Close the Mass Properties Dialog Box and Click Exit Isolate. Save your part.

CHAPTER 11

QUESTION 11 EXAM SCREEN CAPTURES

QUESTION 11 EXAM SCREEN CAPTURE 1

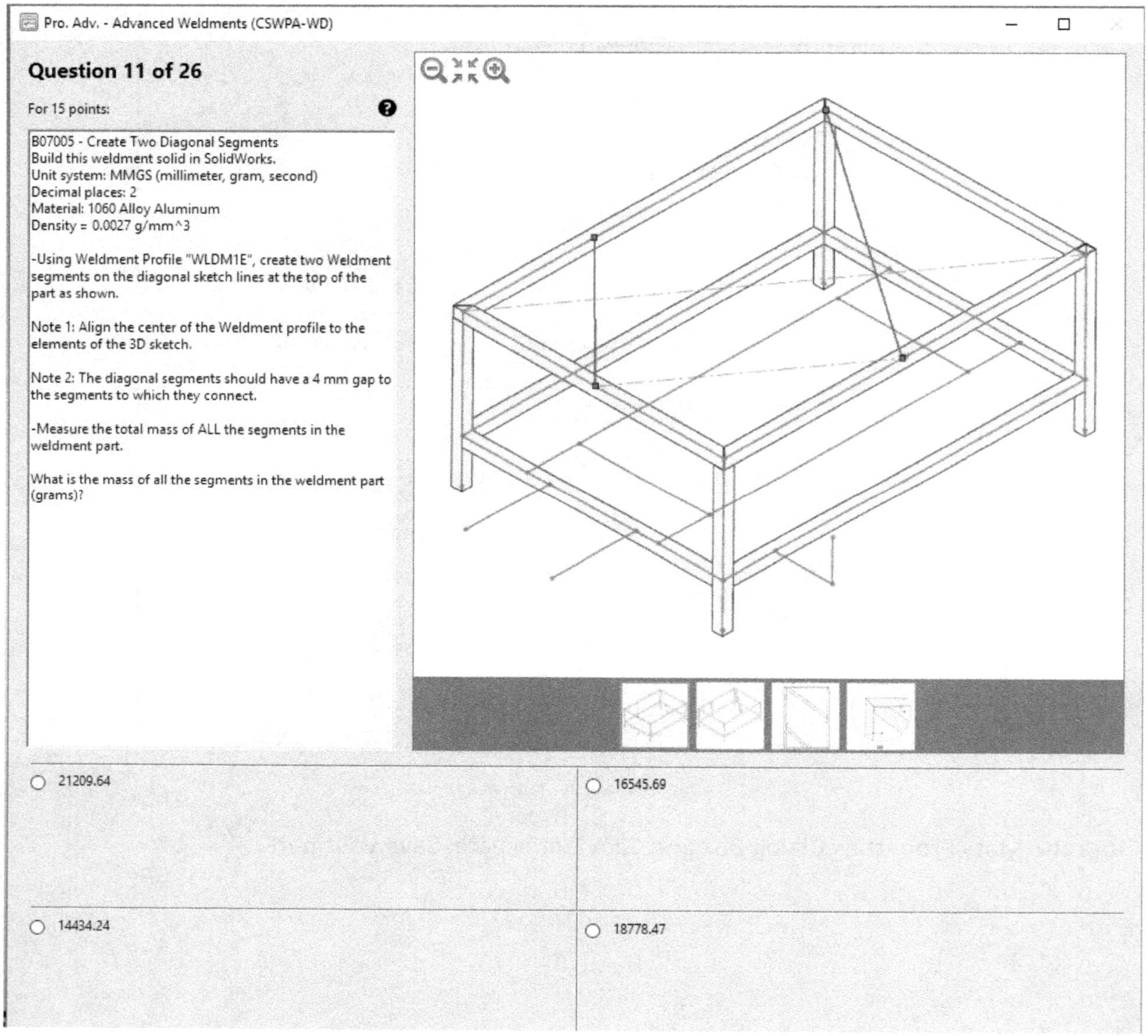

Pro. Adv. - Advanced Weldments (CSWPA-WD)

Question 11 of 26

For 15 points:

B07005 - Create Two Diagonal Segments
Build this weldment solid in SolidWorks.
Unit system: MMGS (millimeter, gram, second)
Decimal places: 2
Material: 1060 Alloy Aluminum
Density = 0.0027 g/mm^3

-Using Weldment Profile "WLDM1E", create two Weldment segments on the diagonal sketch lines at the top of the part as shown.

Note 1: Align the center of the Weldment profile to the elements of the 3D sketch.

Note 2: The diagonal segments should have a 4 mm gap to the segments to which they connect.

-Measure the total mass of ALL the segments in the weldment part.

What is the mass of all the segments in the weldment part (grams)?

○ 21209.64

○ 16545.69

○ 14434.24

○ 18778.47

Figure 116 - Question 11 of 26 - Exam Screen Capture 1

QUESTION 11 EXAM SCREEN CAPTURE 2

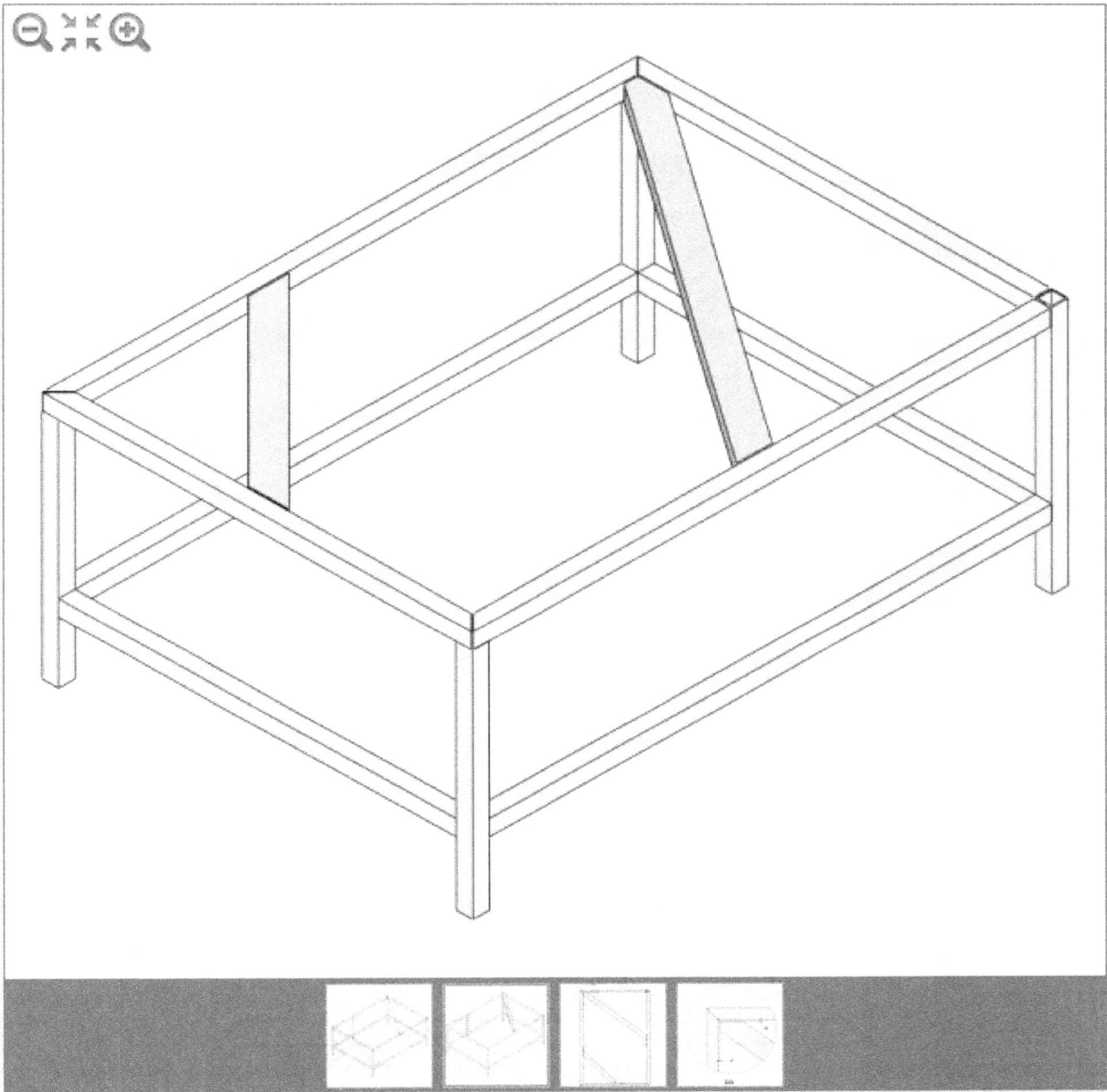

Figure 117 - Question 11 of 26 - Exam Screen Capture 2

QUESTION 11 EXAM SCREEN CAPTURE 3

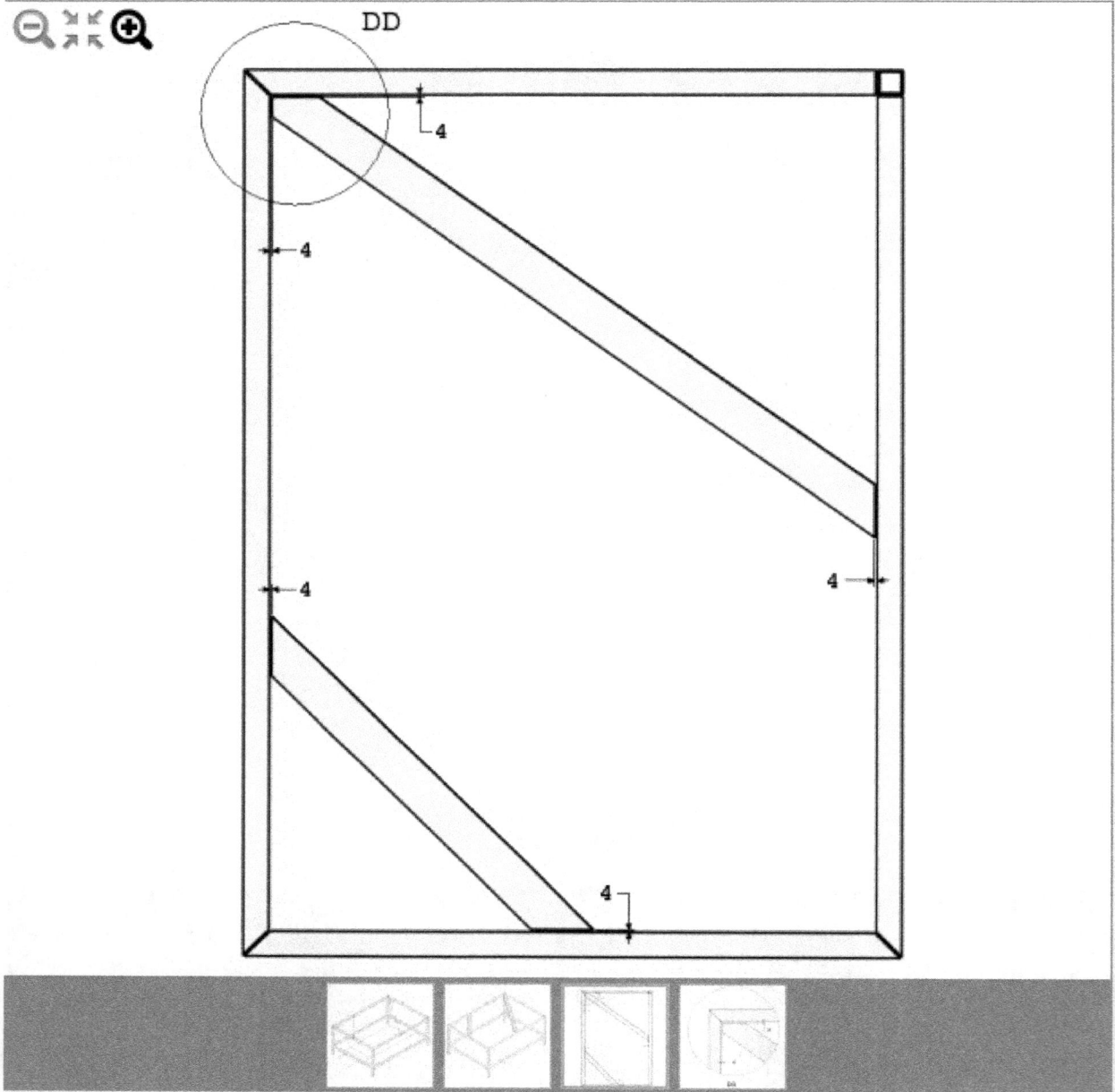

DD

Figure 118 - Question 11 of 26 - Exam Screen Capture 3

DD

Figure 119 - Question 11 of 26 - Exam Screen Capture 4

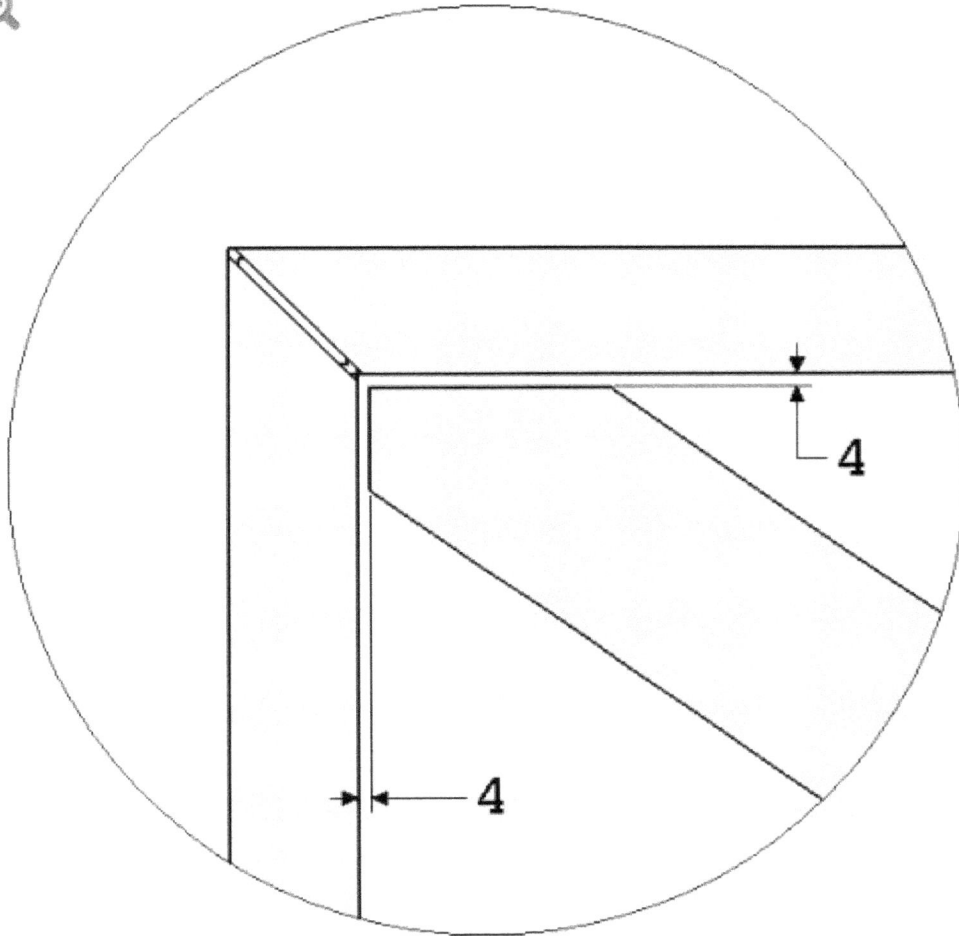

The sketch is already created in this Question. Click Structural Member (Weldments toolbar) or Insert > Weldments > Structural Member.

Make selections in the PropertyManager to define the profile for the structural member as shown in the following image. Under groups, select the first long diagonal line and check the preview in the graphics area to make sure the orientation of the weldment profile is correct. It may be as shown in the following image which is incorrect.

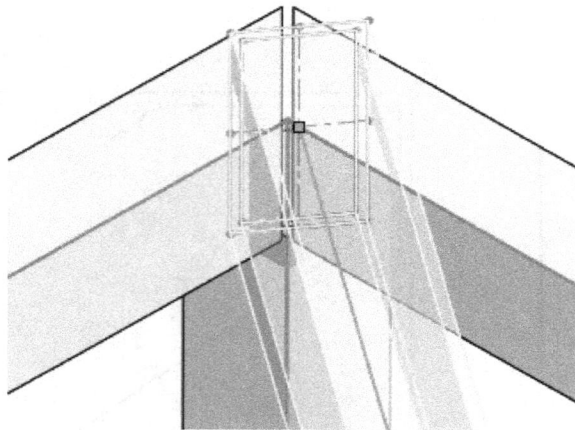

Figure 120 - Question 11 of 26 - Adding a structural member

To change the orientation of the structural member, under Rotation Angle in the PropertyManager, enter 90 Degrees and the structural member sketch profile will rotate by 90 Degrees as shown in the following image.

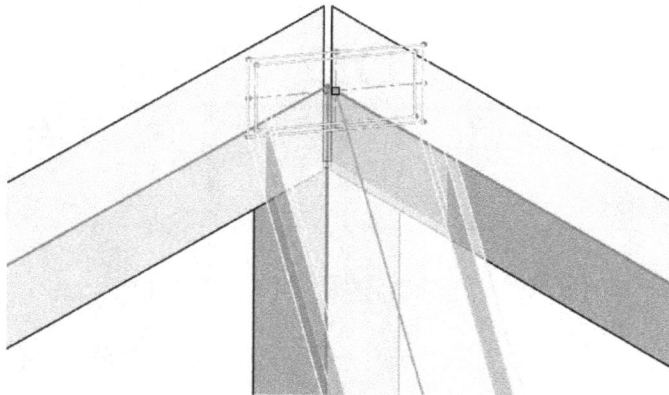

Figure 121 - Question 11 of 26 - Rotation Angle of a structural member

Under Groups, click on the New Group command button and a new group is created - Group 2. With the new Group selected under Groups, select the short diagonal sketch segment and enter 90 Degrees under Rotation Angle to change the orientation of this new segment. The Structural Member PropertyManager and part preview should now look as shown in the following image.

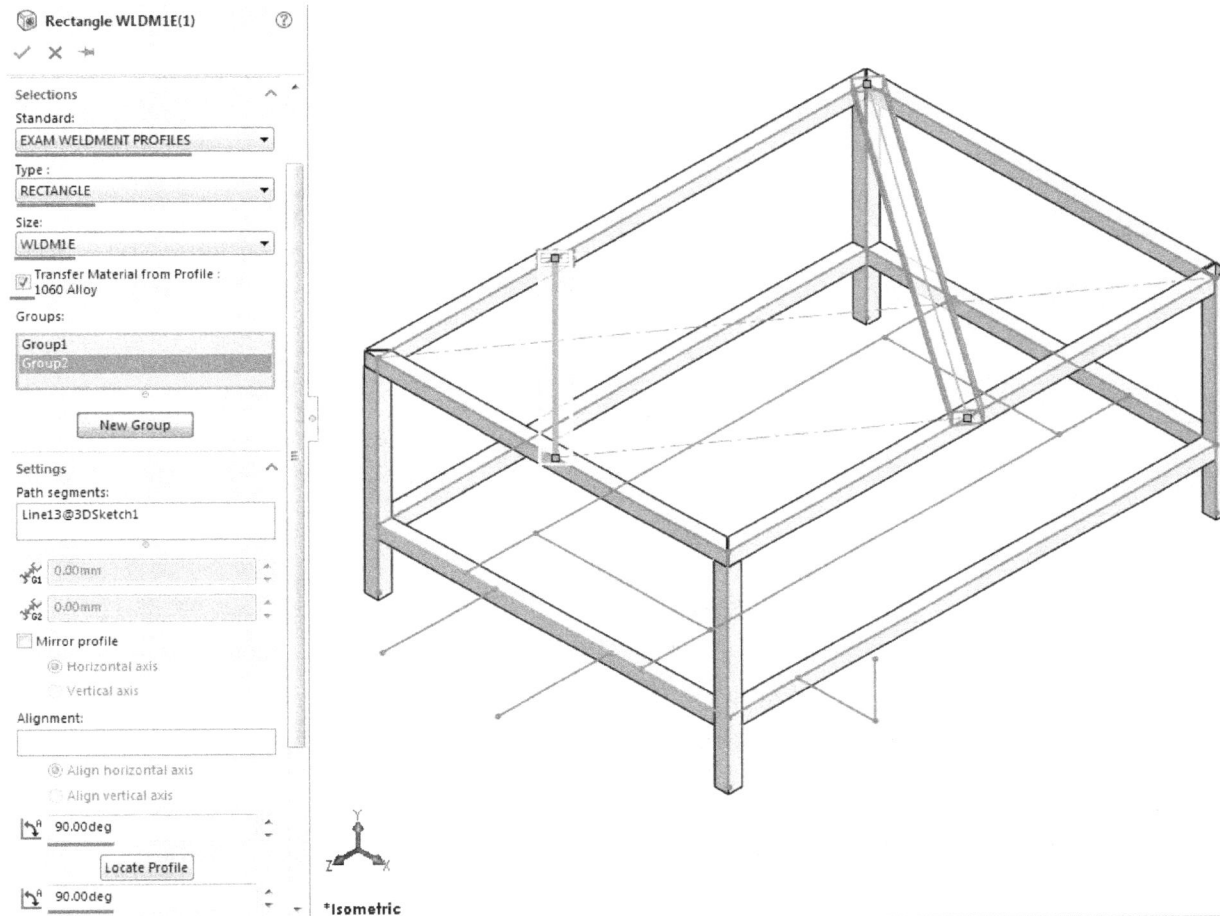

Figure 122 - Question 11 of 26 - Structural Member PropertyManager and Part Preview

Click OK.

WELDEMENTS - TRIM / EXTEND

As you may have noticed, these two new rectangular diagonal segments go through the square weldment segments to which they supposed to connect. Hence we have to trim these new rectangular segments and leave the required 4mm gap to the segments to which the connect - as indicated in question 11 screen capture one. Click Trim/Extend (Weldments toolbar) or Insert > Weldments > Trim/Extend. Under Corner Type, select End Trim. Under Bodies to be Trimmed, select the two rectangular diagonal segments. Clear the Allow Extension Checkbox. Under

Trimming Boundary, select the inner faces of the four top square segments and toggle keep or discard to choose which segments to keep as shown in the following image.

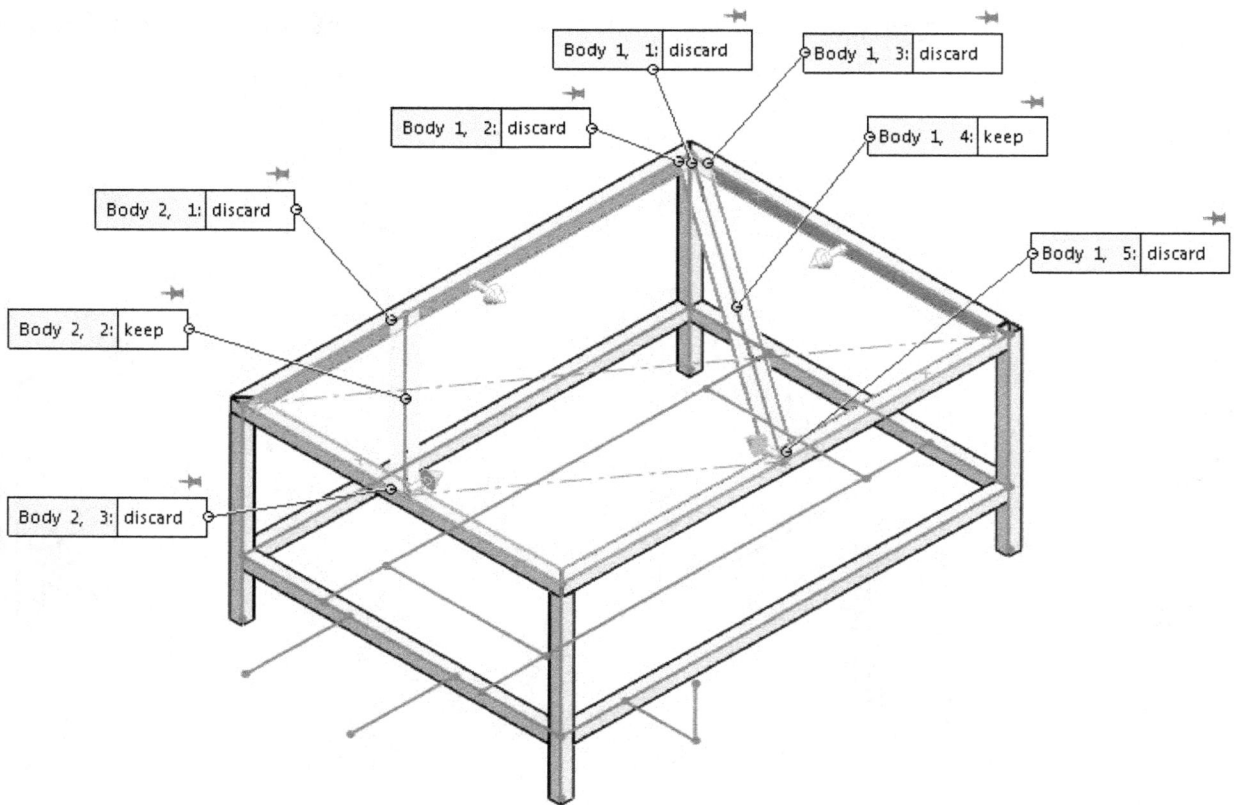

Figure 123 - Question 11 of 26 - Trimming Boundary

Select the Weld Gap Checkbox and enter 4mm. Your PropertyManager should have selections and options as specified in the following image.

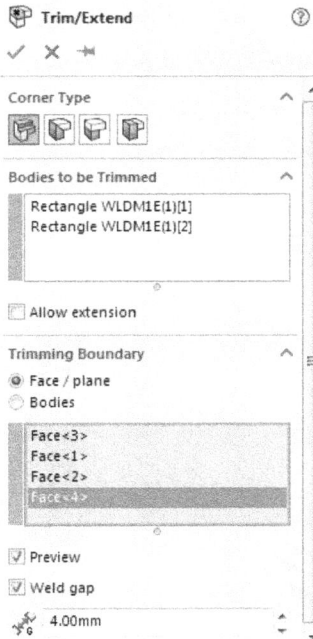

Figure 124 - Question 11 of 26 - Trim / Extend PropertyManager

Click OK. Save your Part.

HIDING AND SHOWING SKETCHES

To hide the 3D Sketch in the model, click on the 3D Sketch in the FeatureManager Design Tree and select Hide as shown in the following image. Follow same procedure to show the sketch.

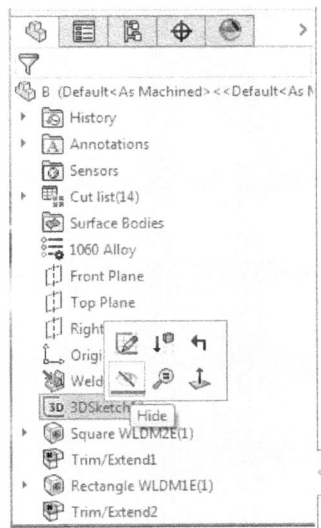

Figure 125 - Question 11 of 26 - Hiding and Showing Sketches

Your part should now look as shown in the following image with the 3D sketch hidden. Remember to always use the Measure Tool to double check the weld gaps between the new rectangular segments and the square segments and this should be 4mm all-round.

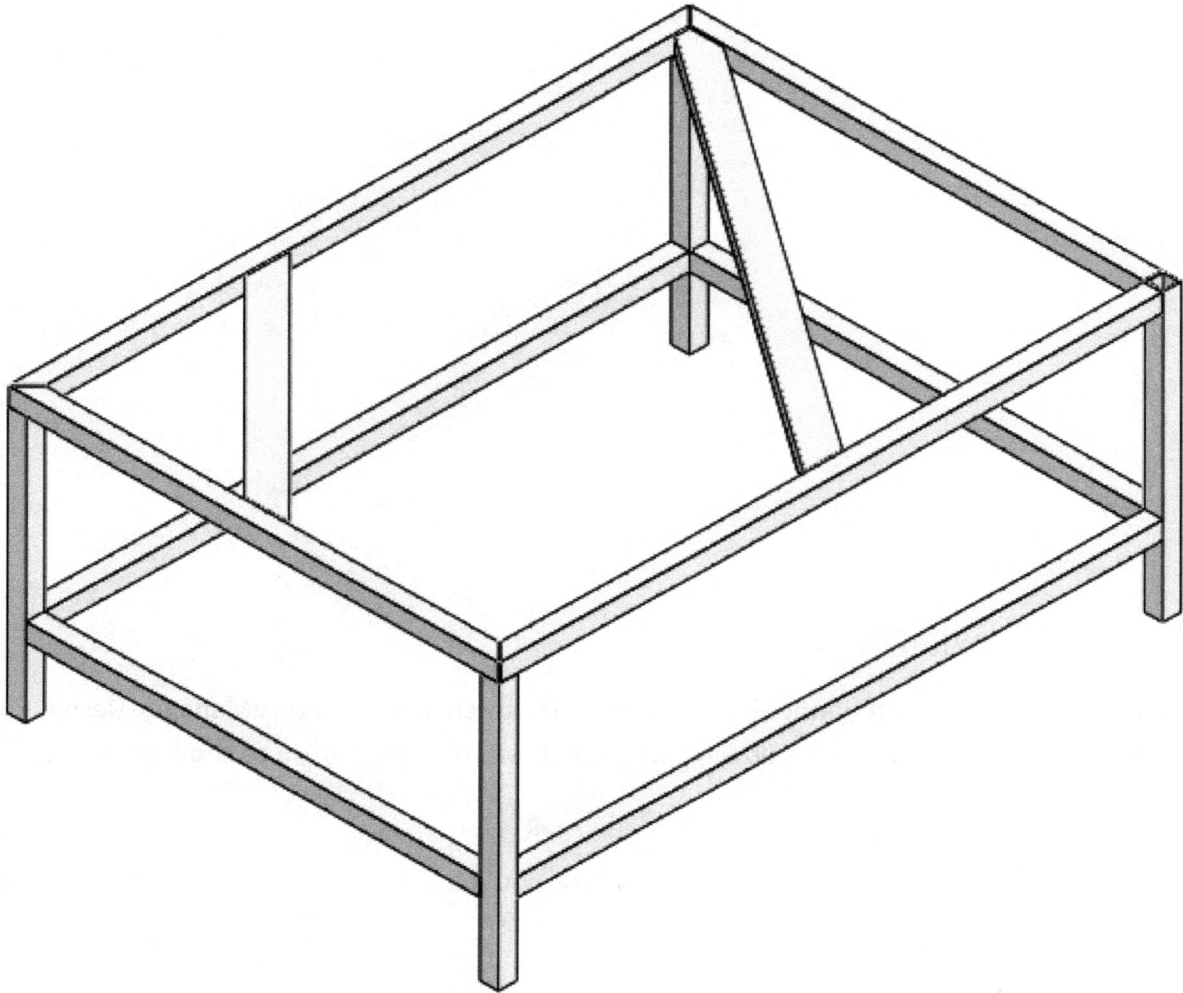

Figure 126 - Question 11 of 26 - Weldment Current Status.

MEASURING THE MASS OF ALL SEGMENTS IN A WELDMENT PART

Under the Evaluate Tab, click on the Mass Properties feature and the Mass Properties Menu appears showing that the total mass of the part is 16545.69 grams.

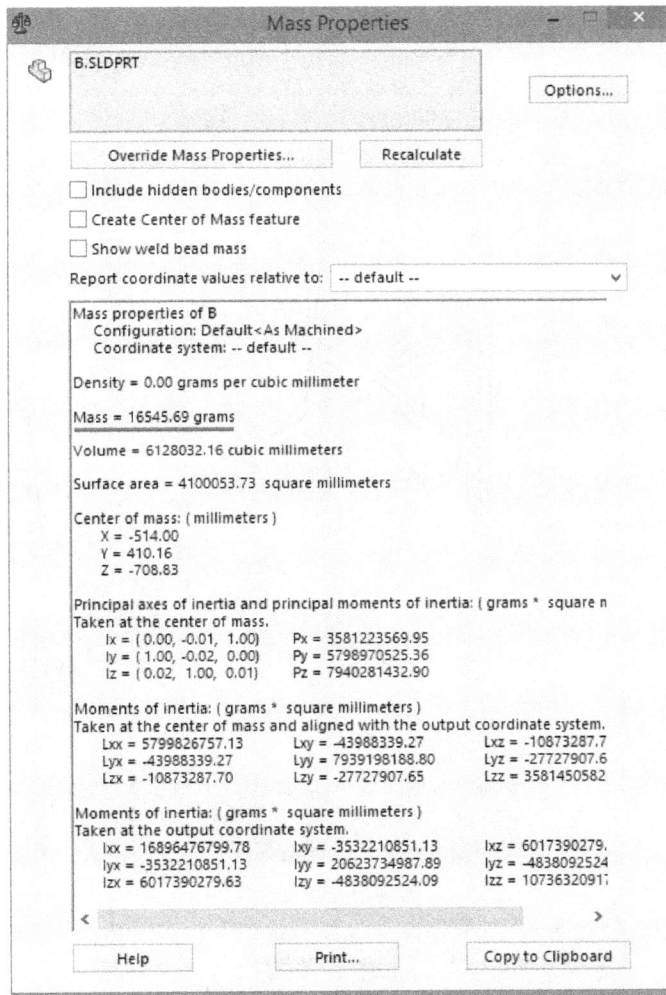

Figure 127 - Question 11 of 26 - Mass Properties

Click save to save the Part.

CHAPTER 12

QUESTION 12 EXAM SCREEN CAPTURES

QUESTION 12 EXAM SCREEN CAPTURE 1

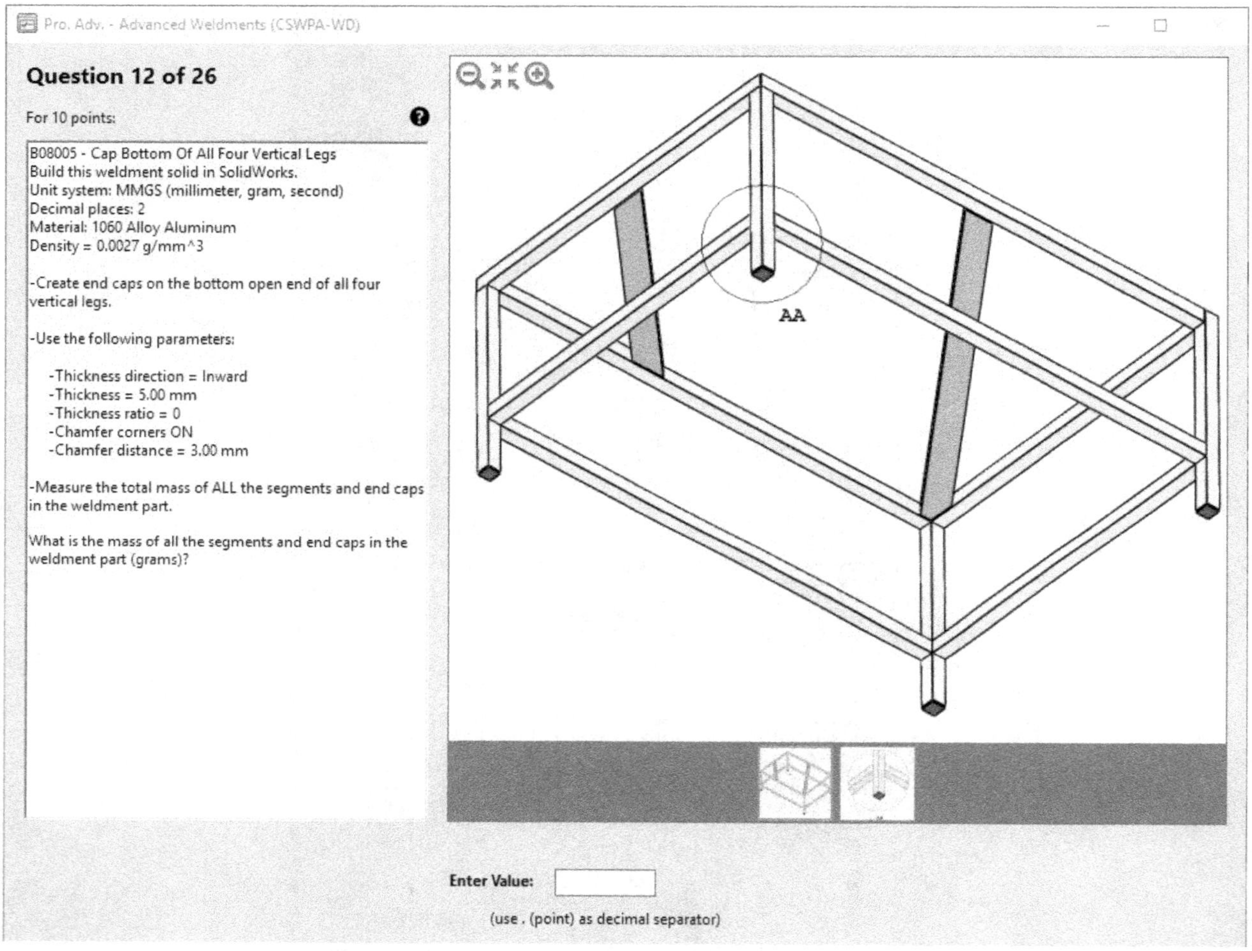

Figure 128 - Question 12 of 26 - Exam Screen Capture 1

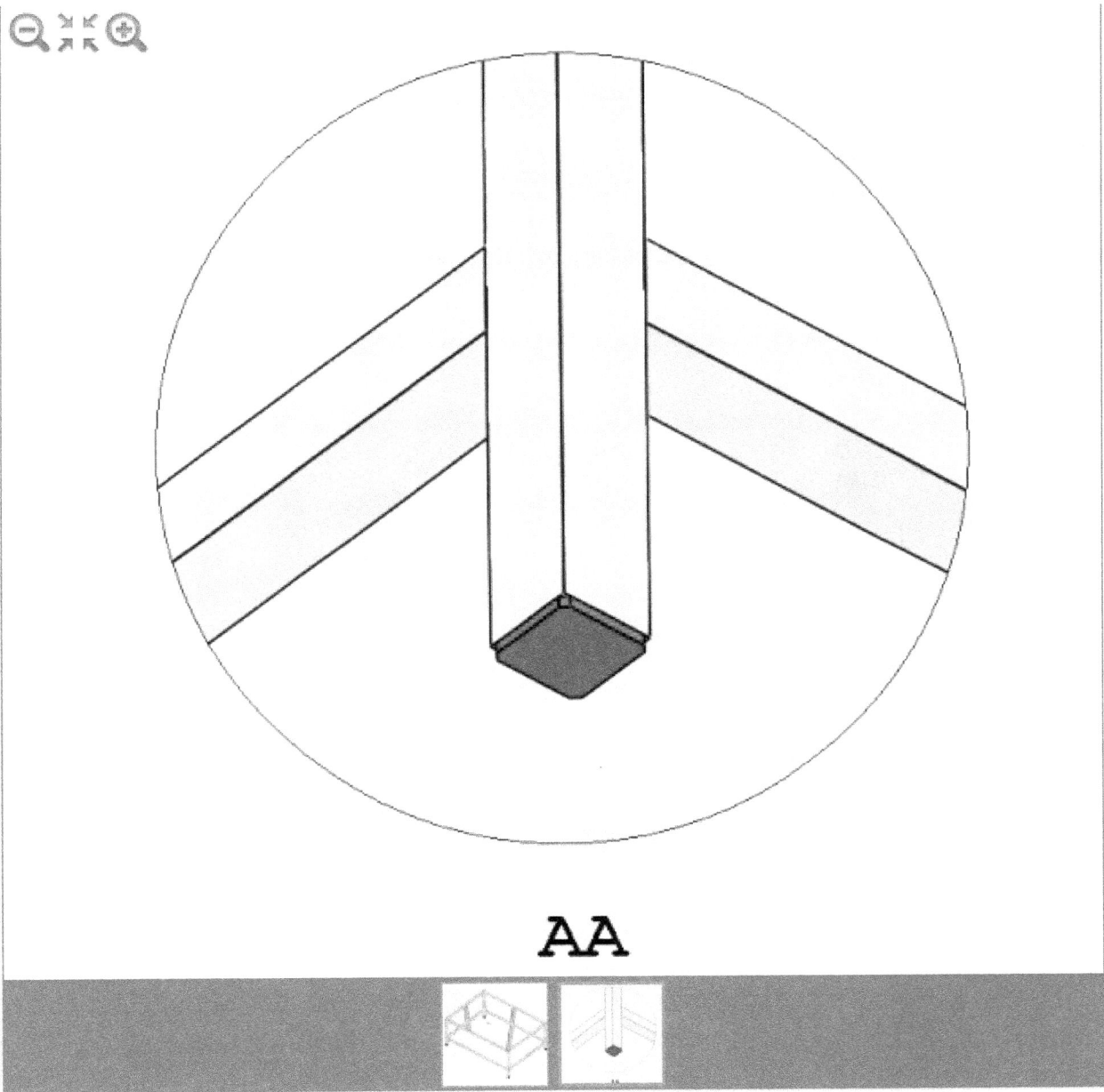

AA

Figure 129 - Question 12 of 26 - Exam Screen Capture 2

Since Question 12 is a continuation from Question 11, we will continue using the same B.sldprt.

ADDING END CAPS

To add End Caps -click End Cap End Cap (Weldments toolbar) or Insert > Weldments > End Cap. In the End Cap PropertyManager, under Parameters:

Select the Four Faces on the bottom end of the four vertical legs as shown in the following image. The preview shows the end cap.

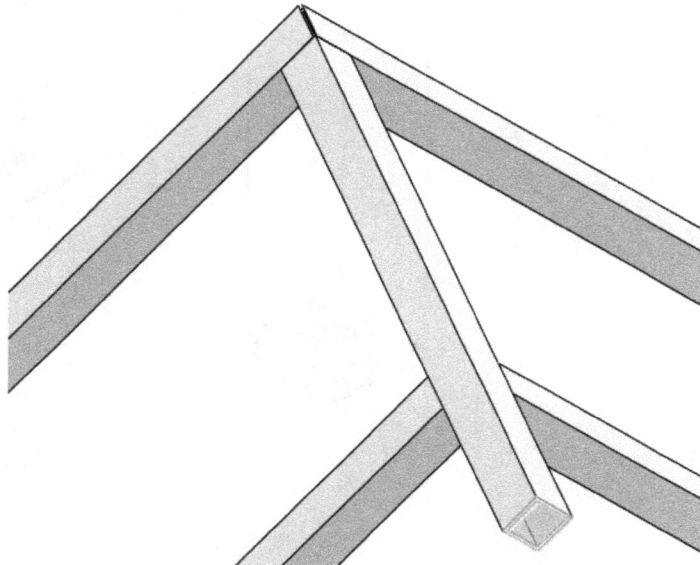

Figure 130 - Question 12 of 26 - Adding End Caps

Select Inward under Thickness direction and enter 5.00mm under end cap Thickness.

Under Offset, select Thickness Ratio as the way to calculate the offset and type a value of zero. You can click to reverse the offset direction but in our case this is not applicable since the thickness ratio is zero.

Select the Corner Treatment Checkbox to specify a corner treatment. Select Chamfer and specify a Chamfer Distance of 3.00mm. If you zoom to the preview of one of the End Caps on one of the vertical legs it should appear as shown in the following image - NB: The Display Style of the part has been changed in the following image for visual clarity.

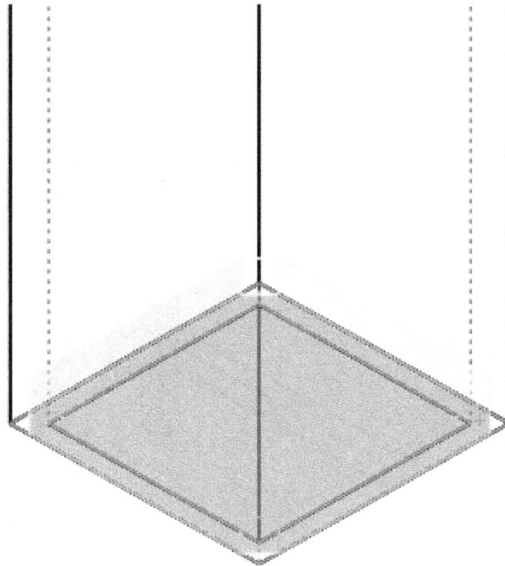

Figure 131 - Question 12 of 26 - End Cap Preview

Options, specifications and selections in your PropertyManager should be as shown in the following image - NB: The part's Display Style has been changed for visual clarity purposes only.

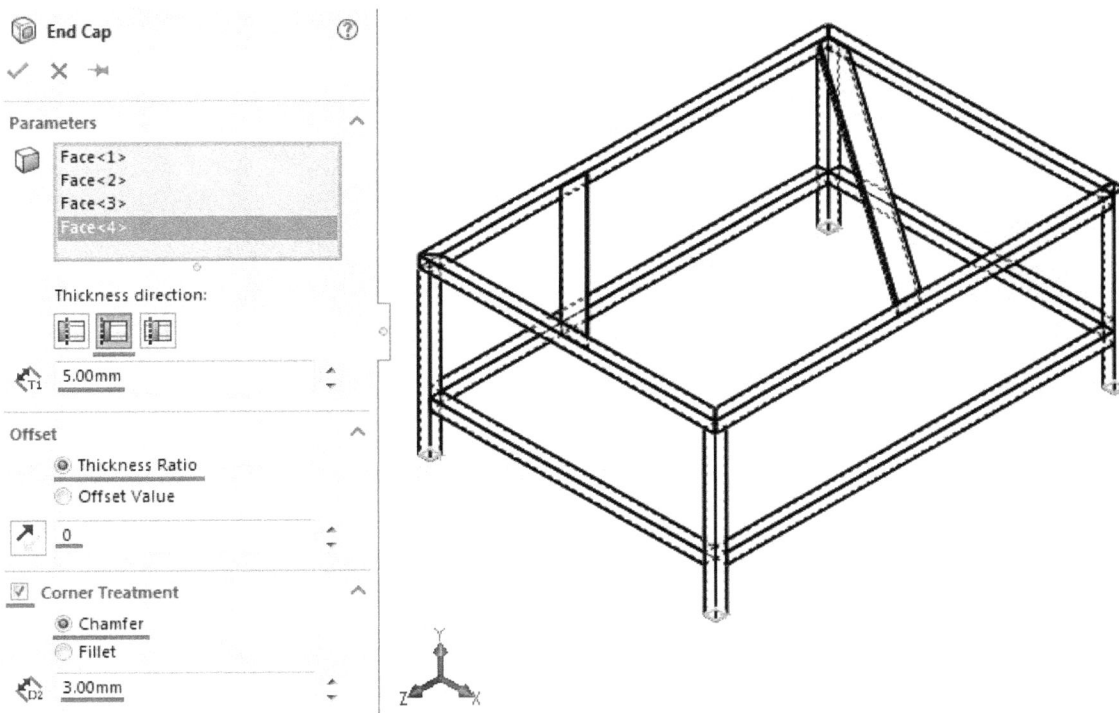

Figure 132 - Question 12 of 26 - End Cap PropertyManager

Click OK. Saves your Part.

Your Part should now look as shown in the following image.

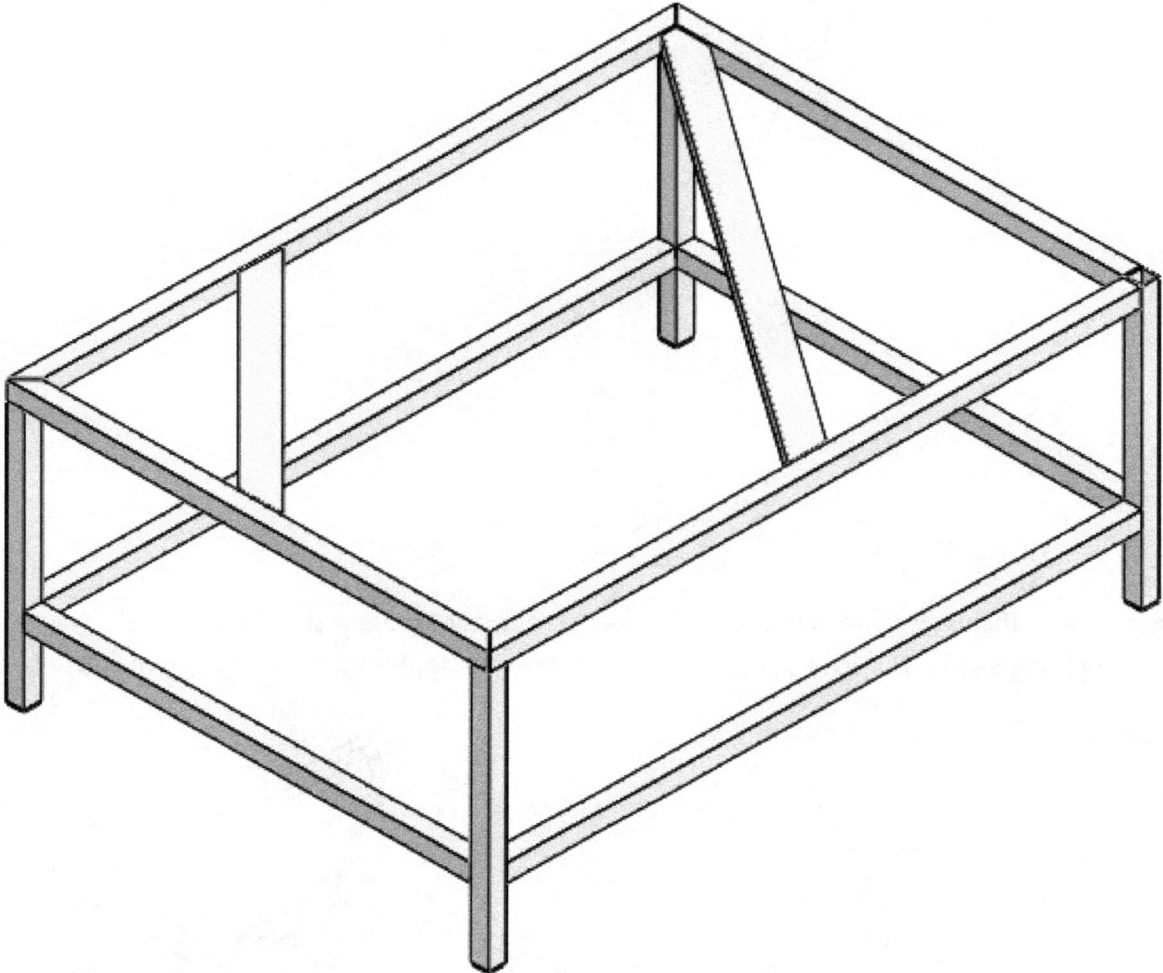

Figure 133 - Question 12 of 26 - Weldment Current Status

MASS PROPERTIES

Under the Evaluate Tab, click on the Mass Properties feature or Tools > Evaluate > Mass Properties. The mass of the part is 16603.52 grams as shown in image below.

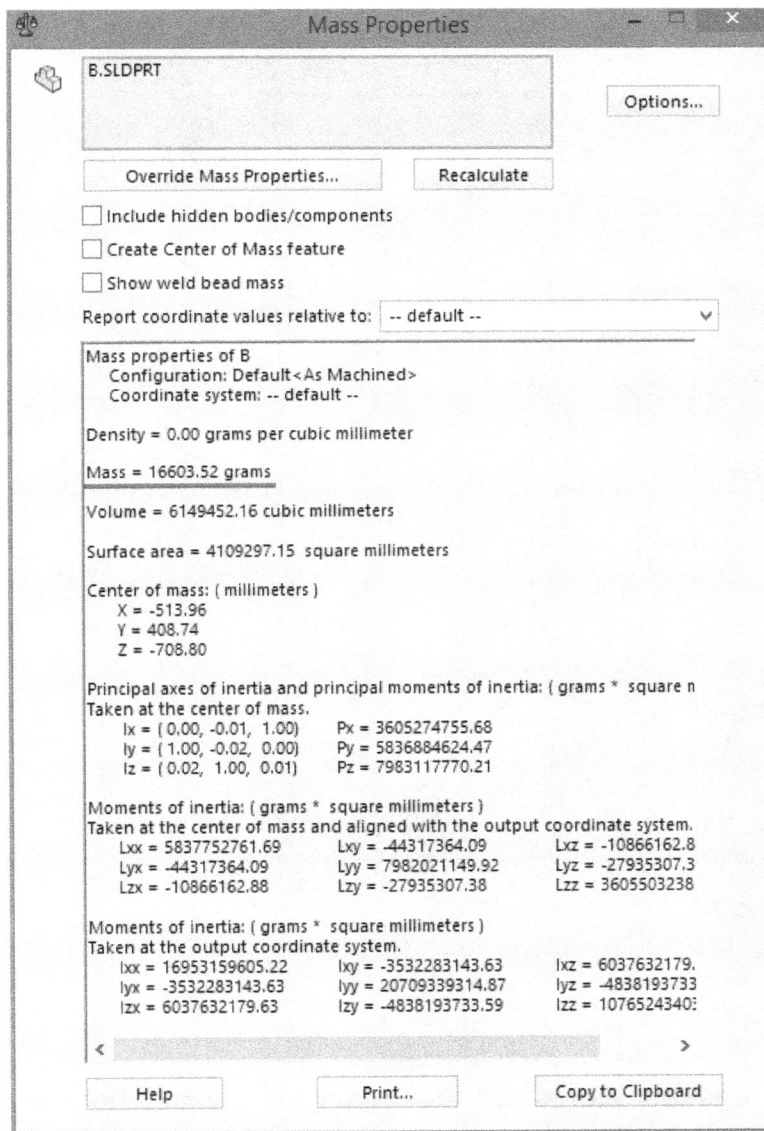

Figure 134 - Question 12 of 26 - Mass Properties

Save and close the part.

CHAPTER 13

QUESTION 13 EXAM SCREEN CAPTURES

QUESTION 13 EXAM SCREEN CAPTURE 1

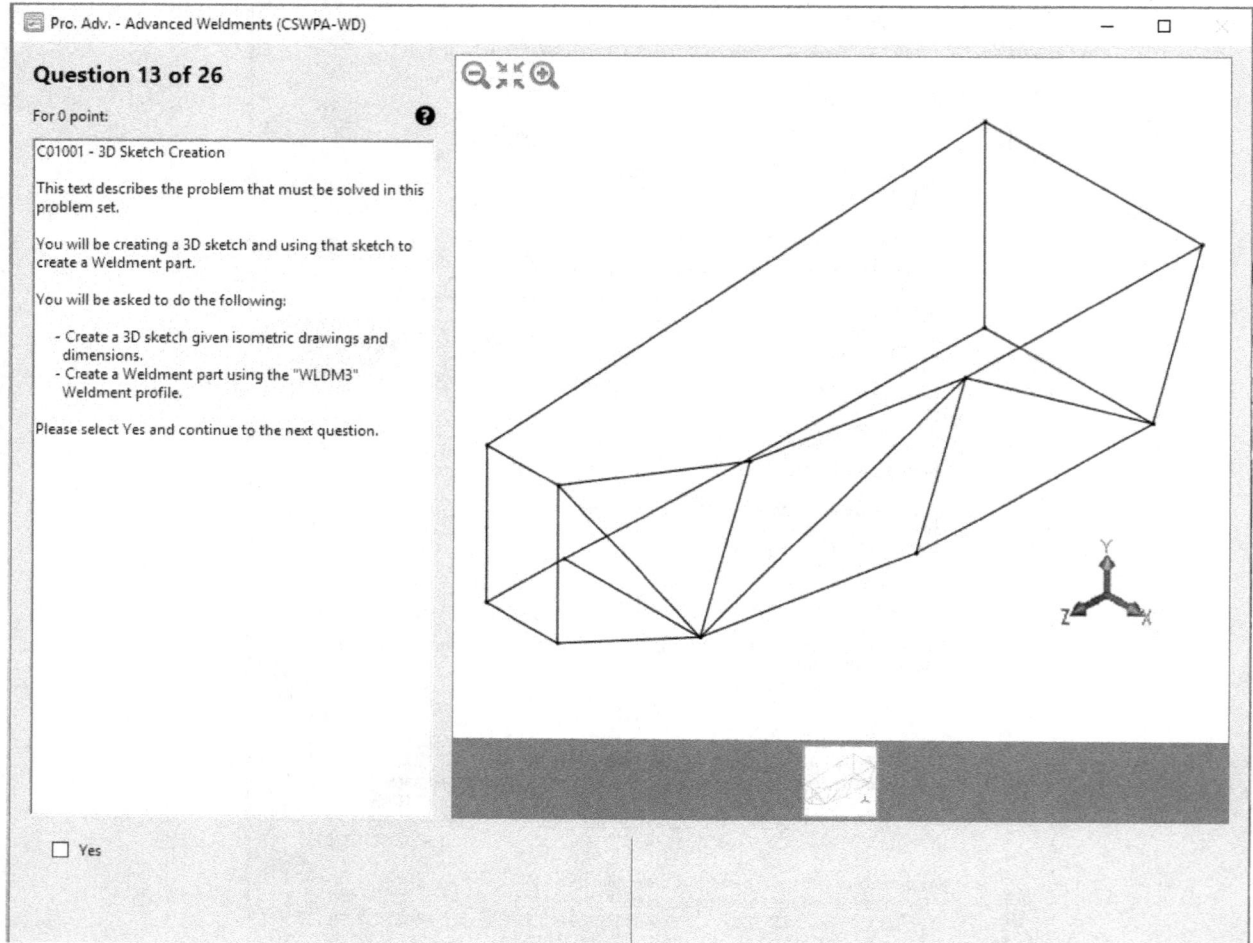

Figure 135 - Question 13 of 26 - Exam Screen Capture 1

Read the instructions under Question 13 then select Yes and click Continue to go to the next question - Question 14.

CHAPTER 14

QUESTION 14 EXAM SCREEN CAPTURES

QUESTION 14 EXAM SCREEN CAPTURE 1

Pro. Adv. - Advanced Weldments (CSWPA-WD)	— ☐

Question 14 of 26

For 20 points: ❓

C02005 - 3D Sketch Creation And Total Length
Build this 3D Sketch in SolidWorks.
Unit system: MMGS (millimeter, gram, second)
Decimal places: 2
Part origin: Arbitrary

-Build this 3D Sketch in SolidWorks using the following parameters:

A = 199
B = 288

Note 1: All the lowest elements can be assumed to lie on the XZ plane.

Note 2: Please refer to both dimensioned isometric images before proceeding to ensure you understand all parameters, dimensions, and relations required.

-After the 3D sketch is created, select all the sketch elements and then click on Tools > Measure to display the total length of the 3D sketch elements.

What is the total length of all the 3D Sketch elements in the 3D Sketch (mm)?

○ 6521 ○ 6735

○ 6869 ○ 6936

Figure 136 - Question 14 of 26 - Exam Screen Capture 1

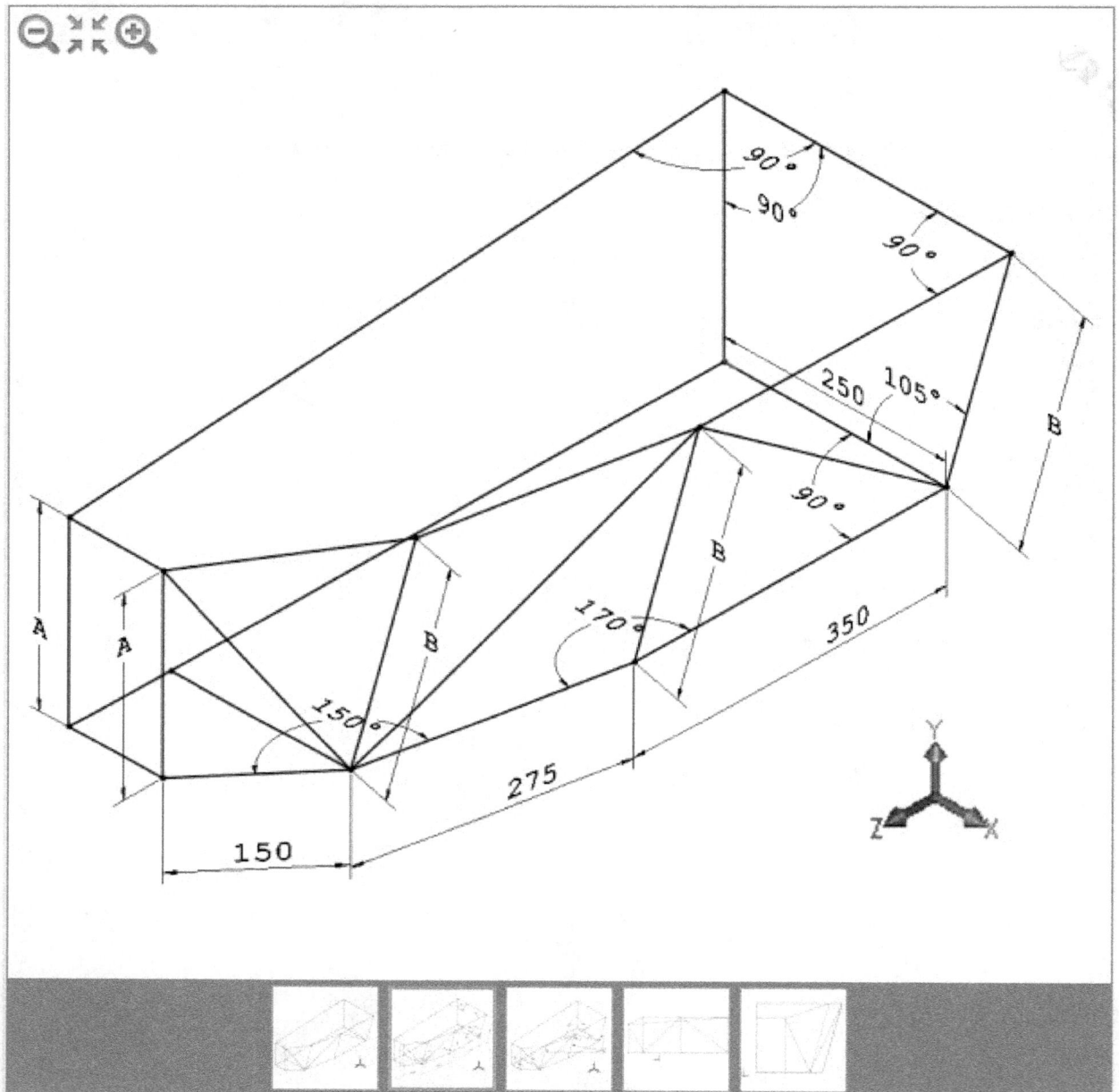

Figure 137 - Question 14 of 26 - Exam Screen Capture 2

Figure 138 - Question 14 of 26 - Exam Screen Capture 3

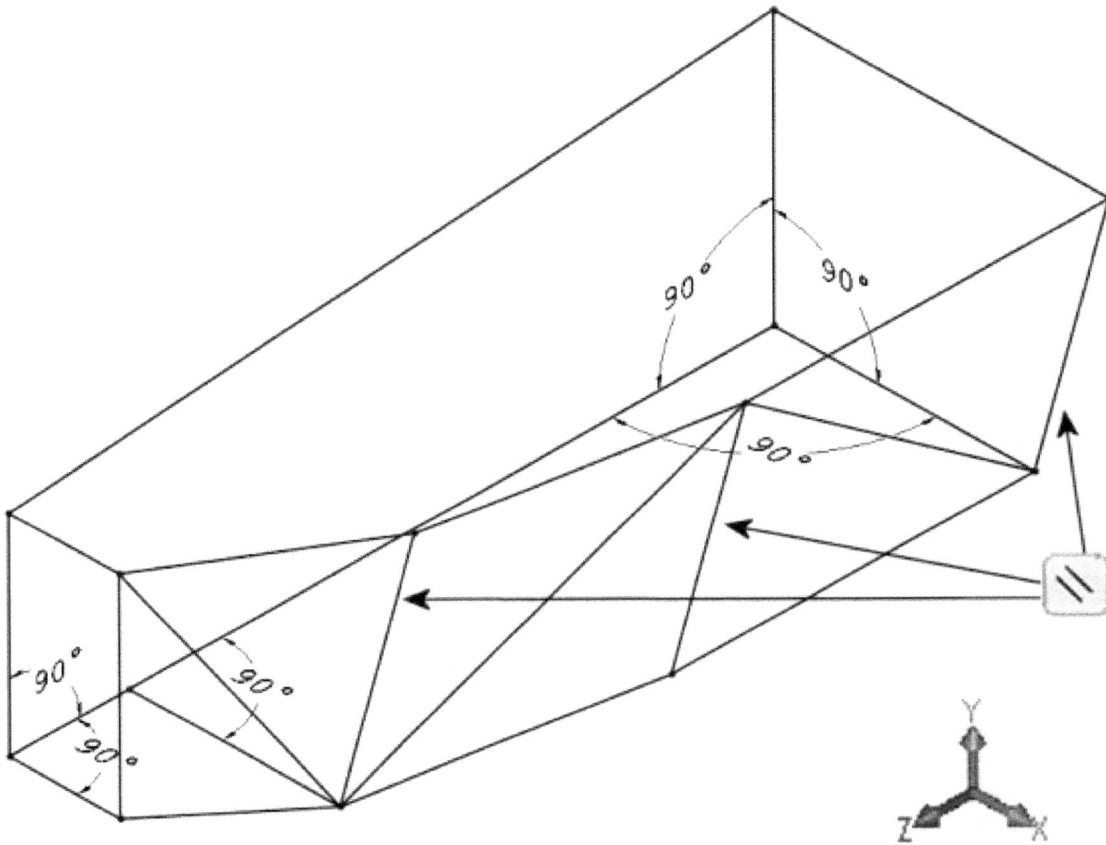

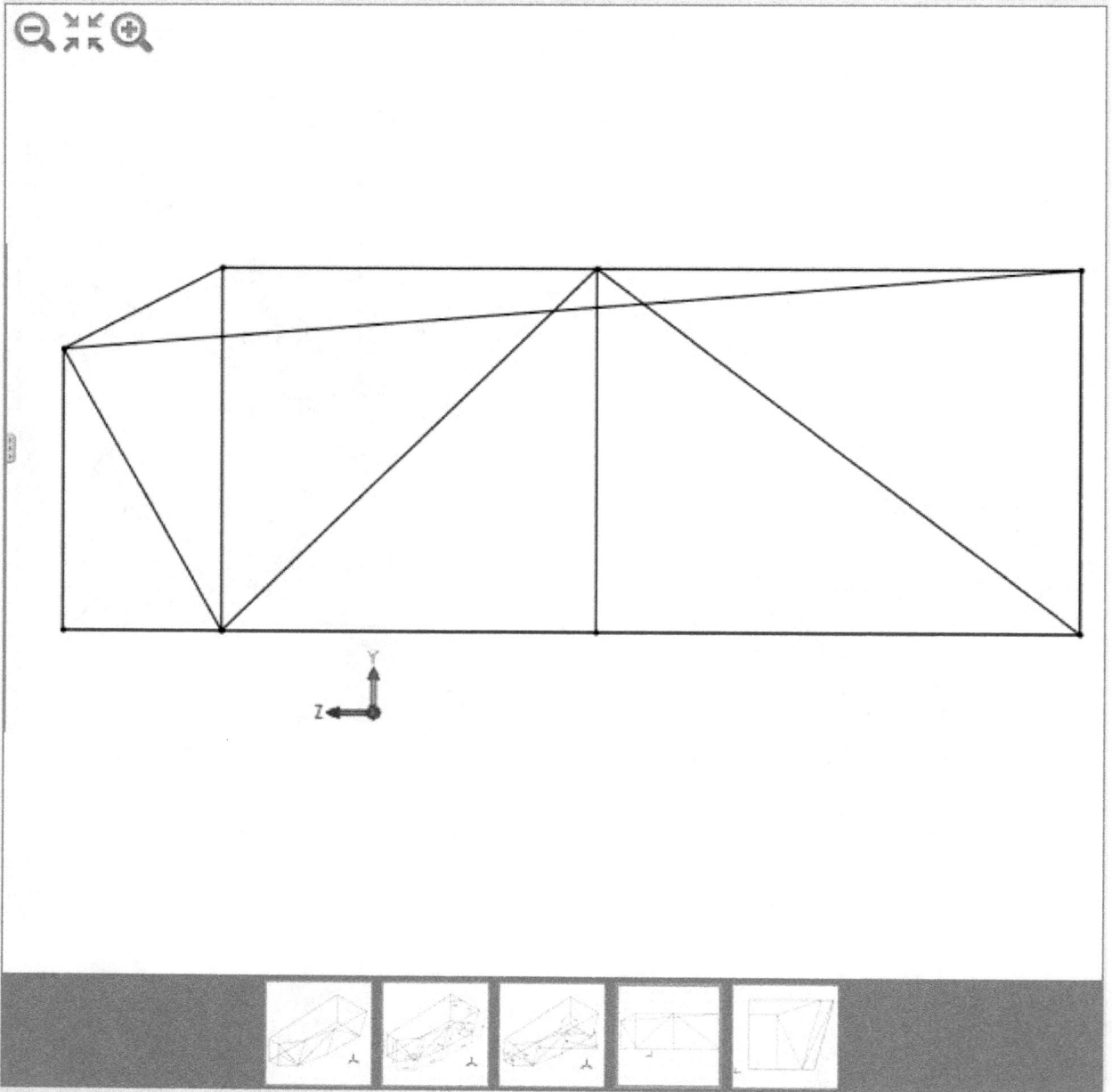

Figure 139 - Question 14 of 26 - Exam Screen Capture 4

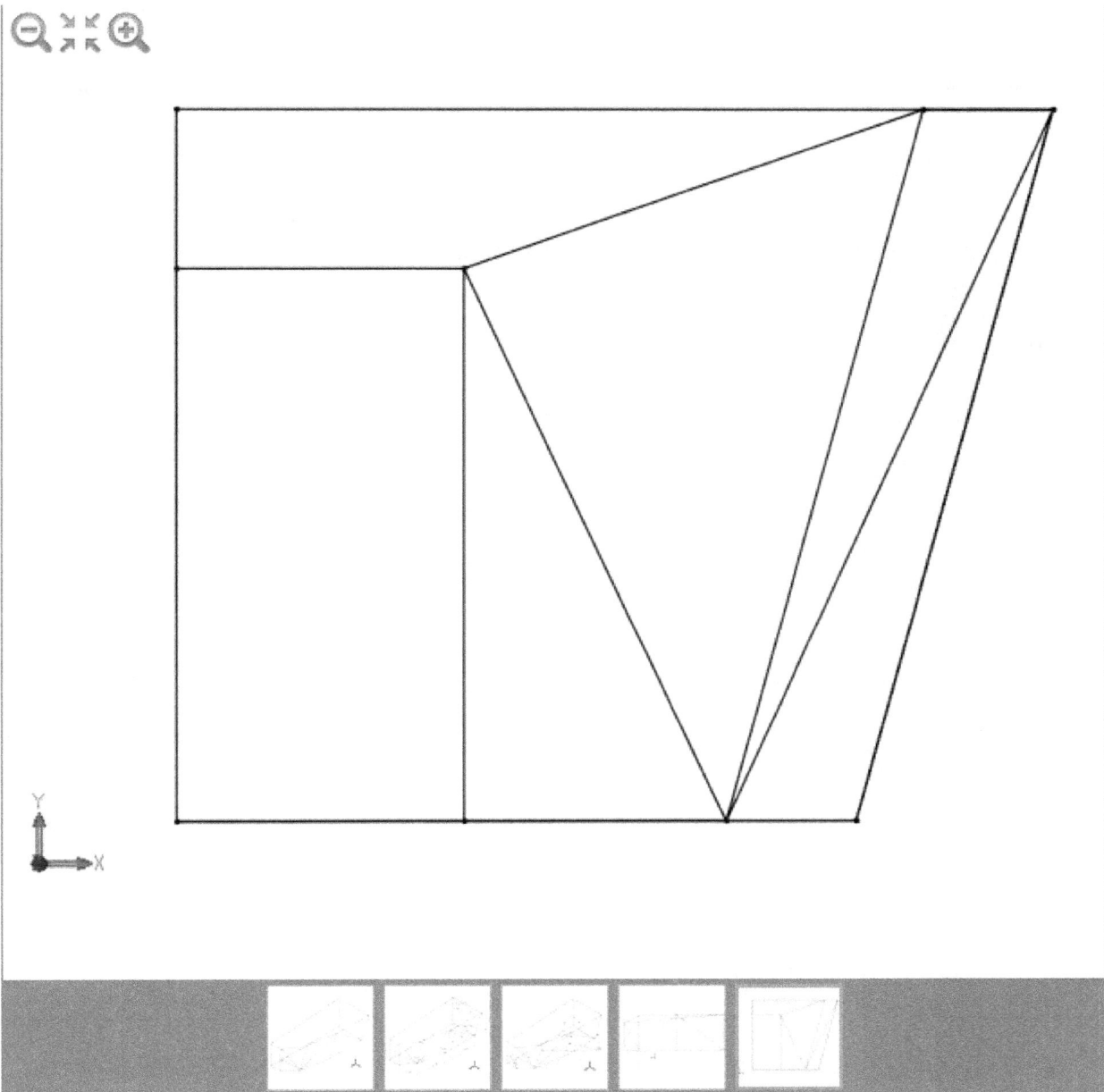

Figure 140 - Question 14 of 26 - Exam Screen Capture 5

Start a new Part in Solidworks and save it as Question 14.

SET A UNIT SYSTEM

Go to Tools > Options > Document Properties > Drafting Standard. From the drop down menu, Select the ISO Standard since our dimensions are given in millimeters.

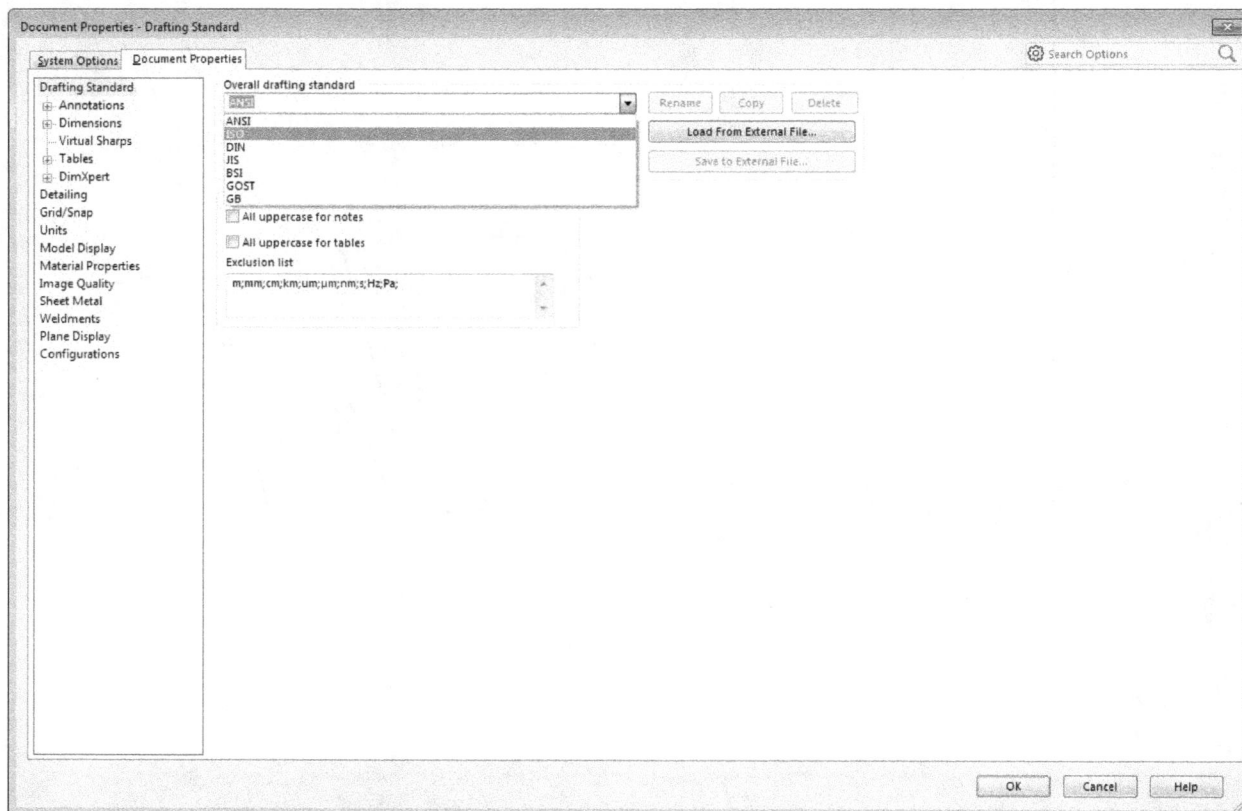

Figure 141 - Question 14 of 26 - Document Properties - Drafting Standard

Go to Tools > Options > Document Properties > Units to change the Unit System to MMGS (millimeter, gram, second) and to also set the number of decimal places to two decimal places.

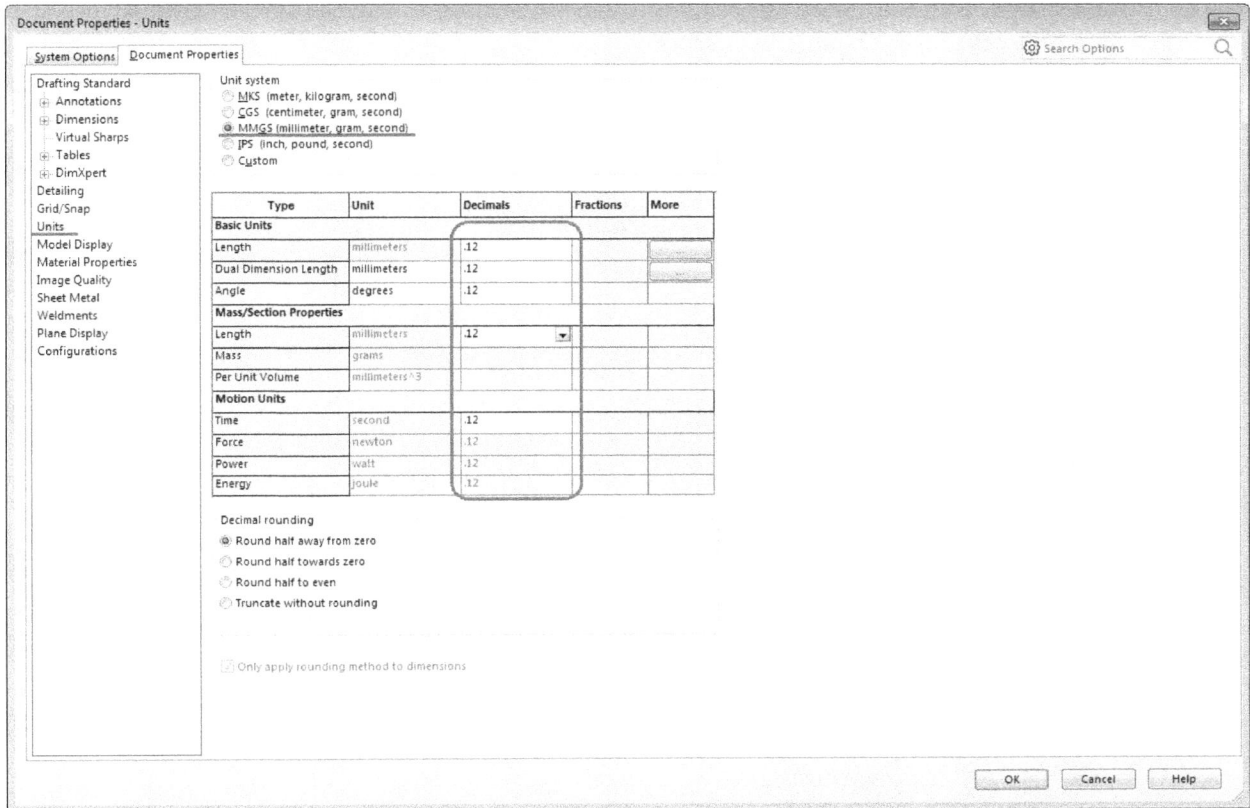

Figure 142 - Question 14 of 26 - Document Properties - Units

Click OK.

3D SKETCHING

Click 3D Sketch (Sketch toolbar) or Insert > 3D Sketch. The view changes to Isometric.

ADDING A 3D SKETCH PLANE

Click Plane (Sketch toolbar) or Tools > Sketch Entities > Plane to display the Sketch Plane PropertyManager. In the Sketch Plane PropertyManager under First Reference - select the Top Plane in the FeatureManager Design Tree and then select the coincident relation as shown in the following image.

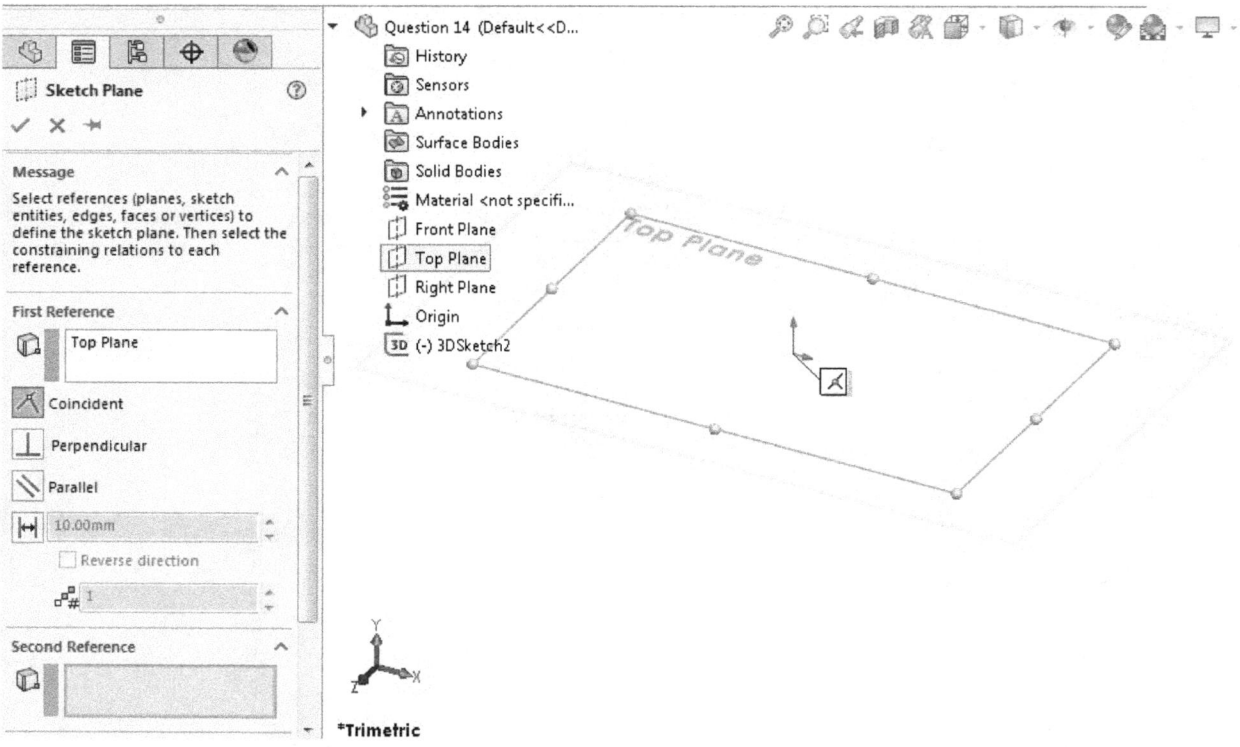

Figure 143 - Question 14 of 26 - Creating a 3D Sketch Plane

Click OK. Create a sketch as shown in the following image using the Line Tool - take note of the automatic sketch relations in the following image.

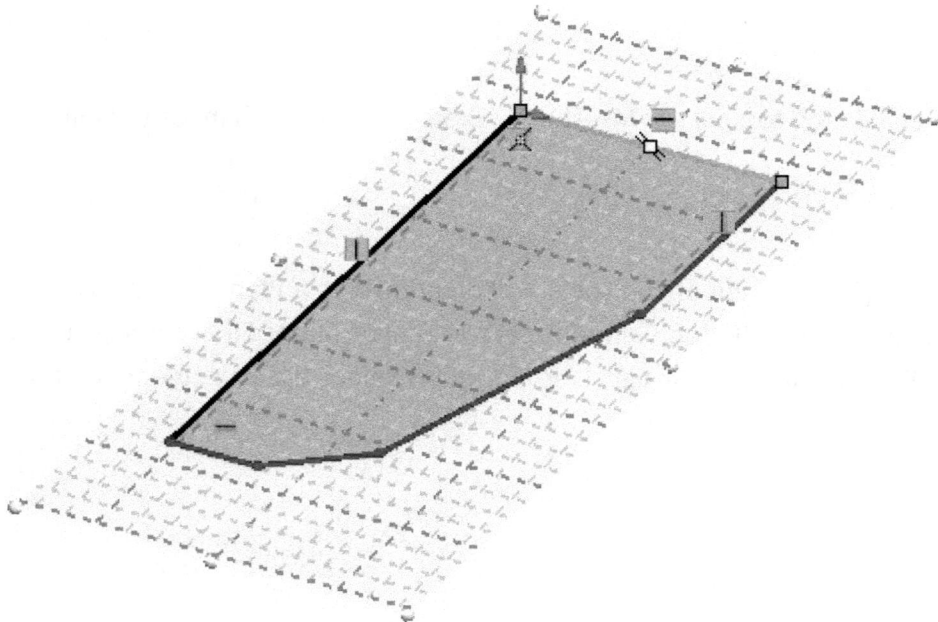

Figure 144 - Question 14 of 26 - Creating a 3D Sketch

Add dimensions to the created sketch as shown in the following image to fully define the sketch - my advice is to add the angular dimensions first - the 150 and 170 Degree Angle Dimensions. To modify the line by dragging, do one of the following: To change the length of the line, select one of the endpoints and drag to lengthen or shorten the line. To move the line, select the line and drag the line to another position - DO THIS TO MAINTAN THE OVERALL SHAPE AS YOU ADD DIMENSIONS ONE AFTER THE OTHER IF THE GENERAL SHAPE DISTORTS.

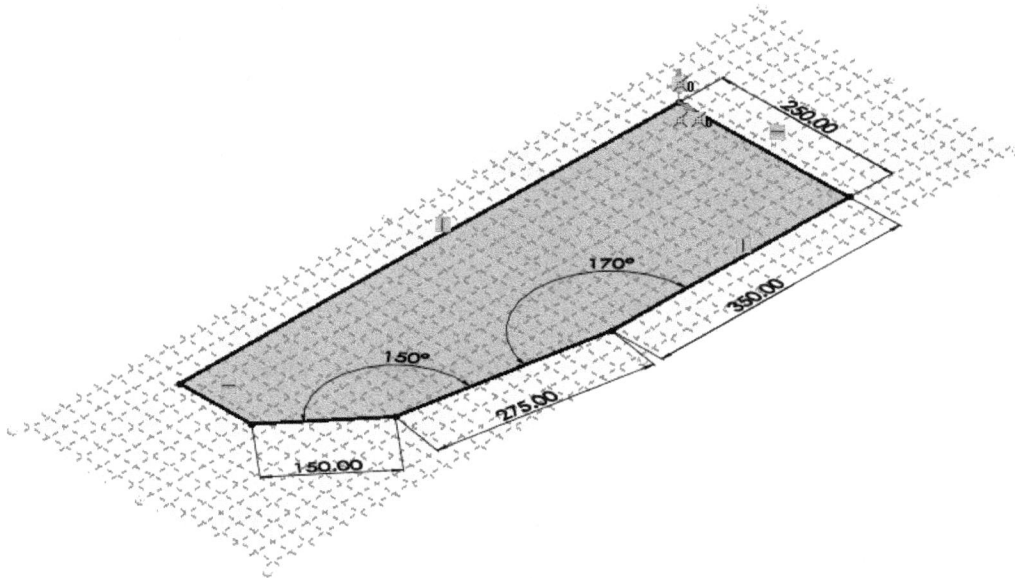

Figure 145 - Question 14 of 26 - Adding dimensions to a Sketch on a 3D Sketch Plane

Exit the 3D Sketch by clicking the Exit Sketch in top left Confirmation Corner (in the Graphics Area). Save Your Part. Click or Right Click on the 3D Sketch in the FeatureManager Design Tree and select Edit Sketch as shown in the following image.

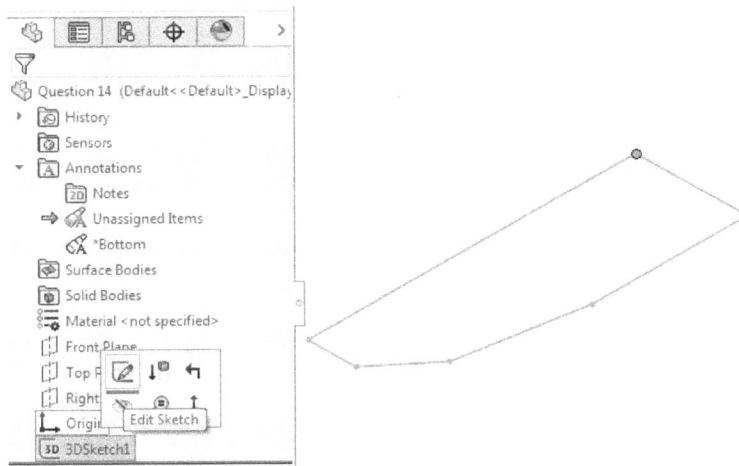

Figure 146 - Question 14 of 26 - Editing a 3D Sketch

117

Your sketch should now appear as shown in the following image.

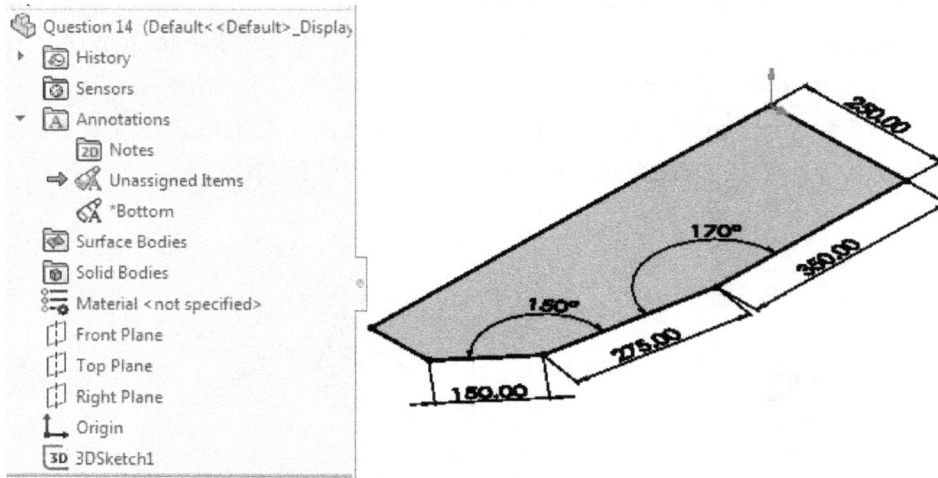

Figure 147 - Question 14 of 26 - 3D Sketch current status

Click Line on the Sketch toolbar, or click Tools > Sketch Entities > Line.

The 3D Line PropertyManager appears and the pointer changes to ✎ᵡʸ . Click the origin in the graphics area to start the line - note the yellow circle to indicate a coincident relation between the origin and the new line you are creating.

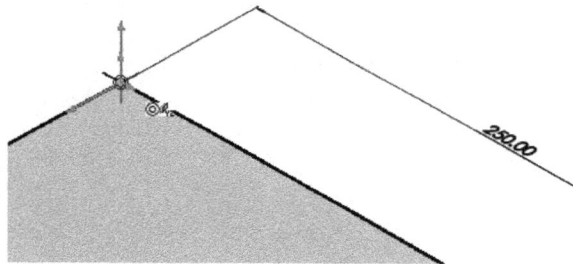

Figure 148 - Question 14 of 26 - 3D Sketch current status

Each time you click, the space handle appears to orient your sketch. If you want to change planes, press Tab.

Drag vertically Along Y to plus or minus 300mm as shown in the following image.

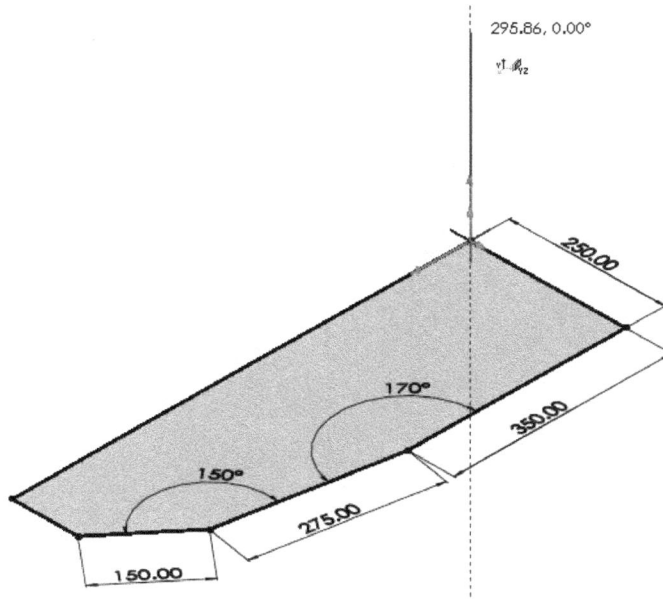

Figure 149 - Question 14 of 26 - 3D Sketch current status

Click to end the line segment then press the Escape key on your keyboard. Your sketch should now appear as shown in the following image.

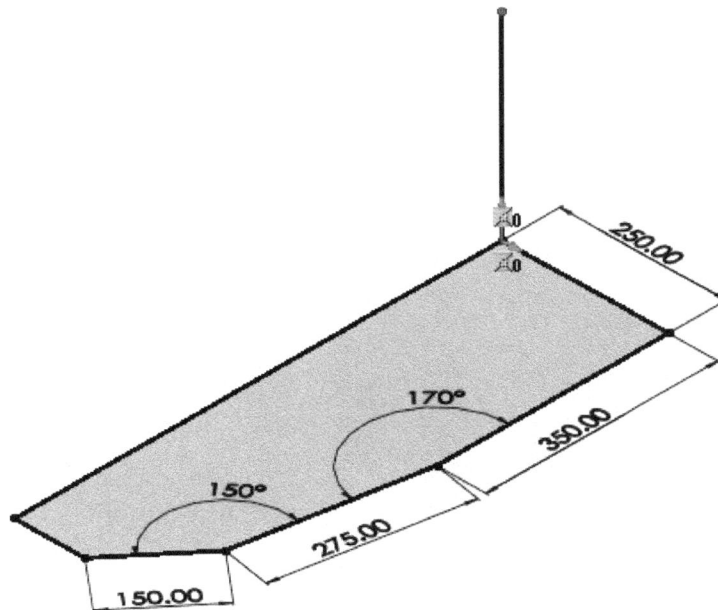

Figure 150 - Question 14 of 26 - 3D Sketch current status

Click Line on the Sketch toolbar, or click Tools > Sketch Entities > Line and add two more line segments as shown in the following image. Note that the vertical line is Along the Y-Axis and

119

the horizontal line in blue has no relations except that its end points are coincident to the two vertical lines which are both along the Y-Axis.

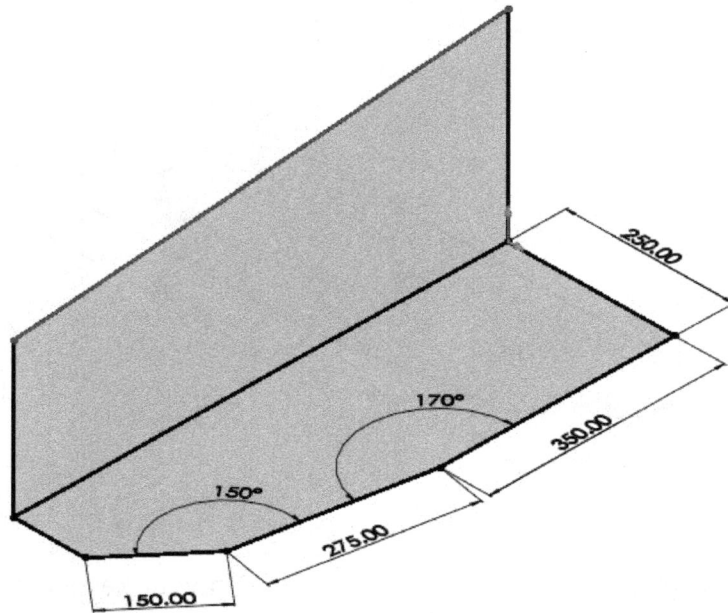

Figure 151 - Question 14 of 26 - 3D Sketch current status

Click Line on the Sketch toolbar, or click Tools > Sketch Entities > Line and add one more line segment as shown in the following image.

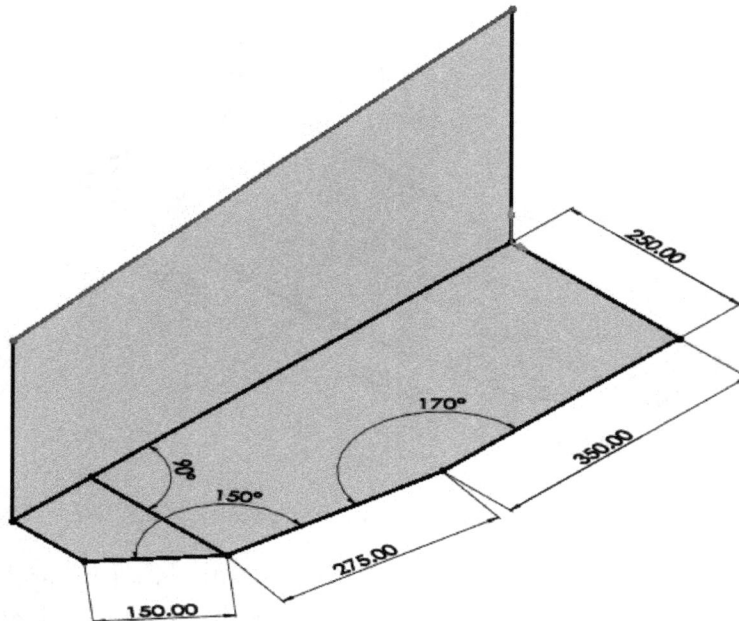

Figure 152 - Question 14 of 26 - 3D Sketch current status

Click Line on the Sketch toolbar, or click Tools > Sketch Entities > Line and add two more line segments as shown in the following image - NB: Also add an equal relation between the two vertical lines which are both Along the Y-Axis as shown in the following image.

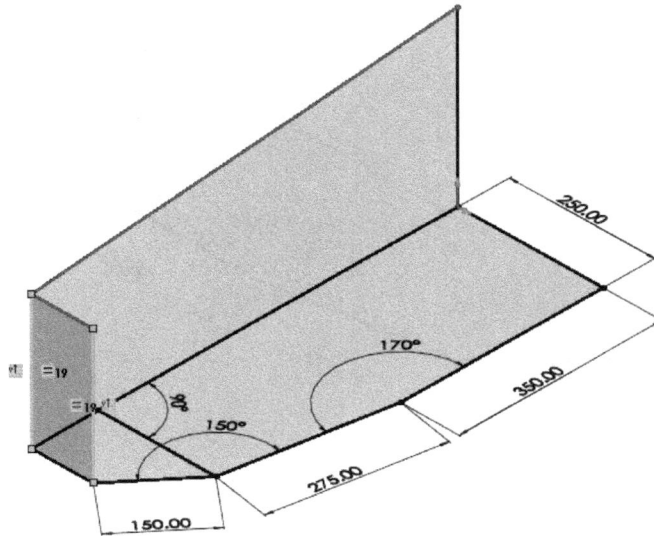

Figure 153 - Question 14 of 26 - 3D Sketch current status

Click Line on the Sketch toolbar, or click Tools > Sketch Entities > Line and add one more line segment as shown in the following two images - press Tab to make sure this new line is on the XY Plane and is Along X.

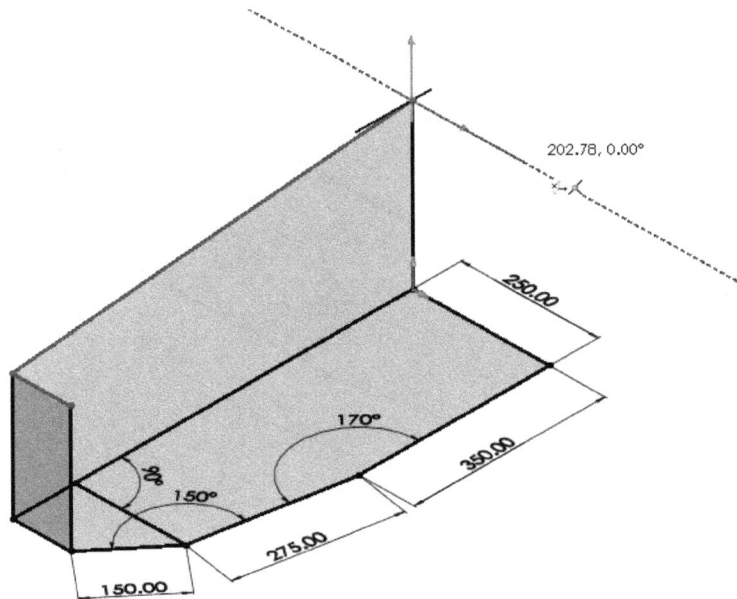

Figure 154 - Question 14 of 26 - 3D Sketch current status

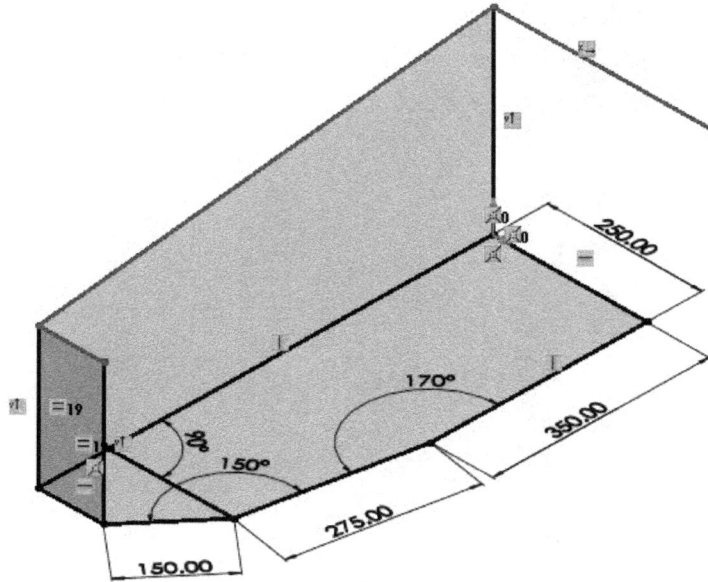

Figure 155 - Question 14 of 26 - 3D Sketch current status

Click Line on the Sketch toolbar, or click Tools > Sketch Entities > Line and add one more line segment as shown in the following image - add the angular dimension of 105 Degrees.

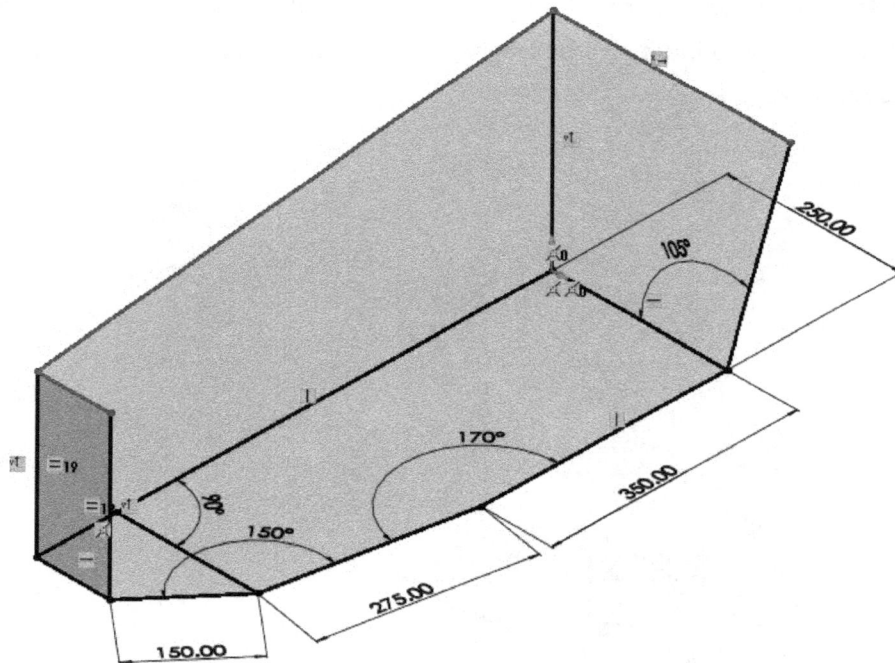

Figure 156 - Question 14 of 26 - 3D Sketch current status

Click Line on the Sketch toolbar, or click Tools > Sketch Entities > Line and add two more line segments as shown in the following image - NB: Make sure no automatic relation is added as you create each of these two line segments.

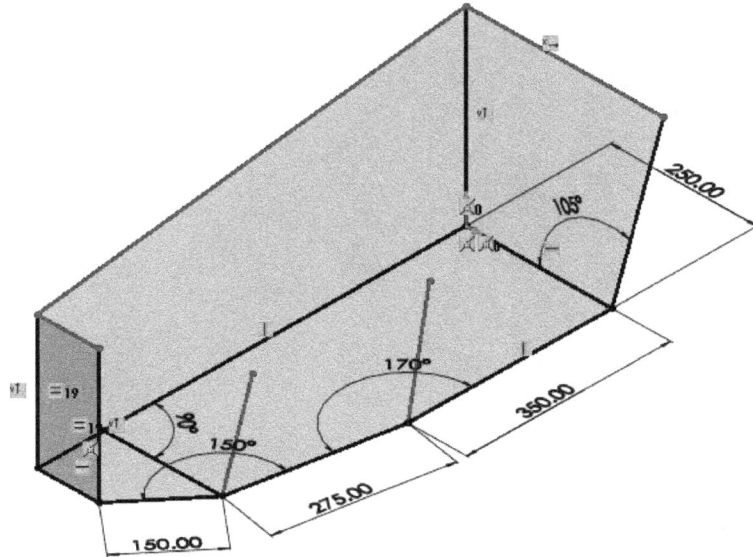

Figure 157 - Question 14 of 26 - 3D Sketch current status

Add equal and parallel relations to the two new vertical lines and the existing line as shown in the following two images.

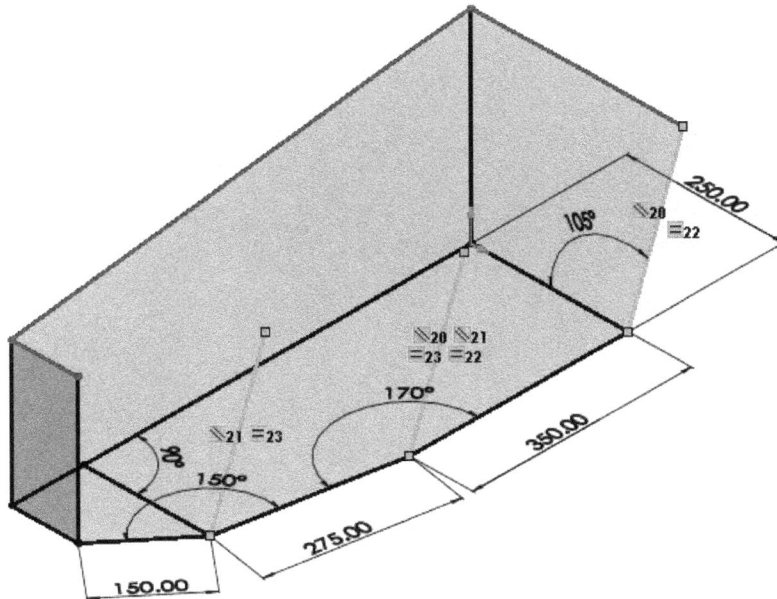

Figure 158 - Question 14 of 26 - 3D Sketch current status

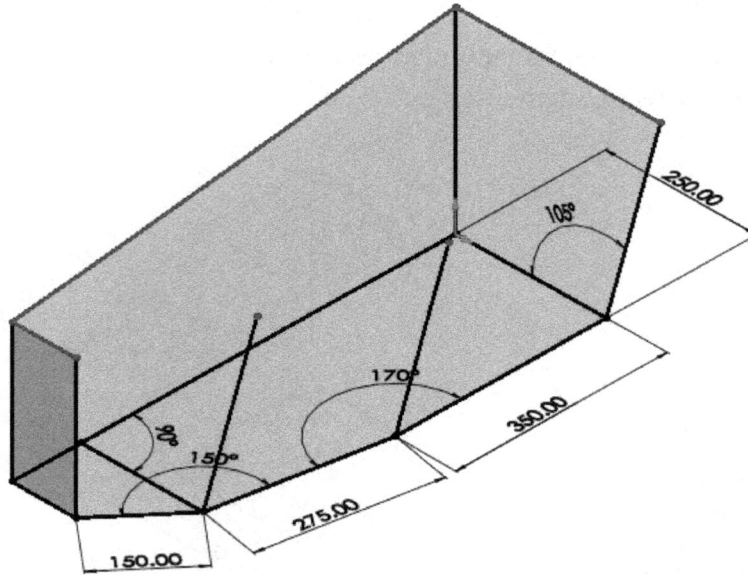

Figure 159 - Question 14 of 26 - 3D Sketch current status

Click Line on the Sketch toolbar, or click Tools > Sketch Entities > Line and add three more line segments as shown in the following two images - NB: Make sure no automatic relation is added other than the coincident relation as you create each of these three line segments.

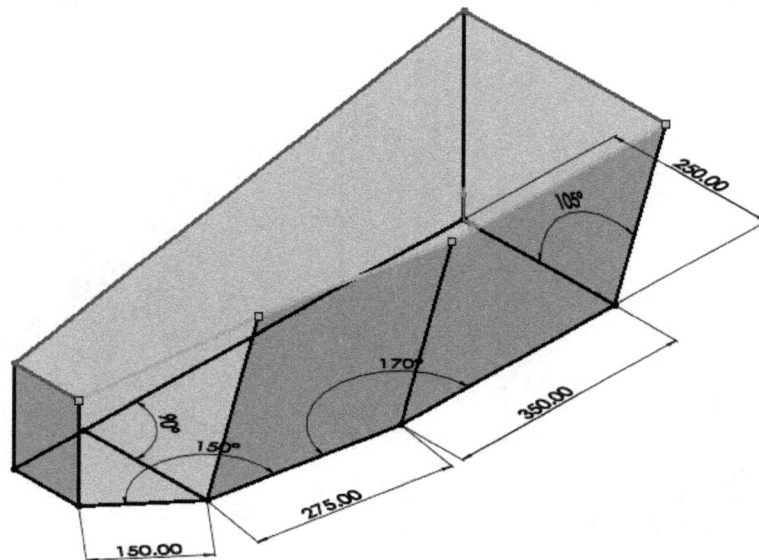

Figure 160 - Question 14 of 26 - 3D Sketch current status

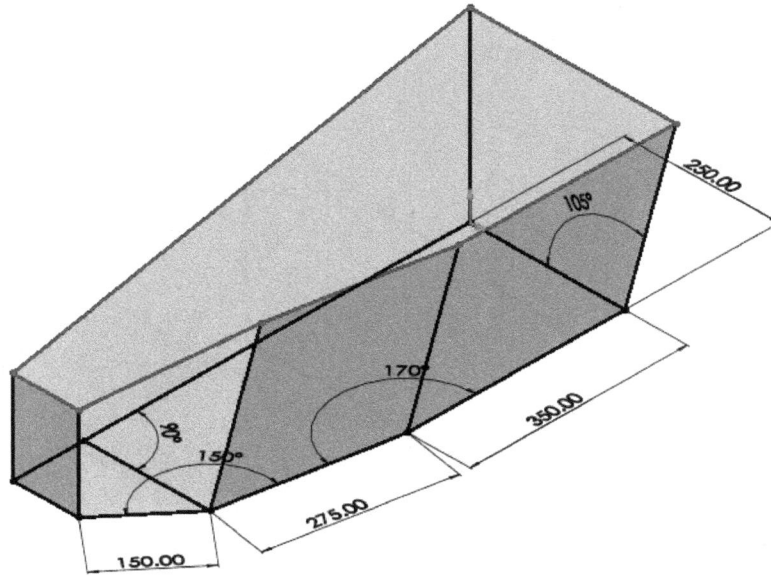

Figure 161 - Question 14 of 26 - 3D Sketch current status

Click Smart Dimension on the Dimensions/Relations toolbar, or click Tools > Dimensions > Smart. Add two dimensions as shown in the image below - 199mm (A) and 288mm (B).

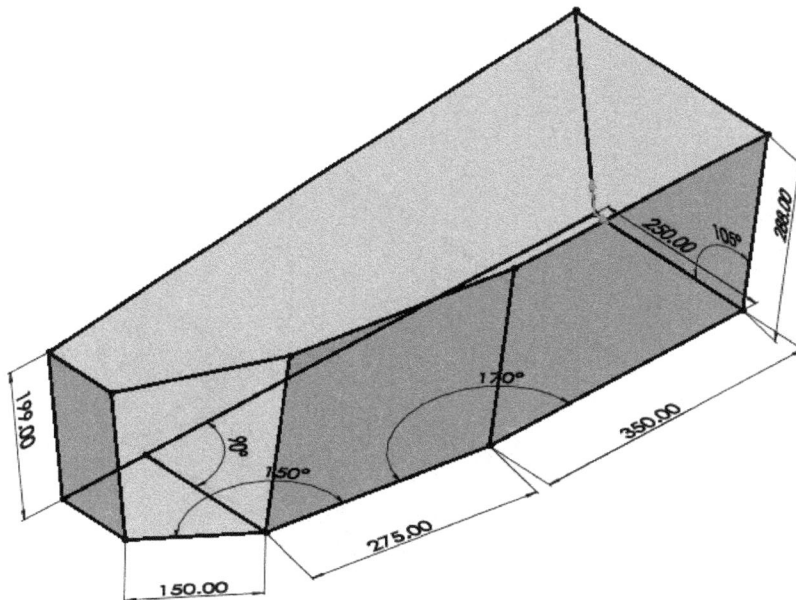

Figure 162 - Question 14 of 26 - Fully Defined 3D Sketch

Click Line on the Sketch toolbar, or click Tools > Sketch Entities > Line and add three more line segments as shown in the following two images - NB: Make sure no automatic relation is added other than the coincident relation *(start and end of each of the three line segments)* as you create each of these three line segments.

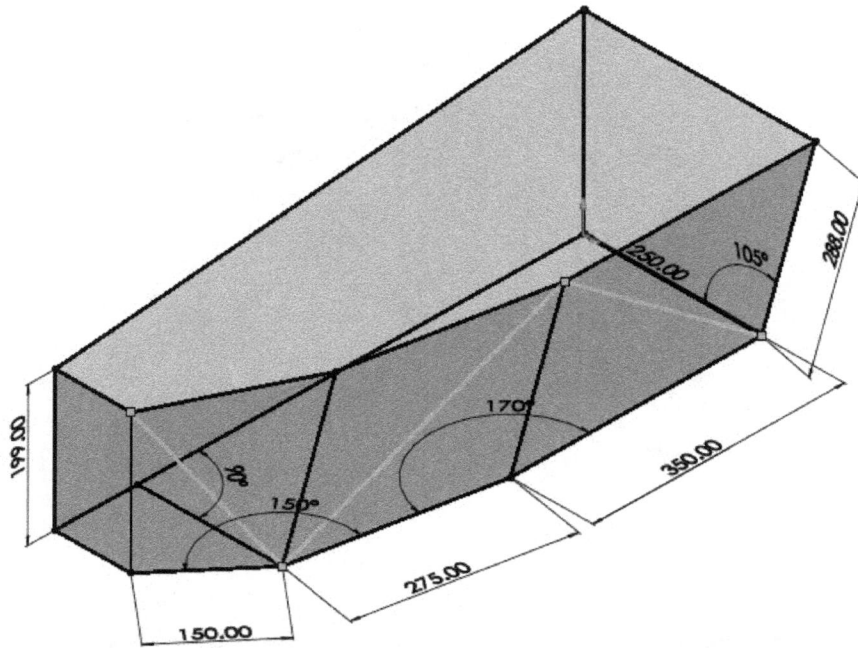

Figure 163 - Question 14 of 26 - Fully Defined 3D Sketch

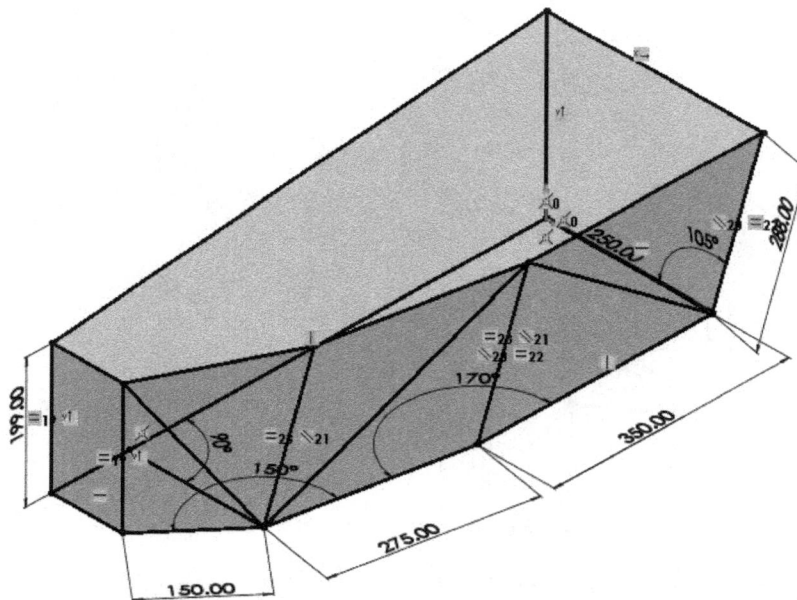

Figure 164 - Question 14 of 26 - Fully Defined 3D Sketch

As a general check *(once satisfied that your isometric view is correct)*, compare the Right View of the 3D Sketch you have created against Exam Screen Capture 4 and the Front View against Exam Screen Capture 5. The following table shows these two comparisons - NB: Checking your work against what is requested in the question is critical throughout this exam - this should be your first priority before you check if you have an answer that is on the list of possible answers since sometimes there is one or two answers consistent with common errors and or omissions that most people will make.

3D SKETCH RIGHT VIEW	EXAM SCREEN CAPTURE 4
3D SKETCH FRONT VIEW	**EXAM SCREEN CAPTURE 5**

TOTAL LENGTH OF 3D SKETCH ELEMENTS

To display the total length of all 3D Sketch elements in this 3D Sketch we have to select all the line segments and then click on Tools > Measure. The quickest way to select all the line segments is to click *(don't click and hold)* on one line segment then press Ctrl + A on your keyboard. All line segments in the 3D Sketch are then selected as shown in the following image.

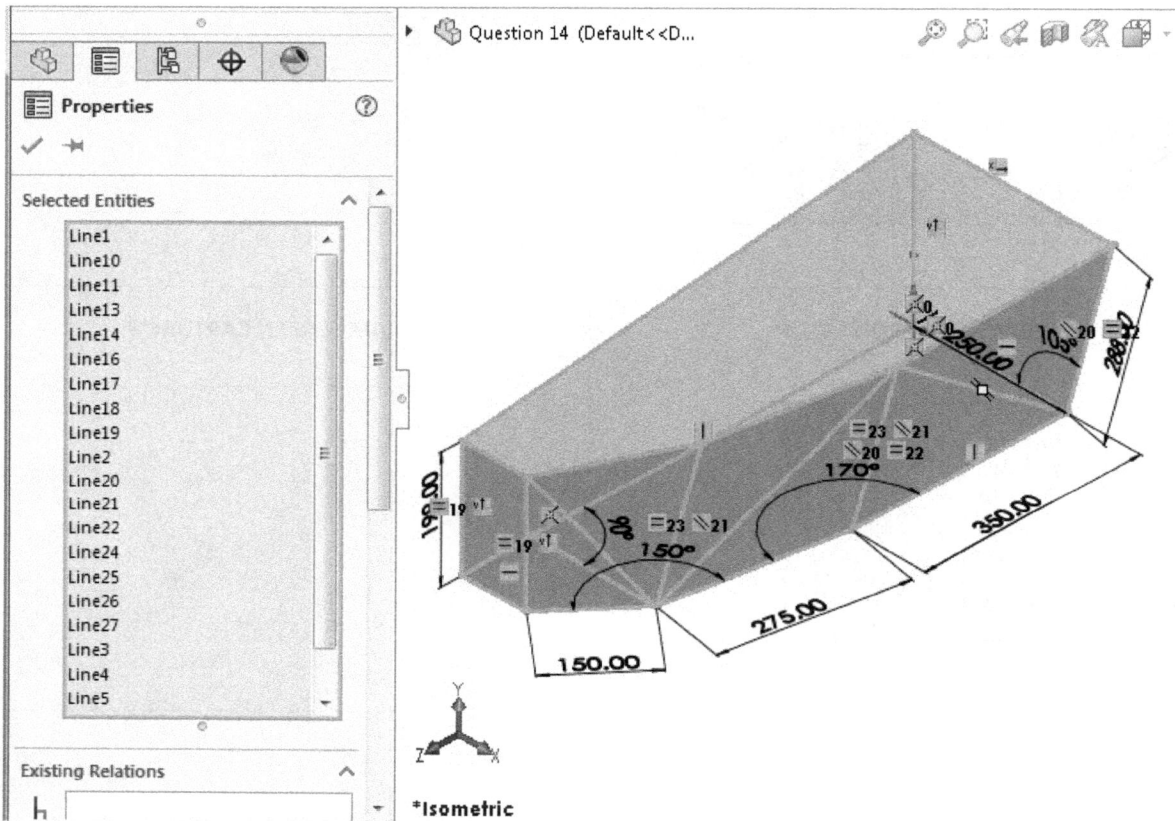

Figure 165 - Question 14 of 26 - Selecting all line segments in a 3D Sketch

To display the total length of all 3D Sketch elements in this 3D Sketch we have to select all the line segments

CHAPTER 15

QUESTION 15 EXAM SCREEN CAPTURES

QUESTION 15 EXAM SCREEN CAPTURE 1

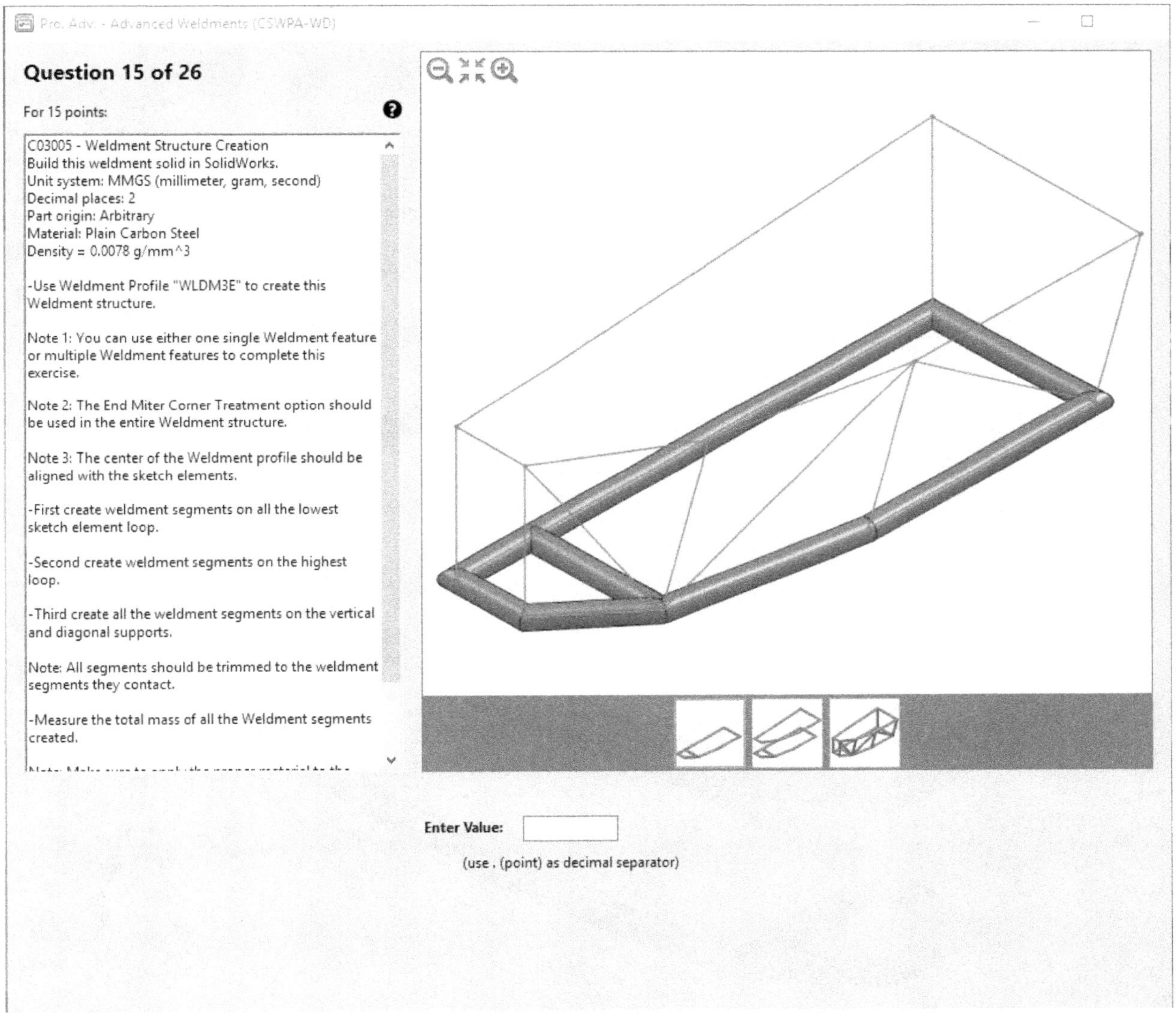

Figure 166 - Question 15 of 26 - Exam Screen Capture 1

QUESTION 15 EXAM SCREEN CAPTURE 2

-Third create all the weldment segments on the vertical and diagonal supports.

Note: All segments should be trimmed to the weldment segments they contact.

-Measure the total mass of all the Weldment segments created.

Note: Make sure to apply the proper material to the part.

What is the total mass of all the Weldment segments (grams)?

Figure 167 - Question 15 of 26 - Exam Screen Capture 2

QUESTION 15 EXAM SCREEN CAPTURE 3

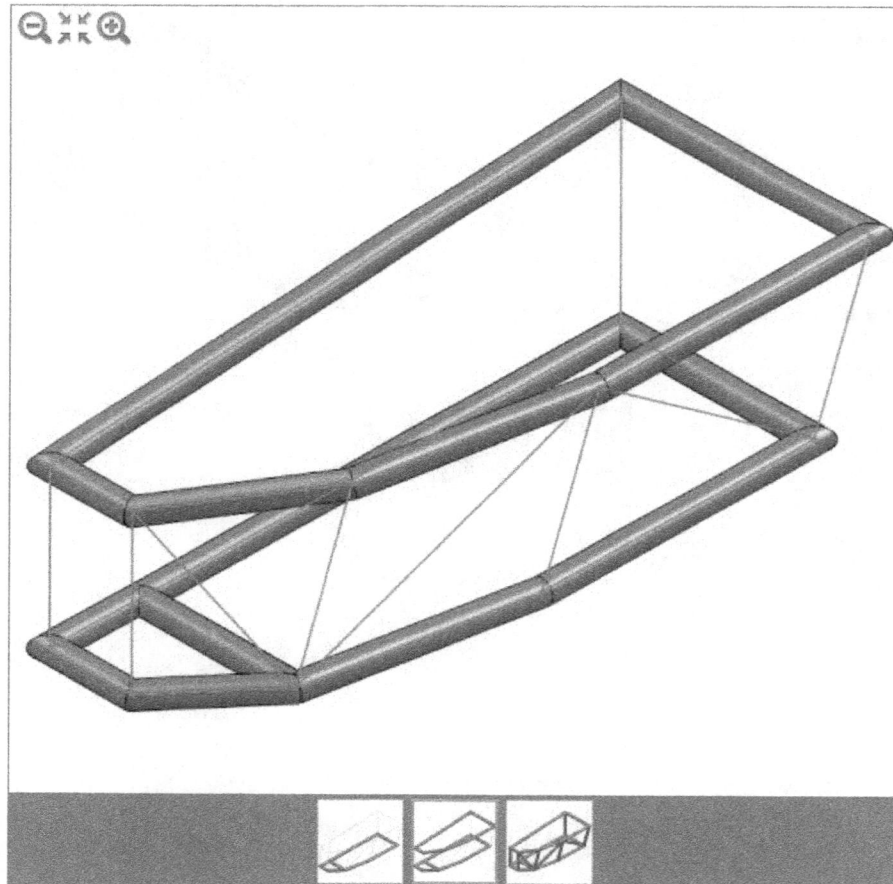

Figure 167 - Question 15 of 26 - Exam Screen Capture 3

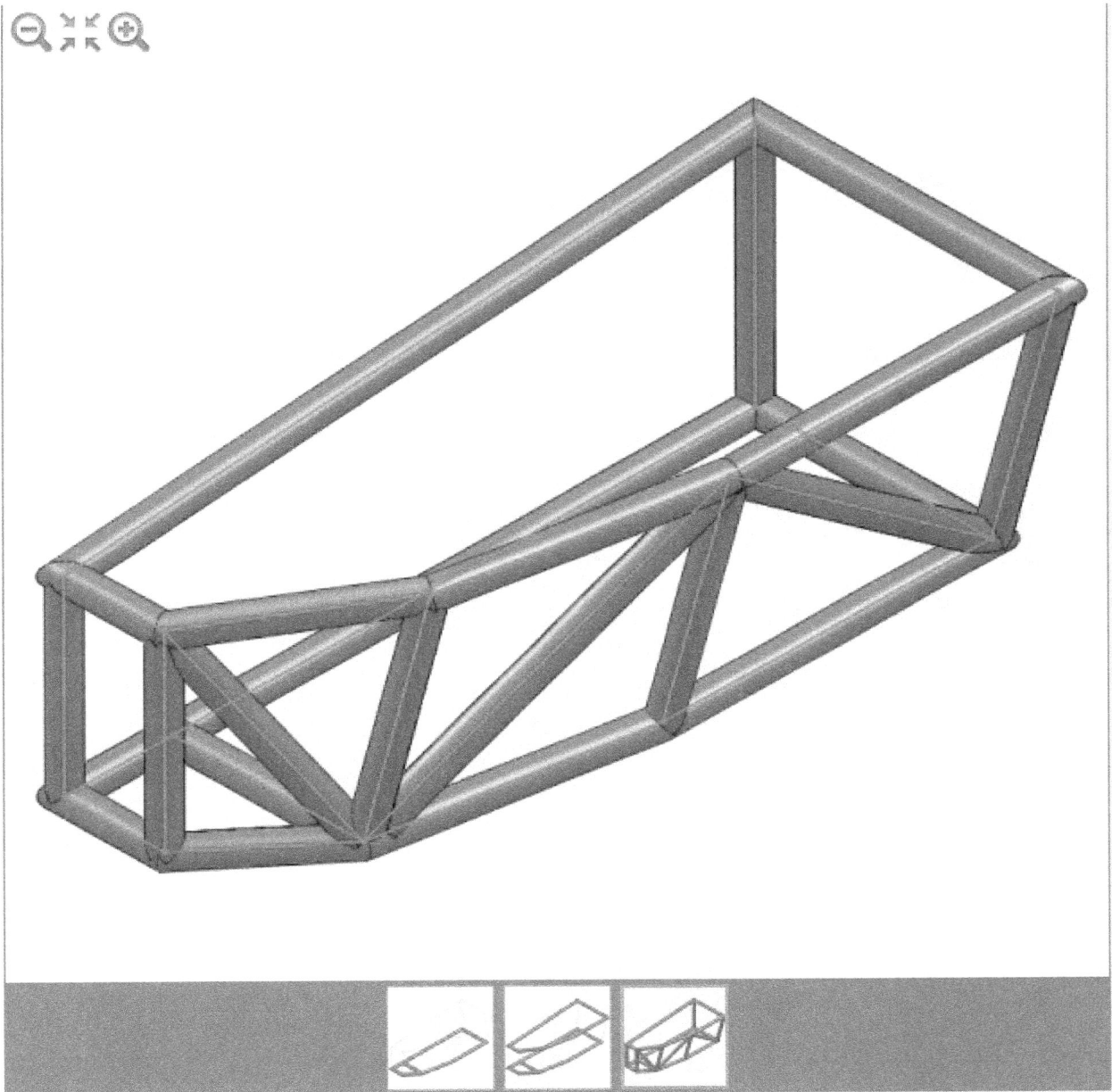

Figure 169 - Question 15 of 26 - Exam Screen Capture 4

USING SAVE AS A COPY

Since Question 15 is a continuation from Question 14, my advice is to use **Save As A Copy** to create a copy of Question 14 that you save as Question 15 then close the Question 14 Part. This is helpful in the event the 3D sketch breaks or an error occurs at least you can always go back to the original 3D Sketch in Question 14 or previous question as you go along with the exam. To do so, Click File, Save As and under the Save Menu select the **Save as copy and open** Radio Button as shown below.

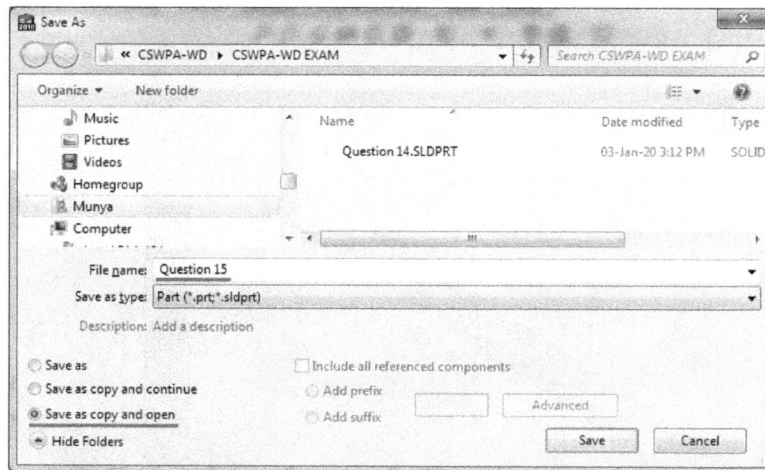

Figure 170 - Question 15 of 26 - Save as a Copy and Open

Click Window and then Close Question 14 as shown below by clicking on the X next to QUESTION1:

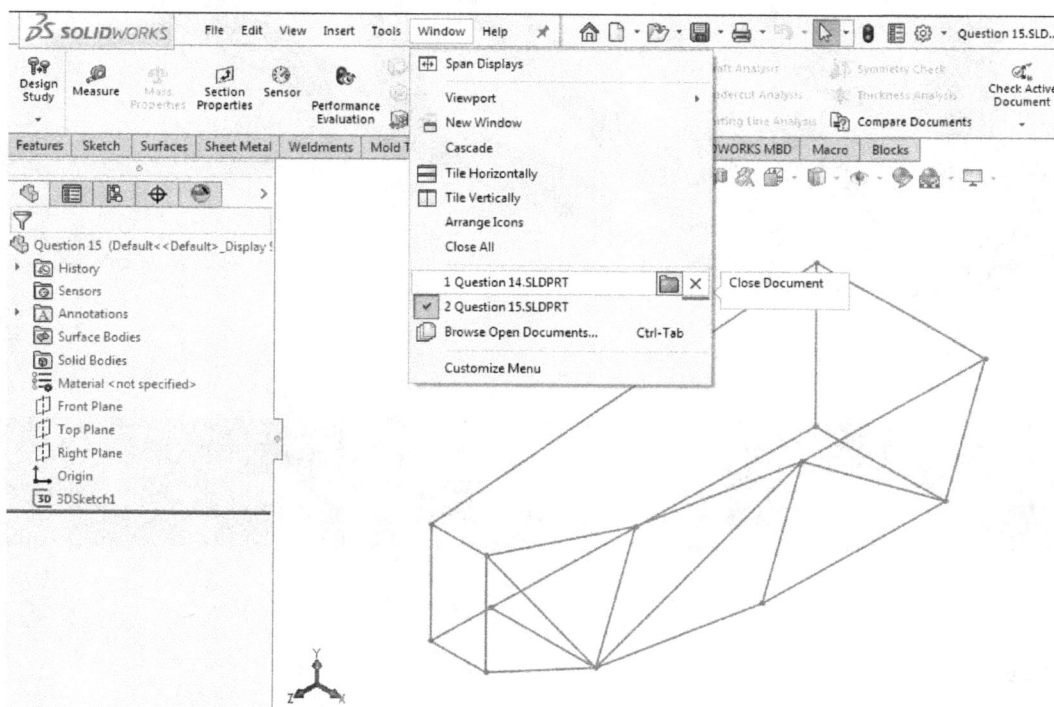

Figure 171 - Question 15 of 26 - Close Question 14

If the Save modified documents Menu comes up, click Save All as shown below. This will save and close the Question 14 Part and leave the new Question 15 Part Open.

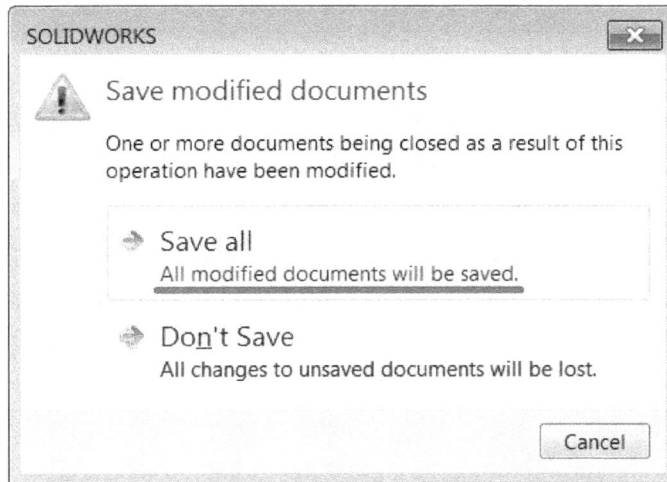

Figure 172 - Question 15 of 26 - Save Modified Documents Menu

ADDING A STRUCTURAL MEMBER

Click Structural Member under the Weldments toolbar or Insert > Weldments > Structural Member. Make selections in the PropertyManager to define the profile for the structural member as shown in the following image. **Deselect** Transfer Material from Profile to prevent the transfer of material from the profile to the weldment you are creating - since the Profile has 1060 Alloy Aluminum as the material but the material is specified as Plain Carbon Steel in this question. However, I made a mistake and left Transfer Material from Profile selected and will show how to resolve this problem at the end of the question and mistakes to avoid if you find yourself in the same situation.

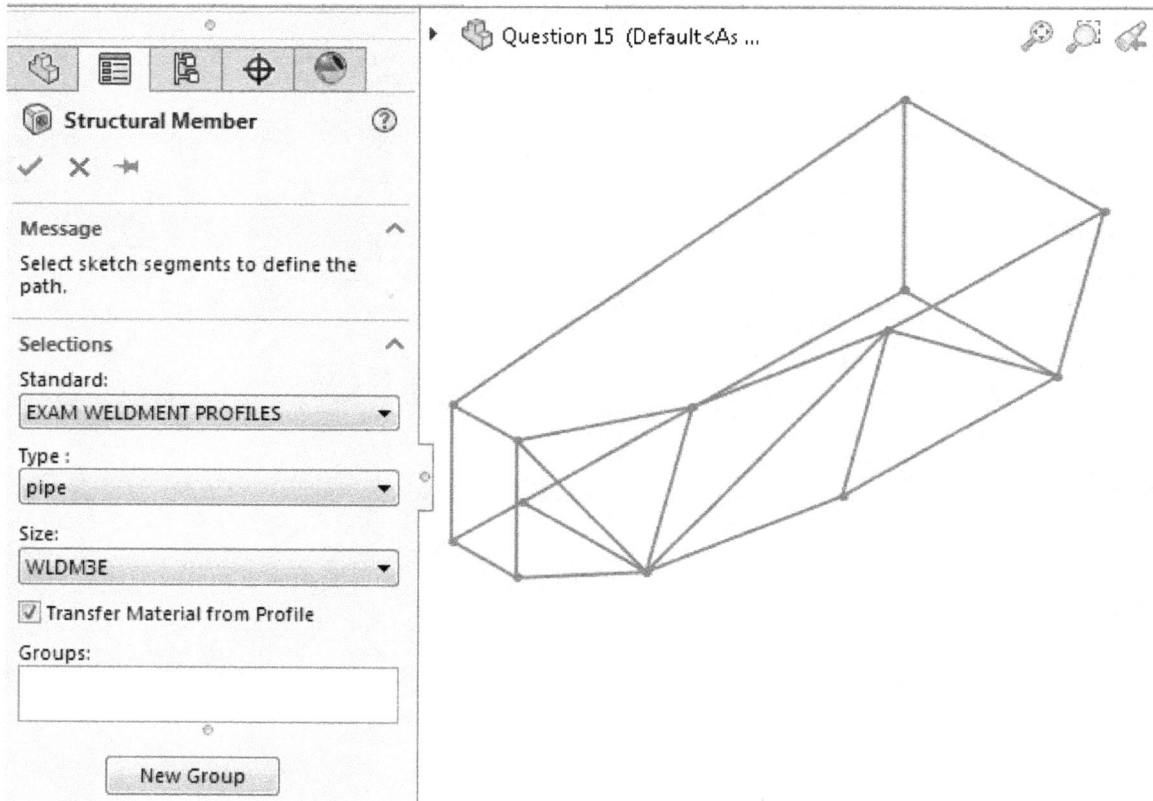

Figure 173 - Question 15 of 26 - Structural Member Property Manager

In the graphics area - select sketch segments shown in the following image to define the path for the structural member.

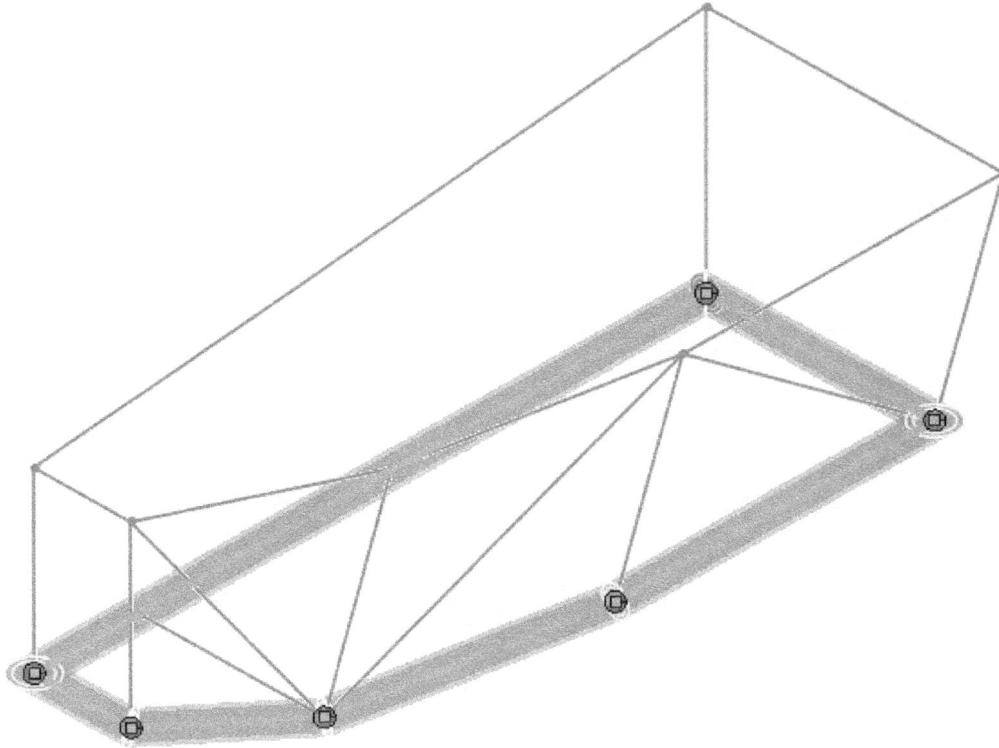

Figure 174 - Question 15 of 26 - Adding a Structural Member

Select Apply Corner Treatment and click on the End Miter option as shown in the following image.

☑ Apply corner treatment

Figure 175 - Question 15 of 26 - Corner Treatment

Click the New Group Command Button on the Structural Member Property Manager and select the sketch segment shown in the following image to define the path for the structural member.

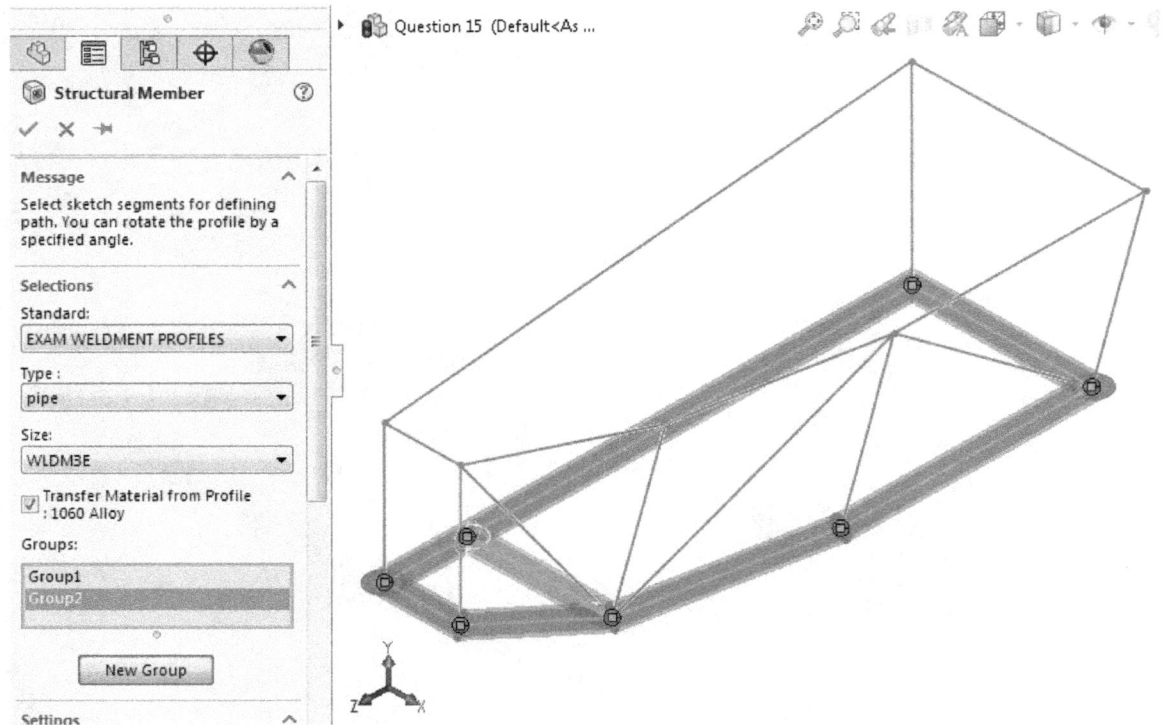

Figure 176 - Question 15 of 26 - Adding a Structural Member

Zoom in to each corner using the Zoom to Area tool on the View toolbar to see the mitered corners and make sure each is trimmed as per the image in this question. If not, you may click to select the point (purple or pink circle). The Corner Treatment dialog box appears, which lets you override the corner treatment for members of each group that meet at this corner. Click Ok to close the Corner Treatment Dialog Box. Click OK to close the Structural Member PropertyManager. Save your part. Your part should now look as shown in the following image.

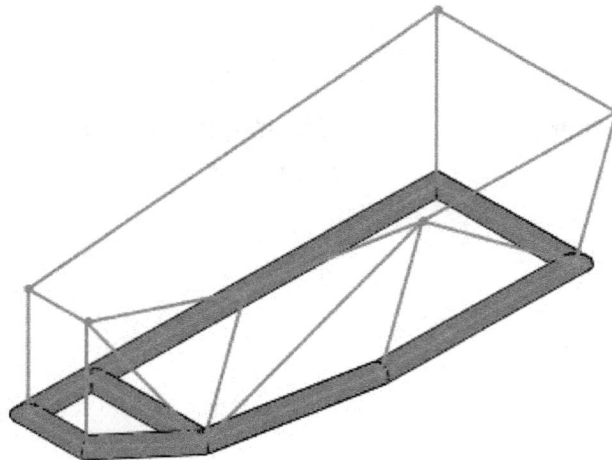

Figure 177 - Question 15 of 26 - Adding a Structural Member

Click or Right Click on the structural member in the Feature Manager Design Tree and select Edit Feature as shown in the following image. The Structural Member Property Manager appears.

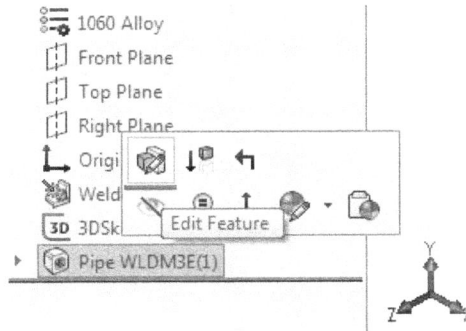

Figure 178 - Question 15 of 26 - Adding a Structural Member

ADDING MORE GROUPS

To create the next group - Group 3, right-click in the graphics area and select Create New Group or under Groups, click New Group. Select the third set of segments as shown in the following image.

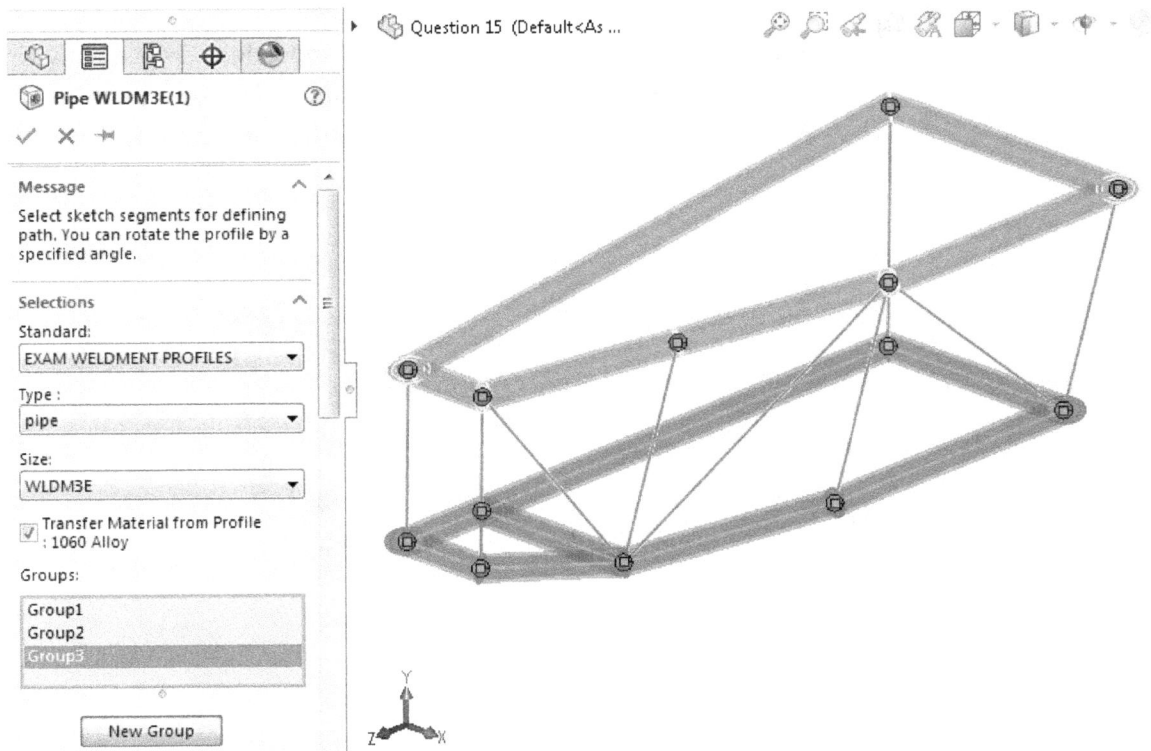

Figure 179 - Question 15 of 26 - Adding a new group

To create the next group - Group 4, right-click in the graphics area and select Create New Group or under Groups, click New Group. Select the fourth set of segments as shown in the following image.

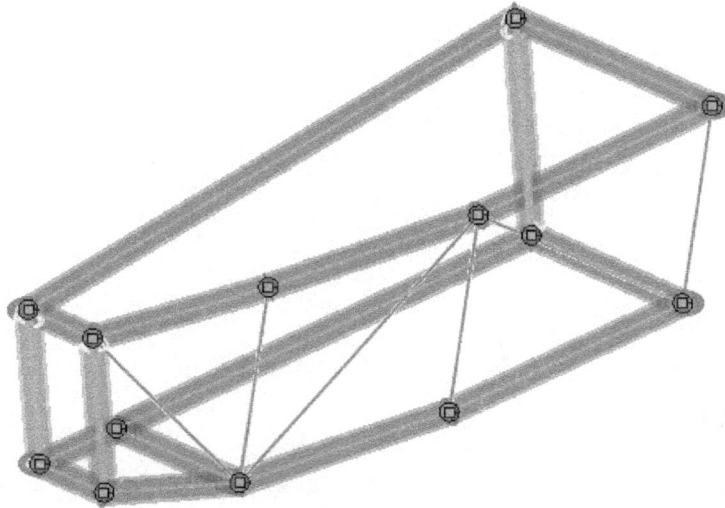

Figure 180 - Question 15 of 26 - Adding a new group

To create the next group - Group 5, right-click in the graphics area and select Create New Group or under Groups, click New Group. Select the fifth set of segments as shown in the following image.

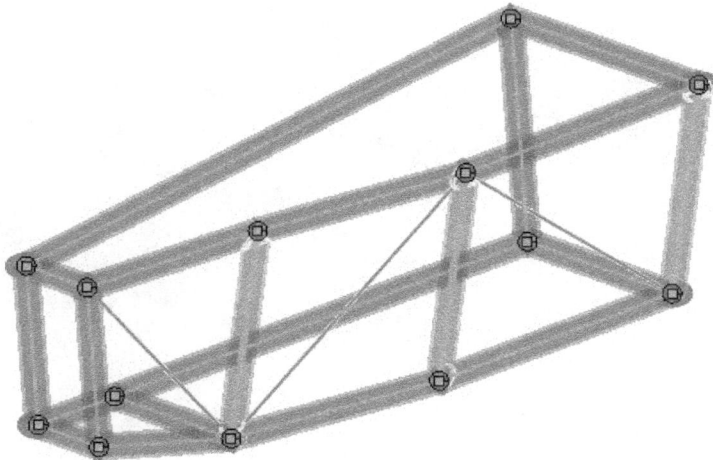

Figure 181 - Question 15 of 26 - Adding a new group

To create the next group - Group 6, right-click in the graphics area and select Create New Group or under Groups, click New Group. Select the sixth set of segments as shown in the following image.

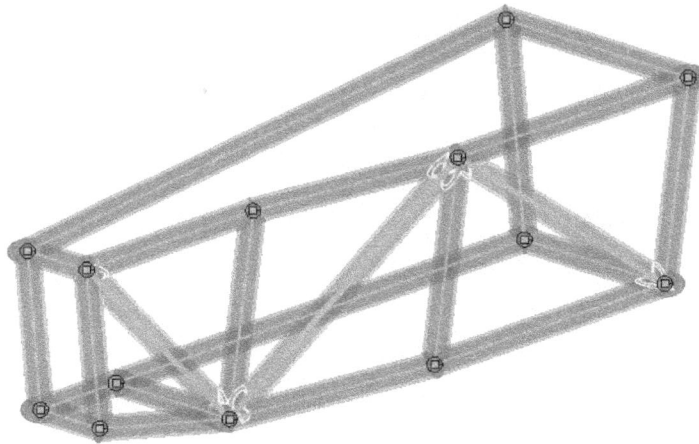

Figure 182 - Question 15 of 26 - Adding a new group

Click OK to close the Structural Member PropertyManager. Save your part. Your part should now look as shown in the following image.

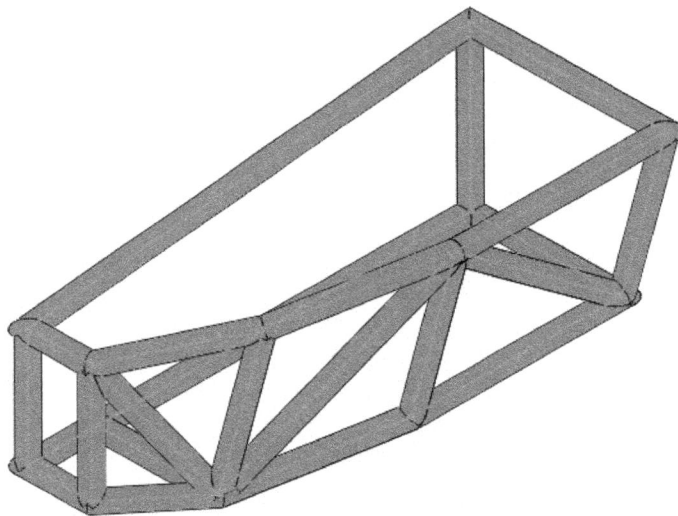

Figure 183 - Question 15 of 26 - Weldment Current Status

Take a closer look at the part and click on each of the vertical structural member to make sure there is no interference or some structural members going through others. You may also change the Display Style to Hidden Lines Visible the zoom into each and every corner or point where two or more structural members meet and rotate to see if there is no interference -

unfortunately this is not an assembly *(but basically a Multibody Part)* hence we cannot use interference detection in Solidworks 2018. If you are using a higher version of Solidworks you may use interference detection under the Evaluate tab or click Tools > Evaluate > Interference Detection. You may notice that there is an interference in the area circled in red in the following image - hence we need to trim the structural member highlighted in light blue in the following image.

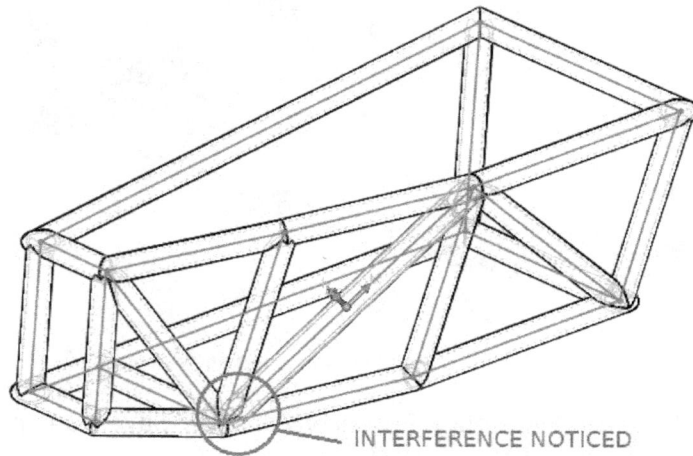

Figure 184 - Question 15 of 26 - Weldment Current Status

TRIM / EXTEND

Click Trim/Extend (Weldments toolbar) or Insert > Weldments > Trim/Extend. In the PropertyManager, set options as shown in the following image then click OK.

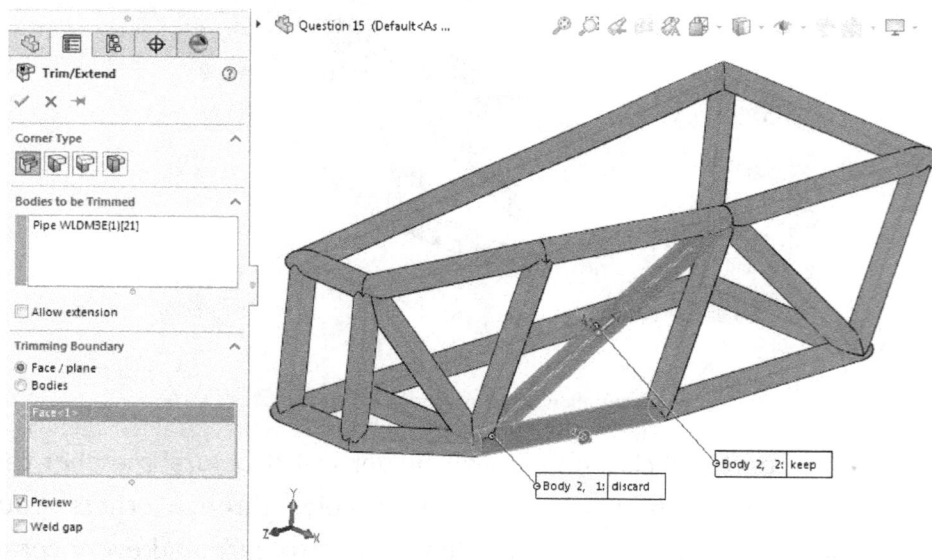

Figure 185 - Question 15 of 26 - Weldments Trim / Extend

140

Click Trim/Extend (Weldments toolbar) or Insert > Weldments > Trim/Extend. In the PropertyManager, set options as shown in the following image then click OK.

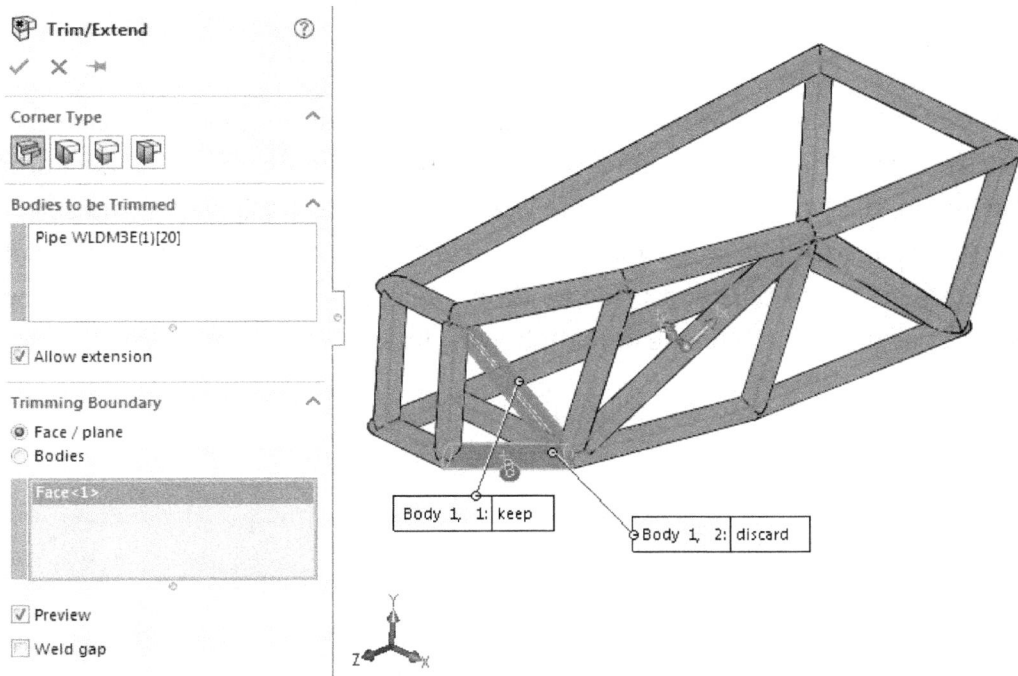

Figure 186 - Question 15 of 26 - Weldments Trim / Extend

Rebuild and save your part once you are happy that there are no other interferences. Your part should now look as shown in the following image.

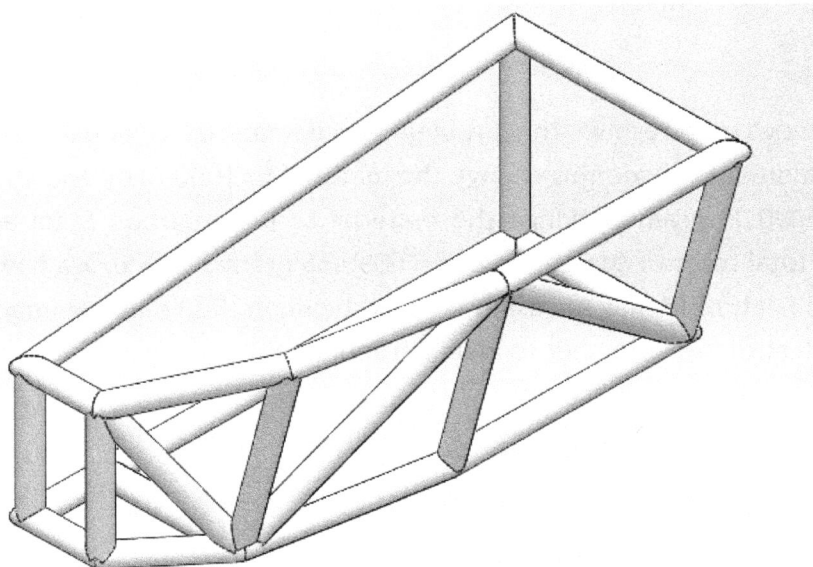

Figure 187 - Question 15 of 26 - Weldment Current Status

MASS PROPERTIES - MEASURING THE TOTAL MASS OF A WELDMENT

NB: The material is specified as Plain Carbon Steel in this question hence we have to override the material transferred from the weldment profile. Change the material to Plain Carbon Steel by right clicking on 1060 Alloy Material in the Feature Manager Design Tree and selecting Edit Material as shown in the following image then select Plain Carbon Steel under Solidworks Materials and Steel - Click Apply and Close or simply select Plain Carbon Steel if it appears on the Favorites List as underlined in blue in the following image.

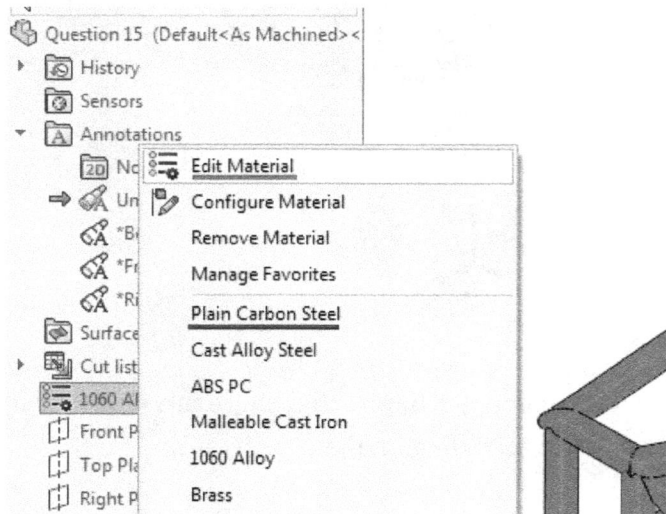

Figure 188 - Question 15 of 26 - Changing a material

However, this does not resolve the problem since material is applied on each structural member in a weldment. To see this change the material to 1060 Alloy and check the total mass of the part = 6090.35 grams. Change the material to Plain Carbon Steel again and you will notice that the total mass of the part is still = 6090.35 grams. Hence, we have to open the Cut list folder in the Feature Manager Design Tree as shown in the following image and change the material on each structural member to Plain Carbon Steel.

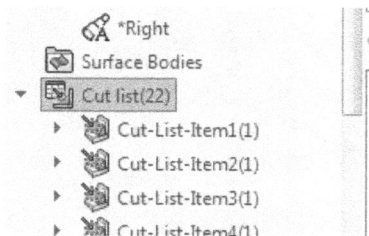

Figure 189 - Question 15 of 26 - Opening the Cut List Folder

Click the LHS Black arrow on each Cut List Item and it appears as shown in the following image.

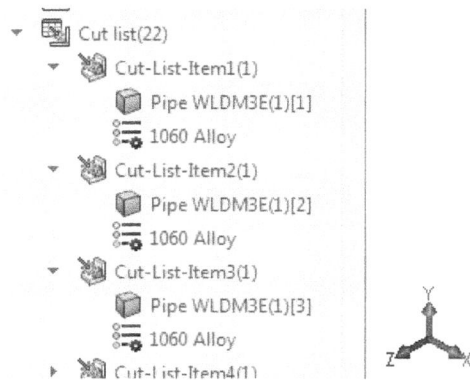

Figure 190 - Question 15 of 26 - Changing material in the cut list folder

Change the material on each Cut List Item to Plain Carbon steel by Right Clicking on Pipe WLDM3E(1)[1] ![icon] Pipe WLDM3E(1)[1] , Selecting Material and following the procedure described in the previous section to change the material to Plain Carbon Steel. Do this for each and every Cut List Item until all are done as shown in the following image.

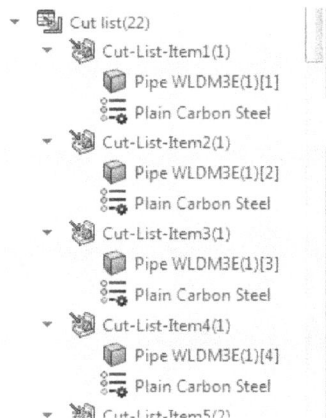

Figure 191 - Question 15 of 26 - Changing material in the cut list folder

Seeing how cumbersome this process is, I hope you will be more careful and or vigilant in using the Transfer material from Profile option when adding structural members. And also always double check the Cut List Folder Items to make sure there is no material applied on the Cut list Items or the applied material is the same as the overall part material or the material specified in the question.

However, the quickest way would have been to Edit the Structural Member as shown in the following image by right clicking on the structural member in the Feature Manager Design Tree and selecting Edit Feature. See the following image.

143

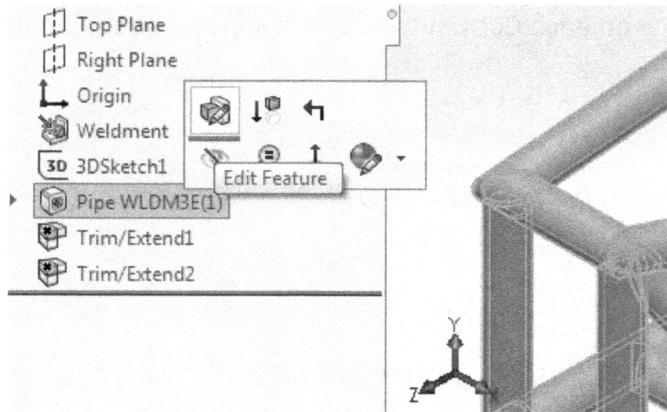

Figure 192 - Question 15 of 26 - Editing a Structural Member

The Structural Member PropertyManager opens and you uncheck the Transfer material from Profile Option as shown in the following image then Click Ok to close the Structural Member PropertyManager.

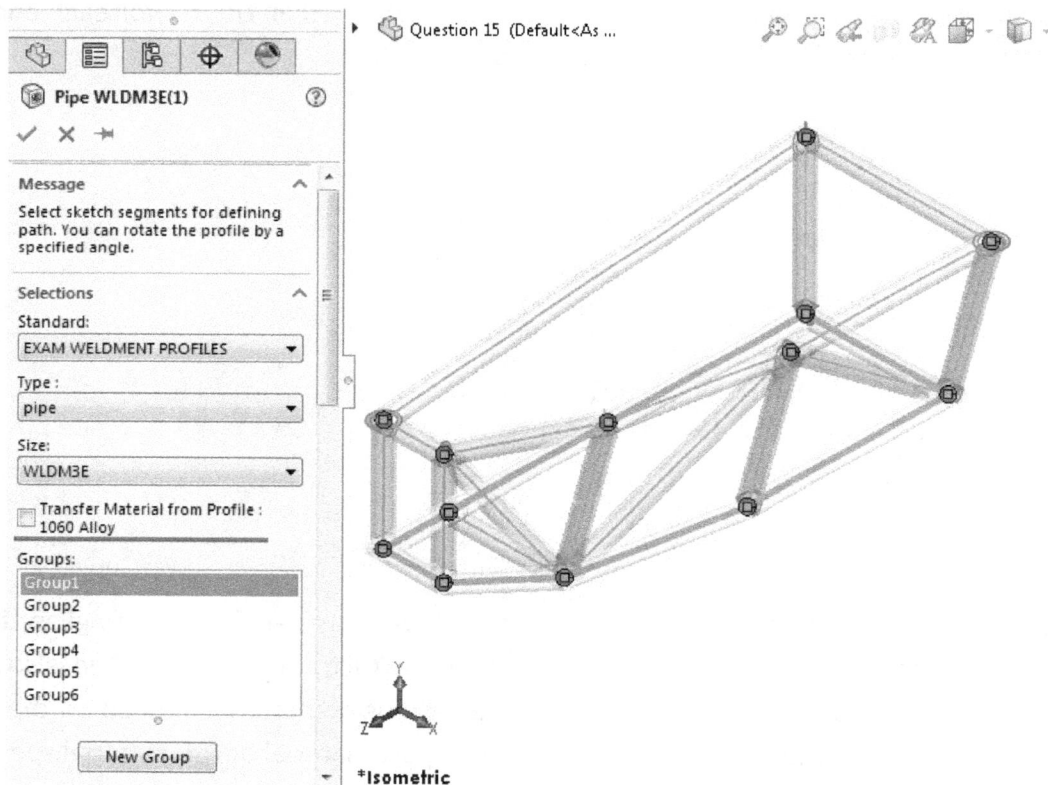

Figure 193 - Question 15 of 26 - Removing the Transfer Material from Profile Option

This automatically removes the applied material on all Cut List Items. If you check the Cut List Items you will notice that no material is applied any more. Now you can leave the overall part applied material as Plain Carbon Steel.

144

TIP: - Cut List Item applied material takes precedence over the overall Part Applied Material.

To measure the mass of the part, under the Evaluate tab, click on the Mass Properties Tool or Tools > Evaluate > Mass Properties - the Mass Properties Dialog Box appears as shown in the following image.

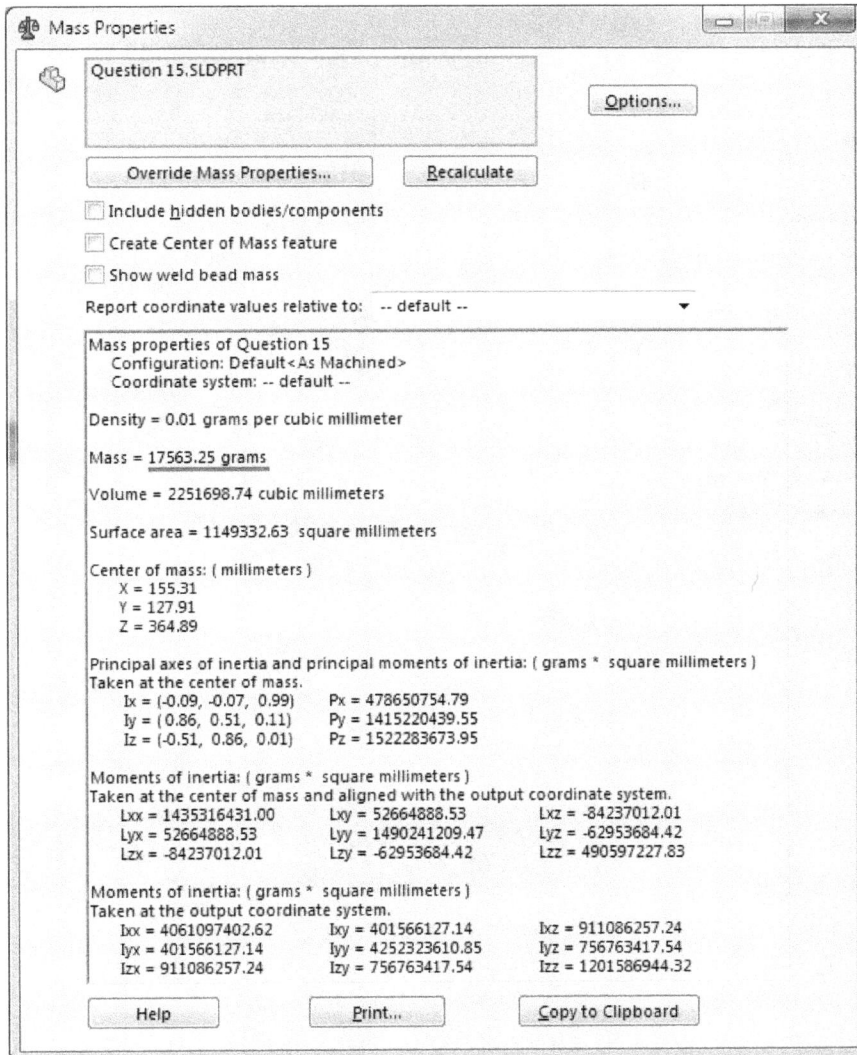

Figure 194 - Question 15 of 26 - Total Mass of a Weldment

Save and close the Part.

CHAPTER 16

QUESTION 16 EXAM SCREEN CAPTURES

QUESTION 16 EXAM SCREEN CAPTURE 1

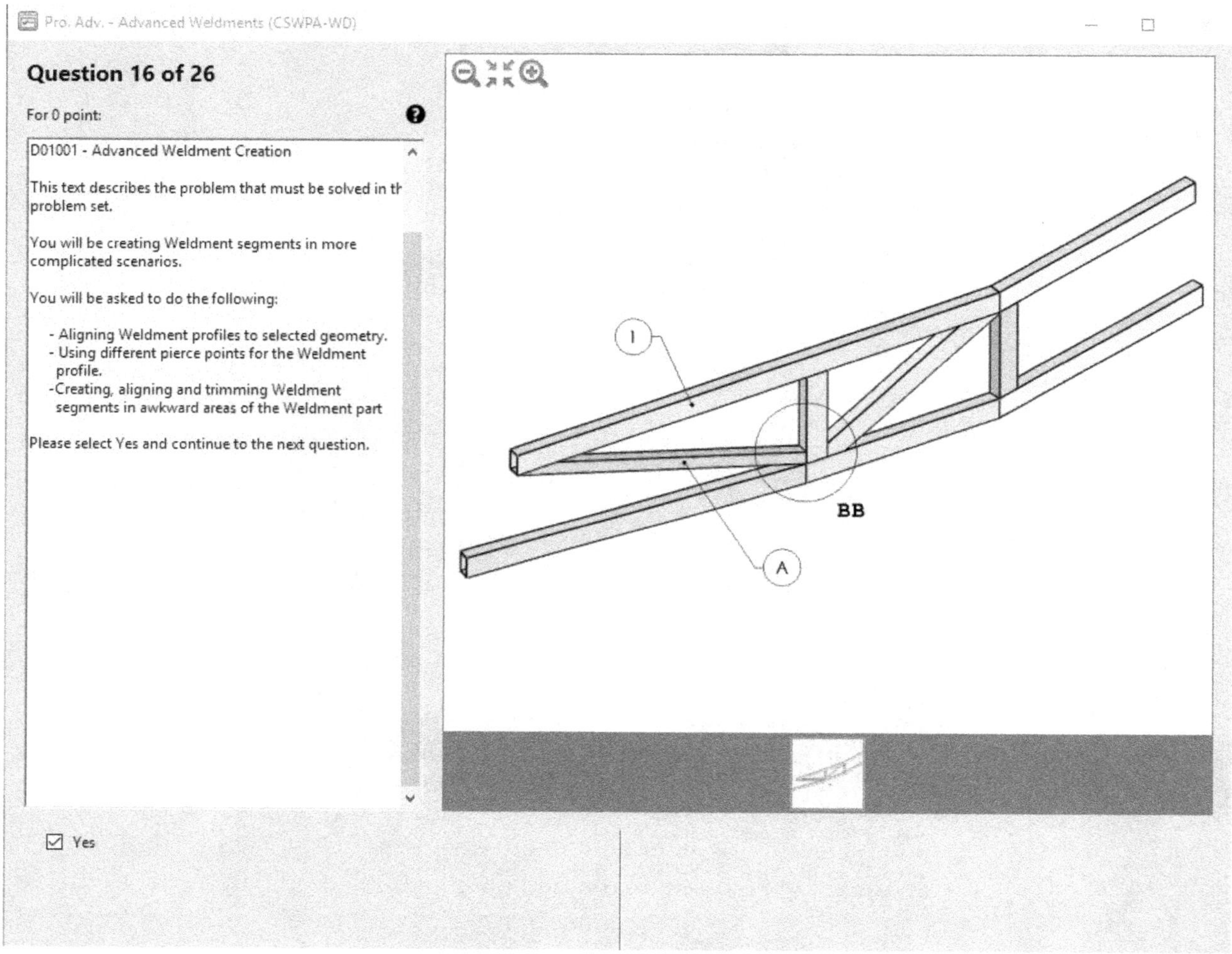

Figure 193 - Question 16 of 26 - Exam Screen Capture 1

Read the instructions in this question then select Yes and continue to Question 17.

CHAPTER 17

QUESTION 17 EXAM SCREEN CAPTURES

QUESTION 17 EXAM SCREEN CAPTURE 1

Pro. Adv. - Advanced Weldments (CSWPA-WD)

Question 17 of 26

For 10 points:

D02005 - Create Vertical Legs
Build this 3D Sketch in SolidWorks.
Unit system: MMGS (millimeter, gram, second)
Decimal places: 2

-Open the attached part.

-Create vertical legs A and B on the two indicated sketch segments using weldment profile "WLDM1E".

Note 1: Align the pierce point of one of the shorter sides of the weldment profile with the vertical sketch line segments (see second image).

Note 2: Align weldment profile so that its longer side lines up with the lower segments 1 and 2 (see third image).

Note 3: Vertical legs A and B must be trimmed to be flush with the segments they contact.

-Measure the center of mass of the entire weldment part.

What is the center of mass of the entire weldment part (mm)?

Attachment to this question

D5.SLDPRT (357.5 kB)

○ X = 453.27 Y = 185.45 Z = -1249.32

○ X = 433.22 Y = 175.44 Z = -1249.36

○ X = 403.22 Y = 165.44 Z = -1249.36

○ X = 413.22 Y = 195.44 Z = -1249.36

Figure 194 - Question 17 of 26 - Exam Screen Capture 1

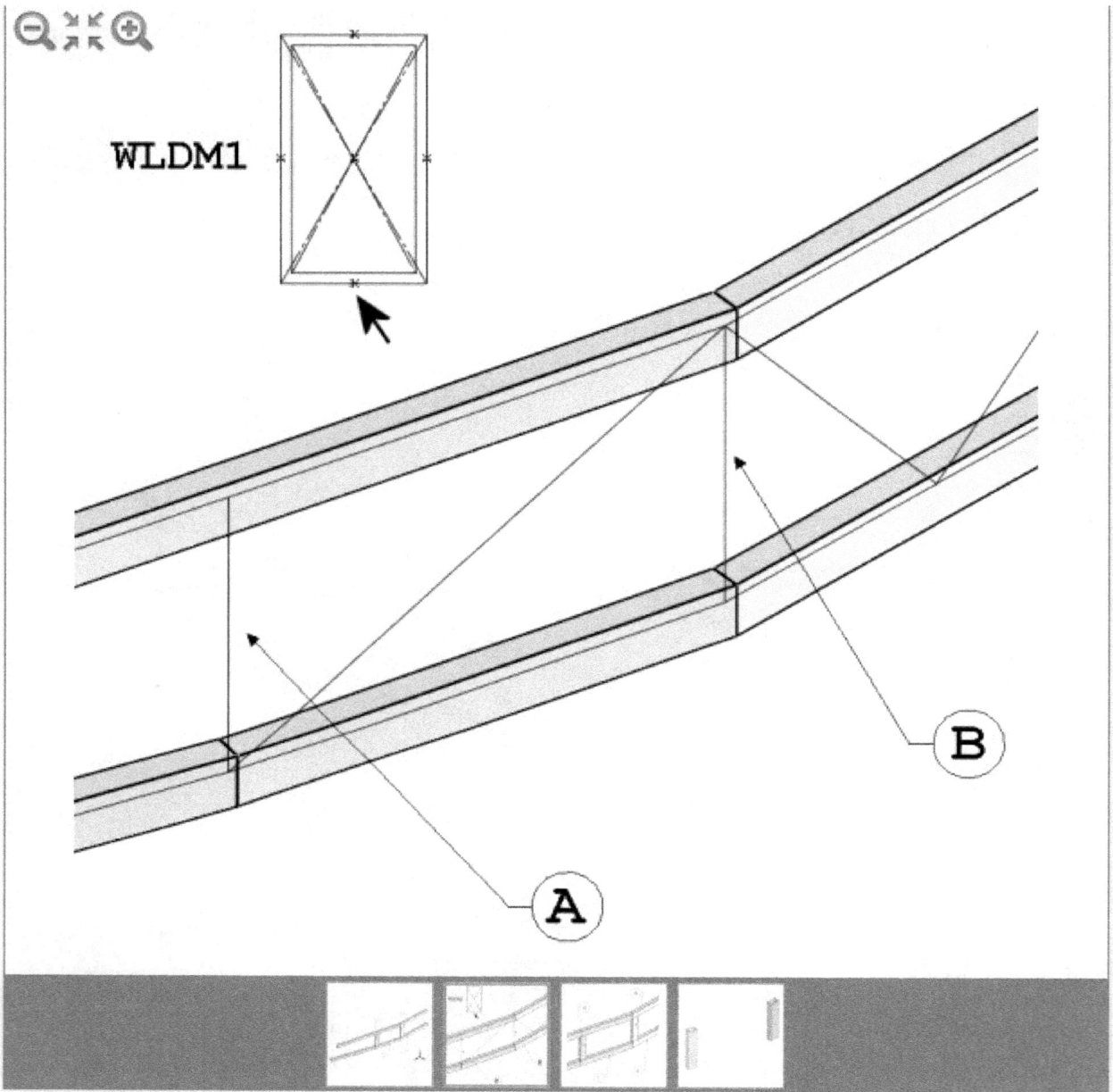

Figure 195 - Question 17 of 26 - Exam Screen Capture 2

Figure 196 - Question 17 of 26 - Exam Screen Capture 3

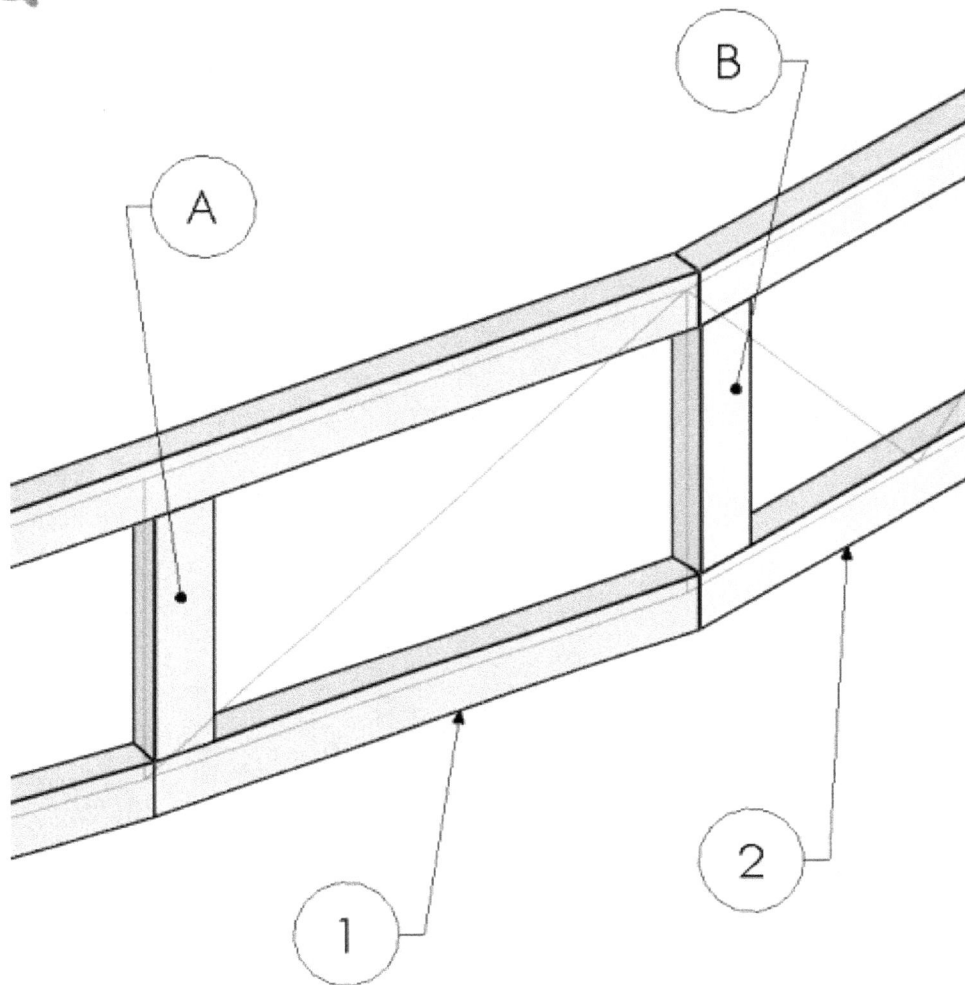

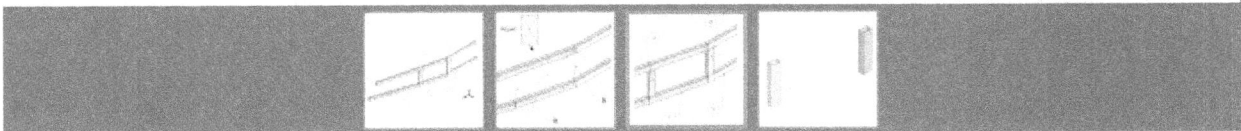

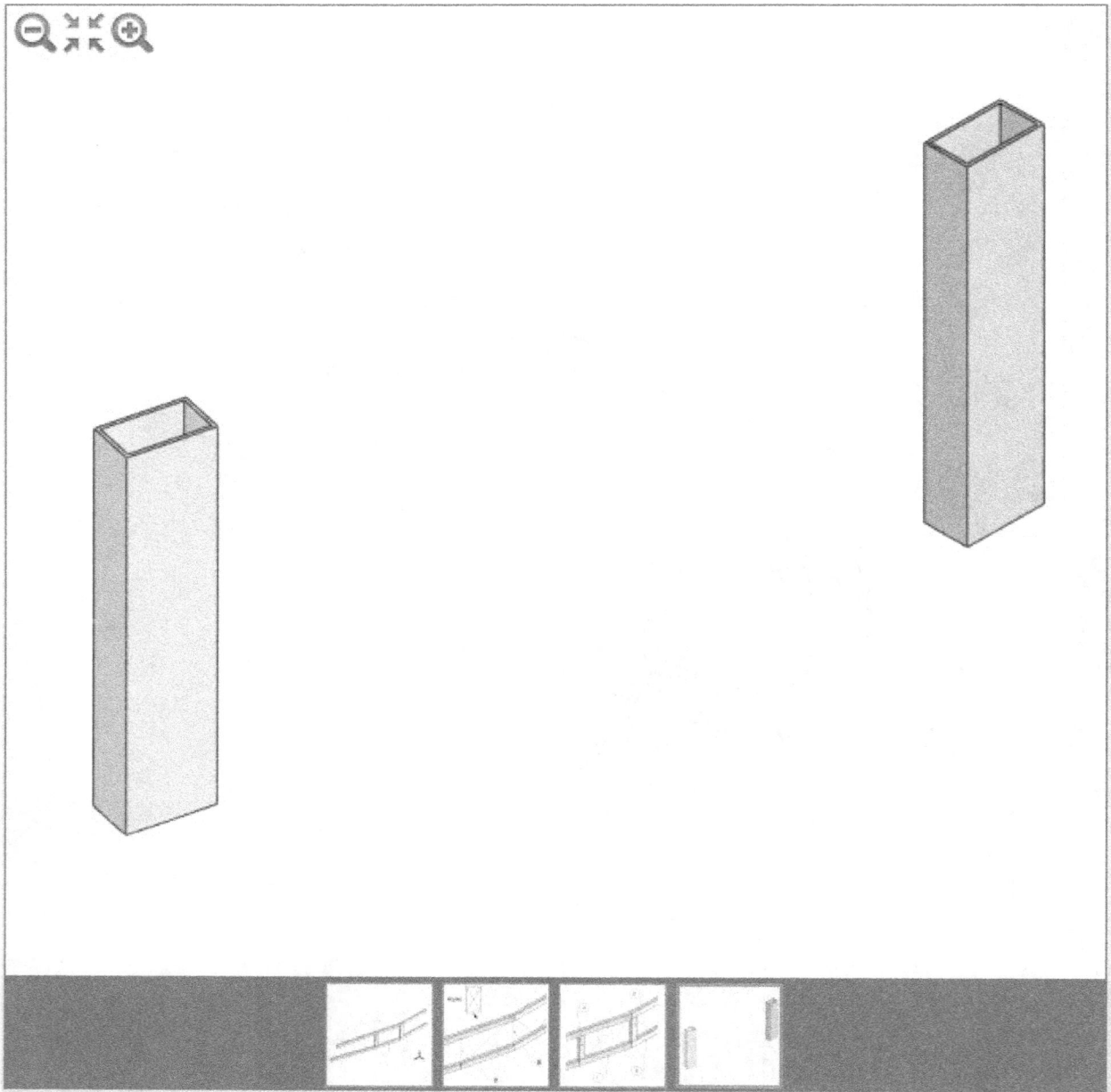

Figure 197 - Question 17 of 26 - Exam Screen Capture 4

Download the part D5.SLDPRT from this Google Drive location *(http://bit.ly/CSWPA-WD)* and save it on your computer.

Open the downloaded part - D5.SLDPRT. The part should look as shown in the following image.

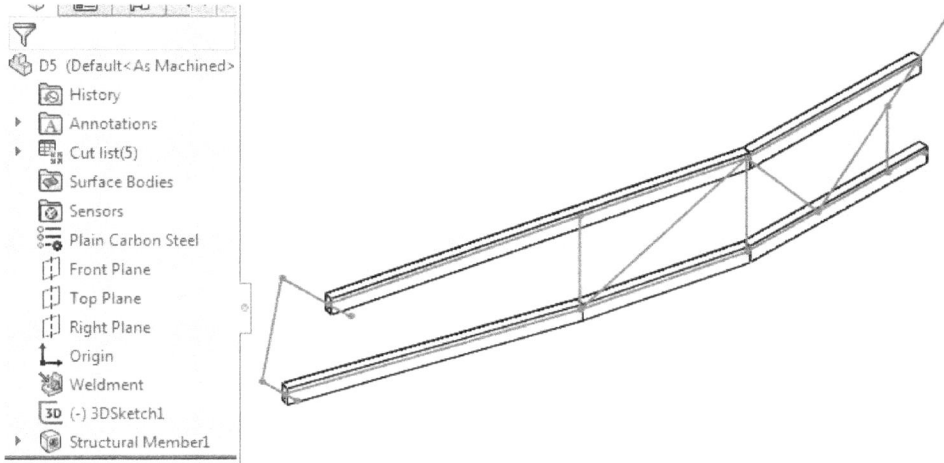

Figure 200 - Question 17 of 26 - D5.SLDPRT - Downloaded Part

ADDING A STRUCTURAL MEMBER

Click Structural Member under the Weldments toolbar or Insert > Weldments > Structural Member. Make selections in the PropertyManager to define the profile for the structural member as shown in the following image. **Deselect** *Transfer Material from Profile* to prevent the transfer of material from the profile to the structural member you are adding - since the Profile has 1060 Alloy Aluminum as the material but the applied material is Plain Carbon Steel in this part.

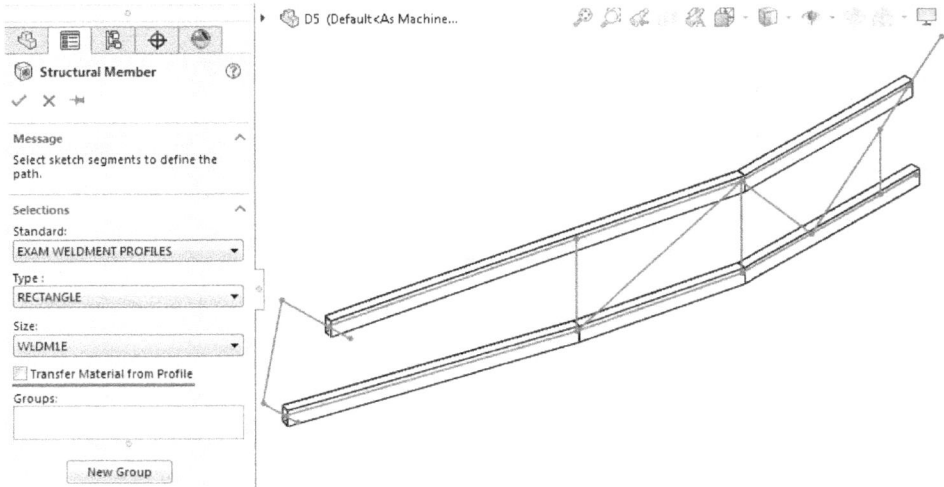

Figure 201 - Question 17 of 26 - Structural Member Property Manager

In the graphics area - select the sketch segment shown in the following image to define the path for the first structural member.

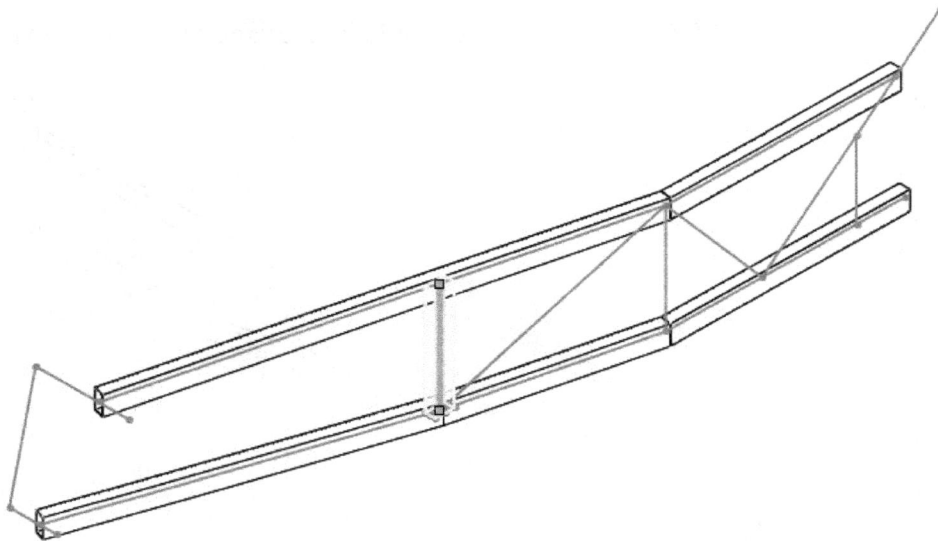

Figure 202 - Question 17 of 26 - Adding a Structural Member

ALIGNING A WELDMENT PROFILE

Click in the Alignment text area in the Structural Member PropertyManager and then in the graphics area click on the horizontal line *(Line15)* below the structural member we are currently adding then select the Align vertical axis Radio Button in the Structural Member PropertyManager under Alignment as shown in the following image .

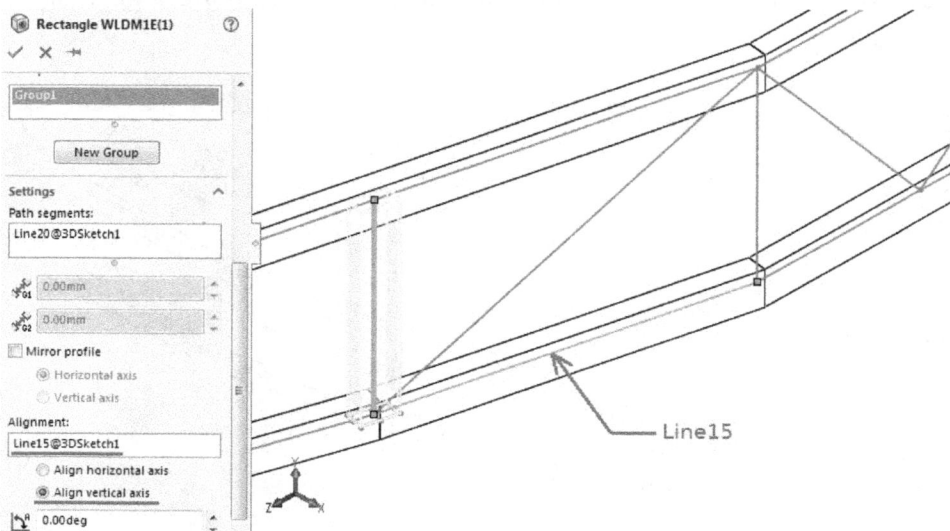

Figure 203 - Question 17 of 26 - Aligning a weldment profile

The pierce point defines the location of the profile, relative to the sketch segment used to create the structural member. The default pierce point is the sketch origin in the profile library feature part. Any vertex or sketch point specified in the profile can also be used as a pierce point.

In the Structural Member PropertyManager, below Alignment and Rotation Angle, click the Locate Profile command button. The display zooms to the profile of the structural member as shown in the following image.

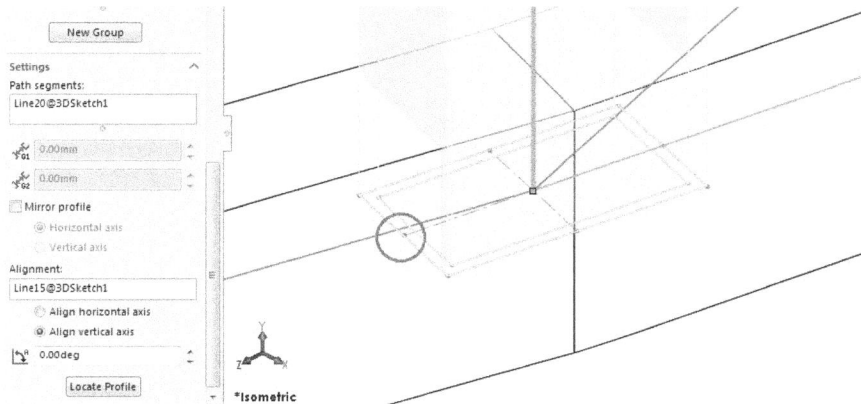

Figure 204 - Question 17 of 26 - Locate profile

Select a vertex point on the profile circled in the red in the previous image. The profile shifts to align the new pierce point with the structural member sketch segment as shown in the following image.

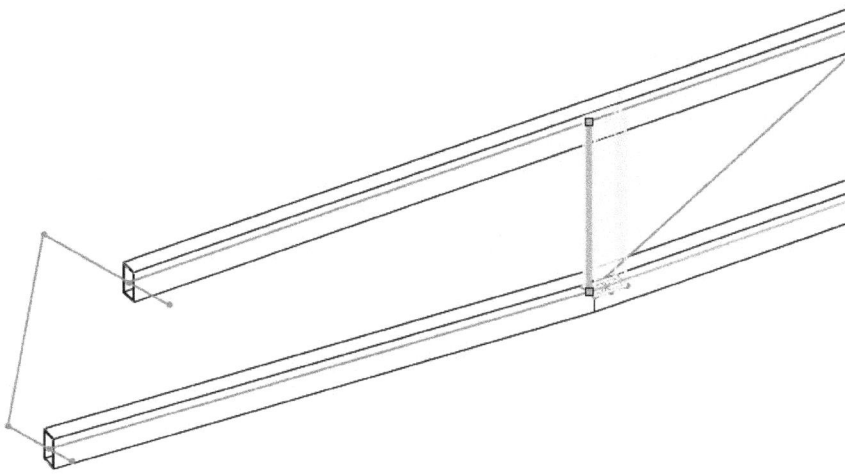

Figure 205 - Question 17 of 26 - Pierce Point

Click Ok to close the Structural Member Property Manager and have a closer look at the newly added structural member. Your part should now look as shown in the following image.

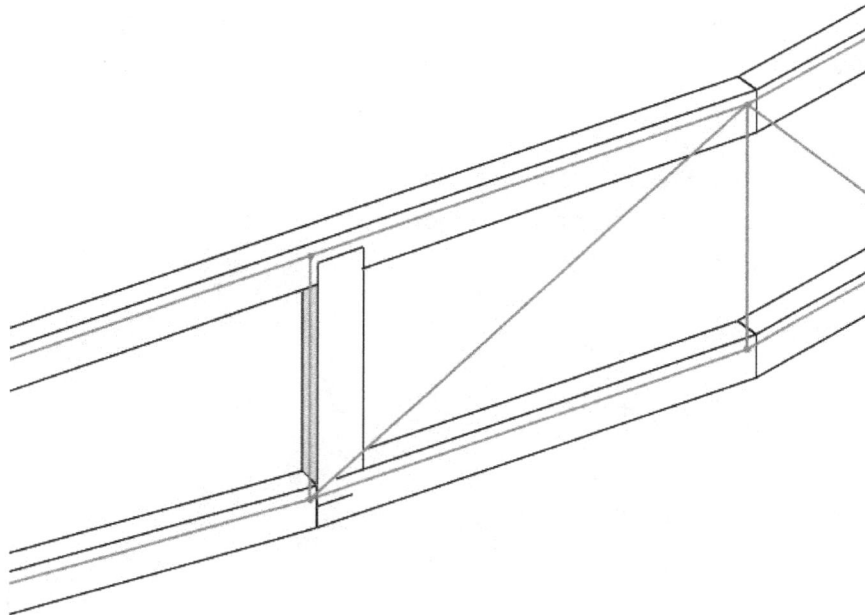

Figure 206 - Question 17 of 26 - Weldment current status

Click or Right Click on the Rectangle WLDM1E(1) structural member in the Feature Manager Design Tree and select Edit Feature as shown in the following image. The Structural Member Property Manager appears.

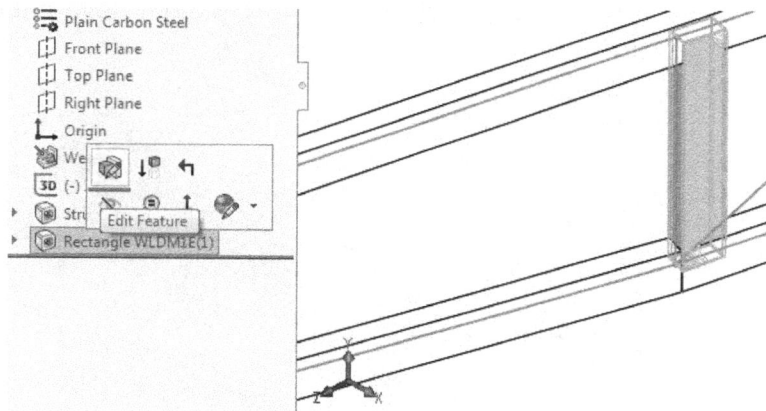

Figure 207 - Question 17 of 26 - Editing a Structural Member

ADDING MORE GROUPS

To create the next group - Group 2, right-click in the graphics area and select Create New Group or under Groups, click New Group. Select the vertical segment shown in the following image.

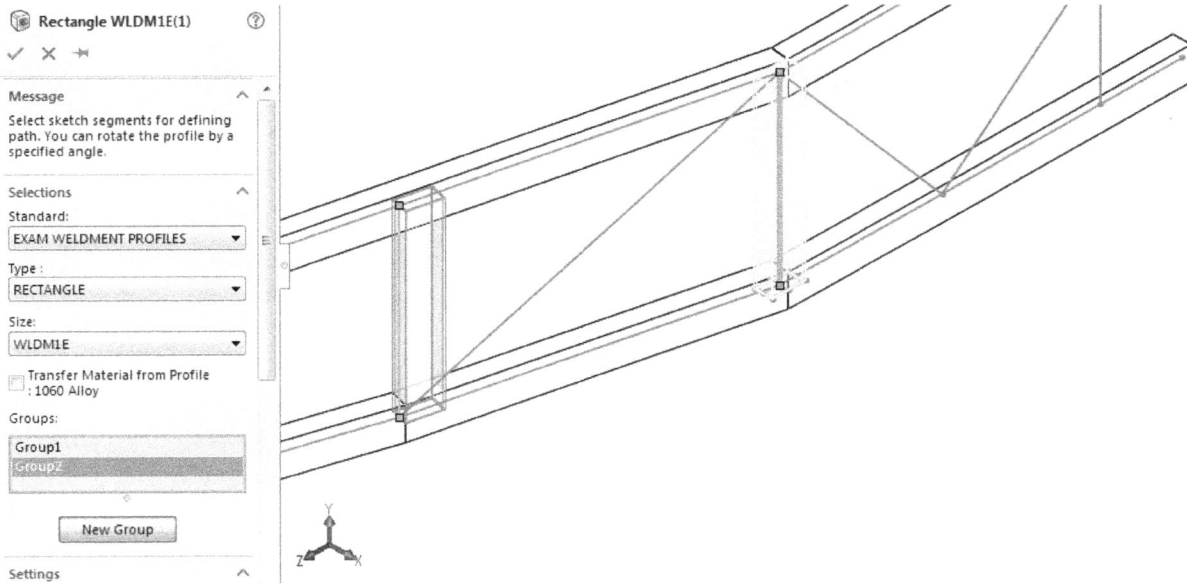

Figure 208 - Question 17 of 26 - Adding a new group

ALIGNING A WELDMENT PROFILE

Click in the Alignment text area in the Structural Member PropertyManager and then in the graphics area click on the horizontal line *(Line8)* below the structural member we are currently adding then select the Align vertical axis Radio Button in the Structural Member PropertyManager under Alignment as shown in the following image .

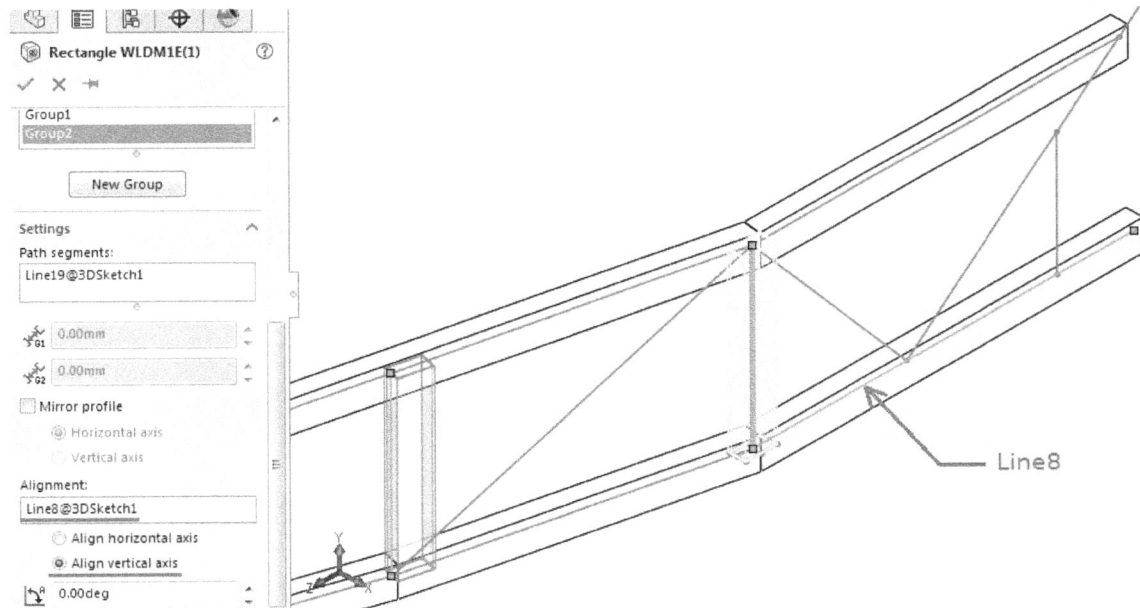

Figure 209 - Question 17 of 26 - Aligning a weldment profile

In the Structural Member PropertyManager, below Alignment and Rotation Angle, click the Locate Profile command button. The display zooms to the profile of the structural member as shown in the following image.

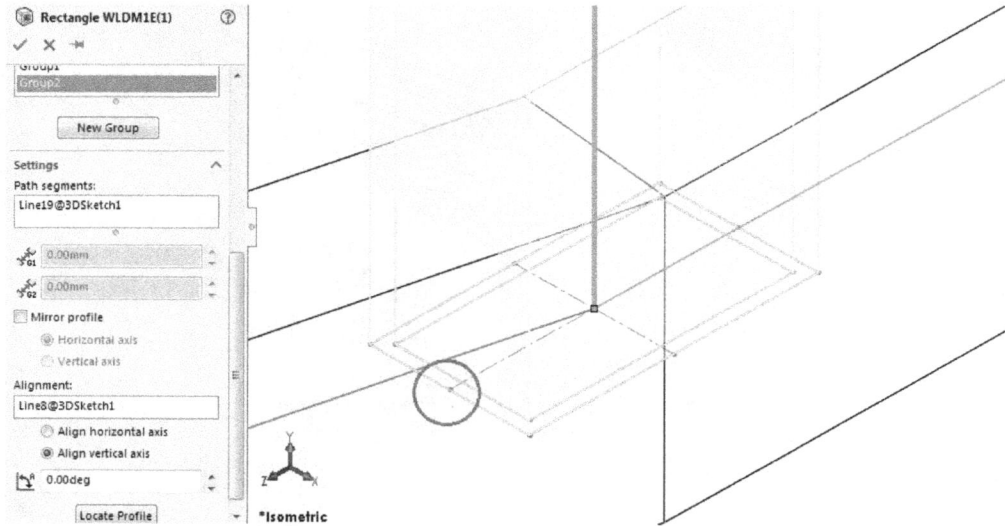

Figure 210 - Question 17 of 26 - Locate profile

Select a vertex point on the profile circled in the red in the previous image. The profile shifts to align the new pierce point with the structural member sketch segment as shown in the following image.

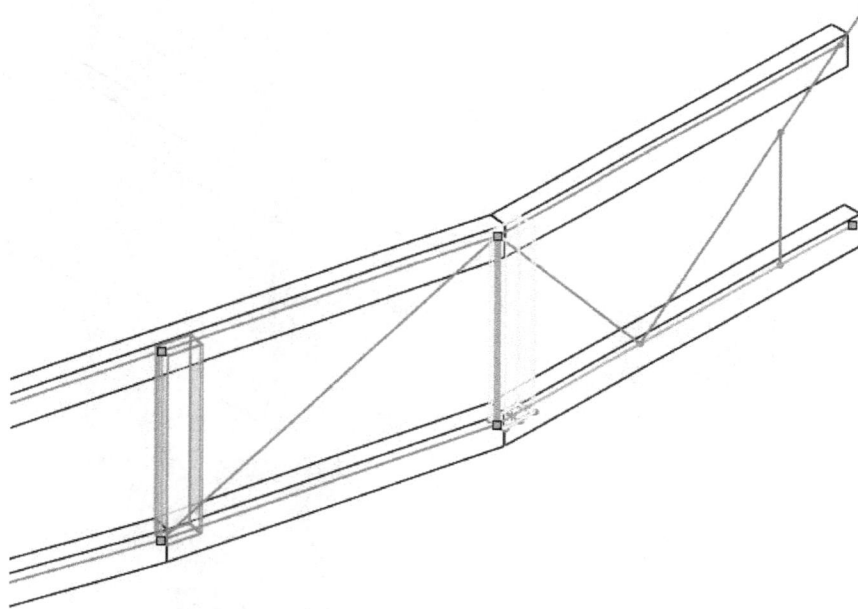

Figure 211 - Question 17 of 26 - Pierce Point

Click Ok to close the Structural Member Property Manager and have a closer look at the newly added structural member. Your part should now look as shown in the following image.

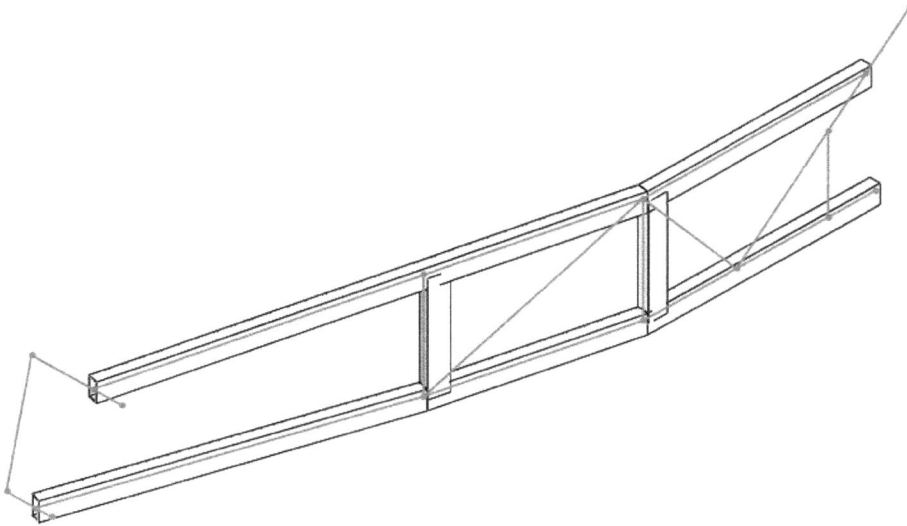

Figure 212 - Question 17 of 26 - Weldment current status

TRIMMING SEGMENTS

Click Trim/Extend (Weldments toolbar) or Insert > Weldments > Trim/Extend. In the PropertyManager, set options as shown in the following image.

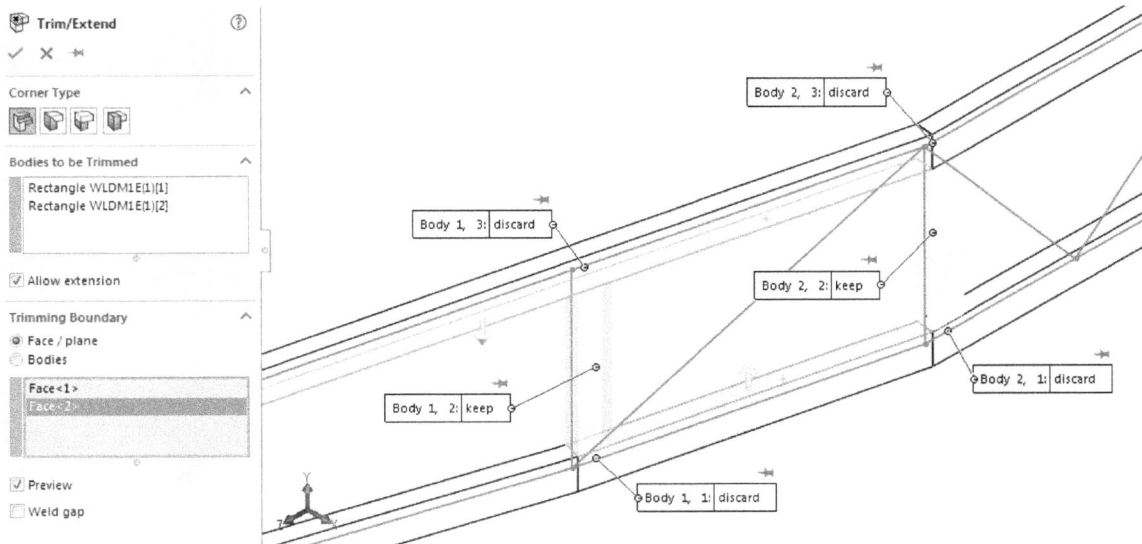

Figure 213 - Question 17 of 26 - Trimming Segments

Click Ok and save you part. Your part should now look as shown in the following image.

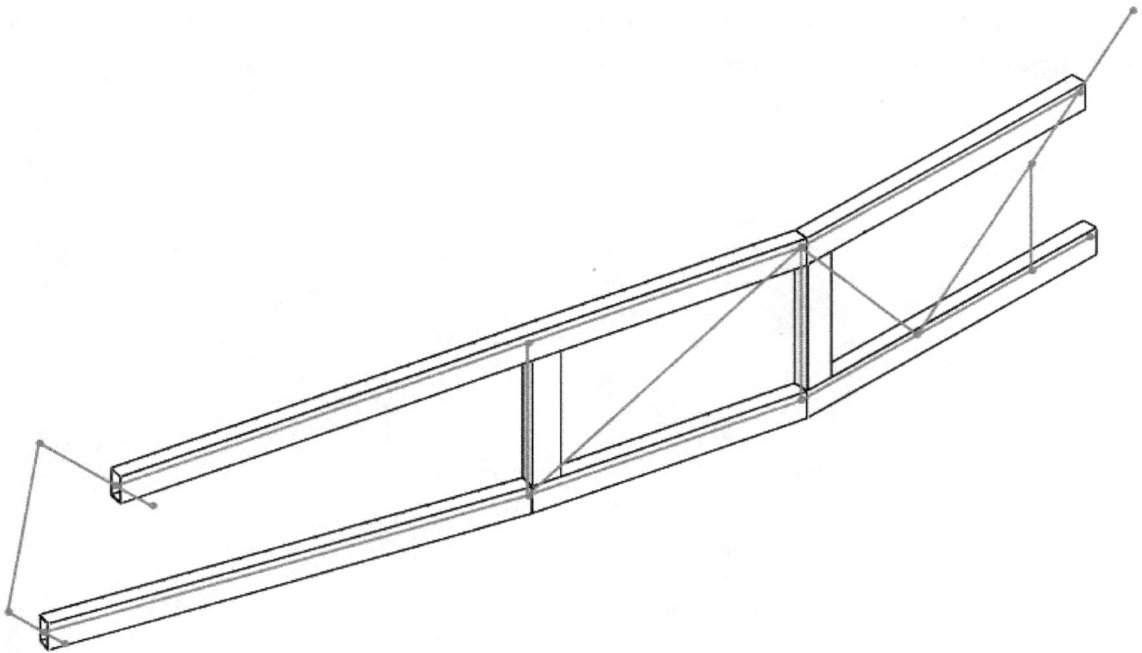

Figure 214 - Question 17 of 26 - Weldment Current Status

DISPLAYING THE CENTER OF MASS

Click Mass Properties (Tools toolbar) or Tools > Evaluate > Mass Properties. The calculated mass properties appear in the dialog box including the center of mass as shown in the following image - center of mass in underlined in red - thus X = 453.27, Y = 185.45 and Z = 1249.32.

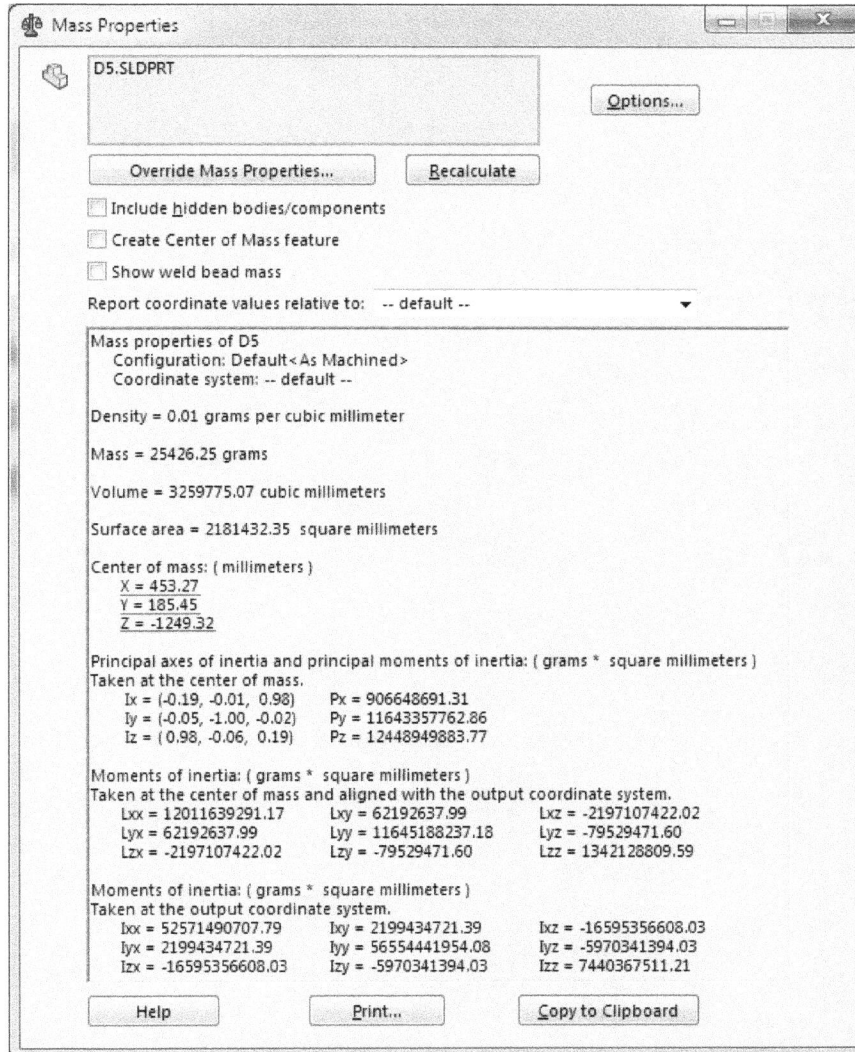

Figure 215 - Question 17 of 26 - Center of Mass

CHAPTER 18

QUESTION 18 EXAM SCREEN CAPTURES

QUESTION 18 EXAM SCREEN CAPTURE 1

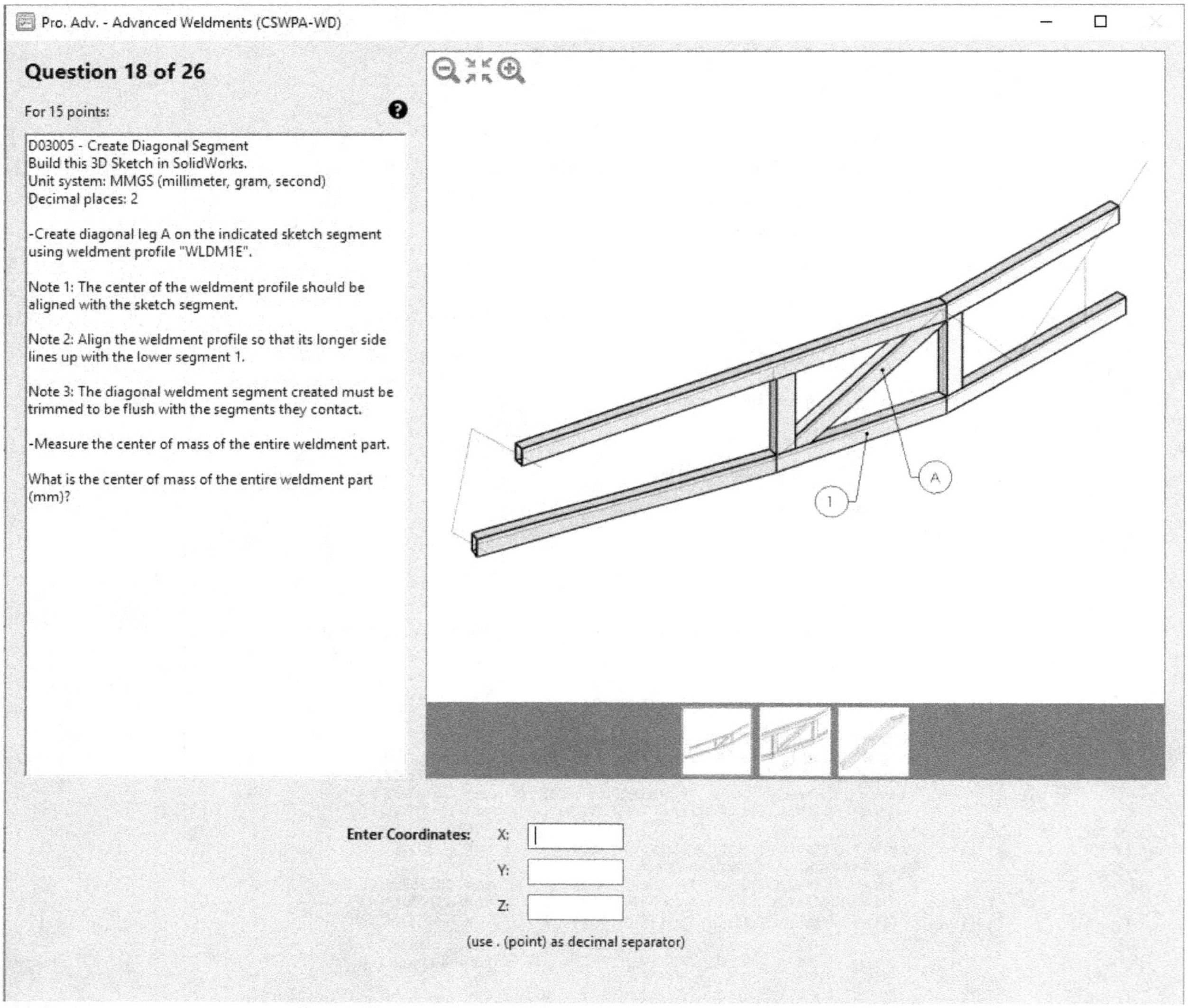

Pro. Adv. - Advanced Weldments (CSWPA-WD)

Question 18 of 26

For 15 points:

D03005 - Create Diagonal Segment
Build this 3D Sketch in SolidWorks.
Unit system: MMGS (millimeter, gram, second)
Decimal places: 2

-Create diagonal leg A on the indicated sketch segment using weldment profile "WLDM1E".

Note 1: The center of the weldment profile should be aligned with the sketch segment.

Note 2: Align the weldment profile so that its longer side lines up with the lower segment 1.

Note 3: The diagonal weldment segment created must be trimmed to be flush with the segments they contact.

-Measure the center of mass of the entire weldment part.

What is the center of mass of the entire weldment part (mm)?

Enter Coordinates: X:
Y:
Z:

(use . (point) as decimal separator)

Figure 216 - Question 18 of 26 - Exam Screen Capture 1

Figure 217 - Question 18 of 26 - Exam Screen Capture 2

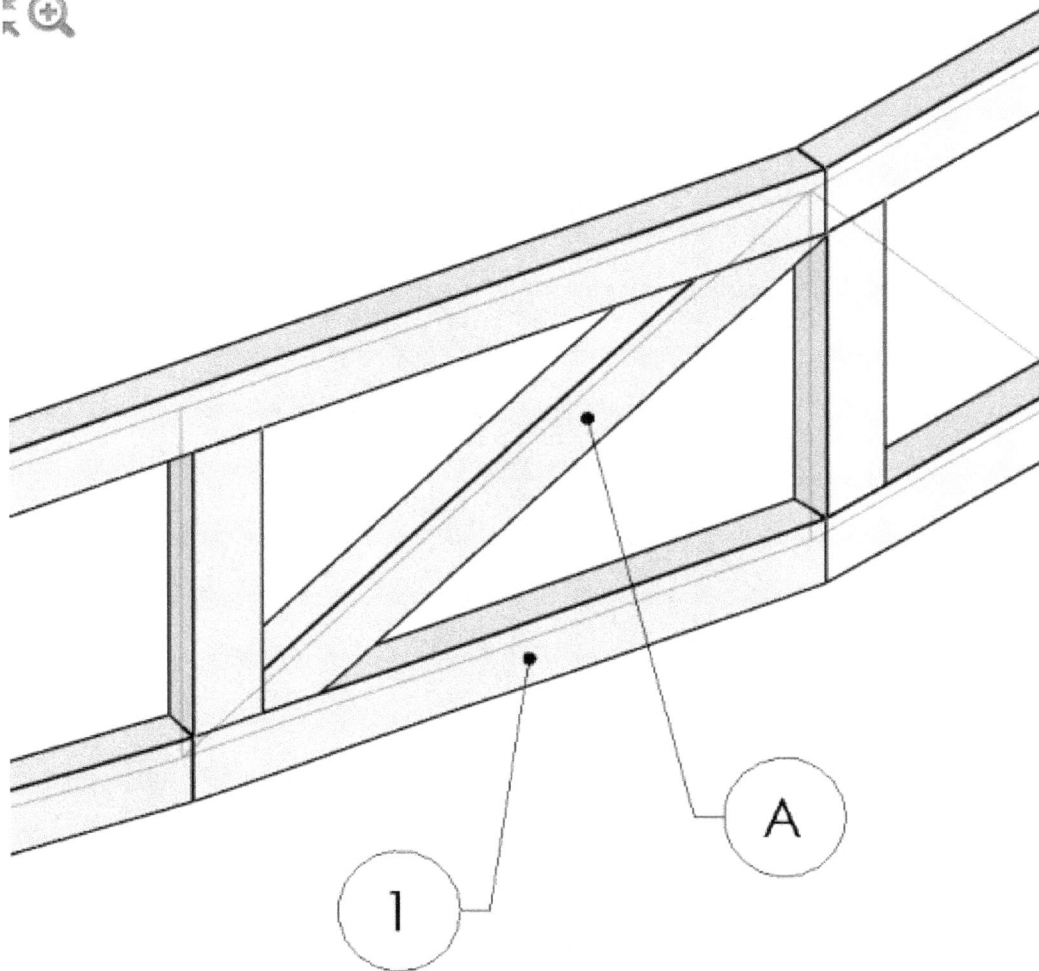

Figure 218 - Question 18 of 26 - Exam Screen Capture 3

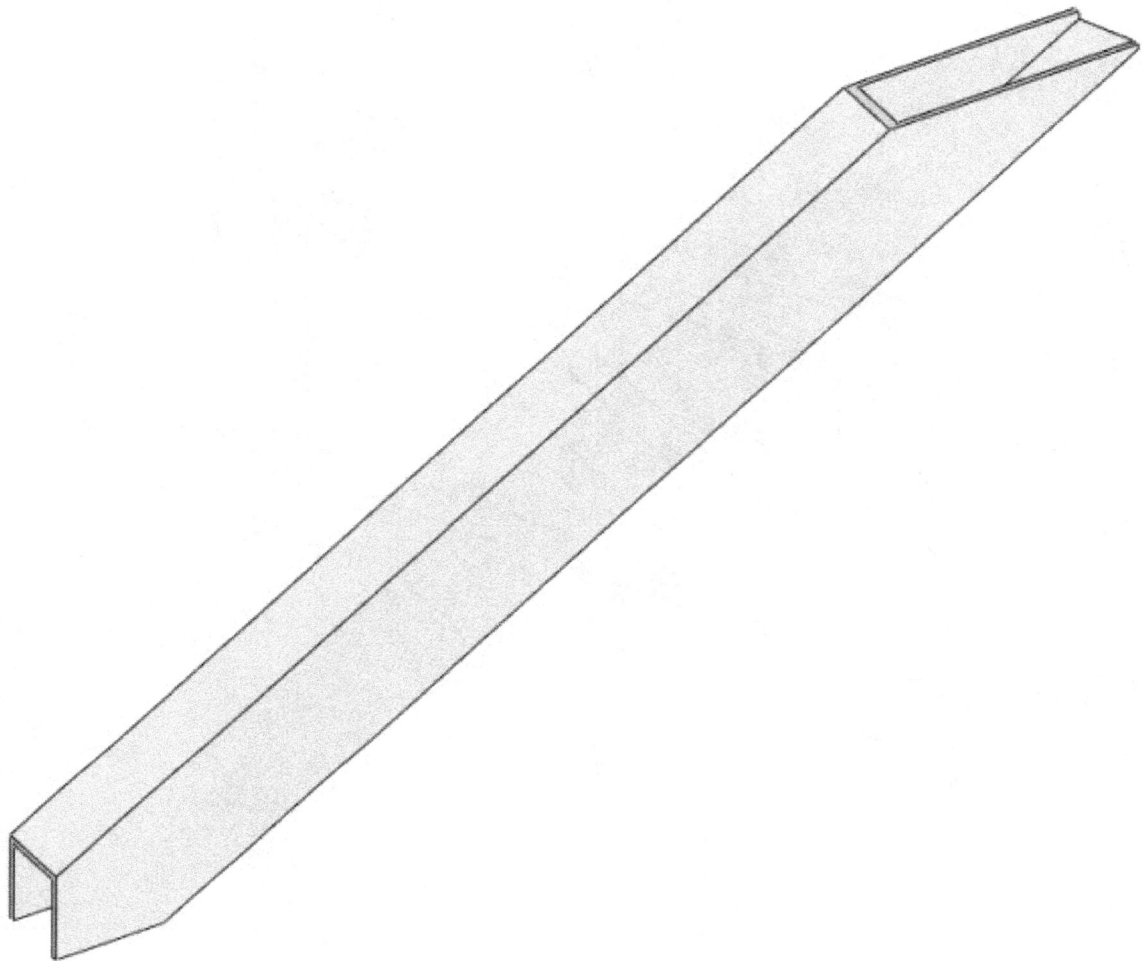

ADDING A STRUCTURAL MEMBER

Click Structural Member under the Weldments toolbar or Insert > Weldments > Structural Member. Make selections in the PropertyManager to define the profile for the structural member as shown in the following image. **Deselect** *Transfer Material from Profile* to prevent the transfer of material from the profile to the structural member you are adding - since the Profile has 1060 Alloy Aluminum as the material but the applied material is Plain Carbon Steel in this part.

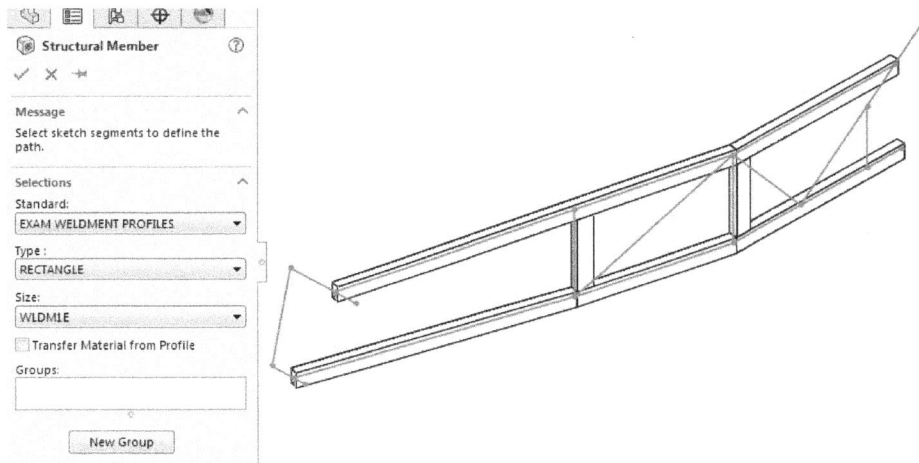

Figure 219 - Question 18 of 26 - Structural Member Property Manager

In the graphics area - select the sketch segment shown in the following image to define the path for the structural member.

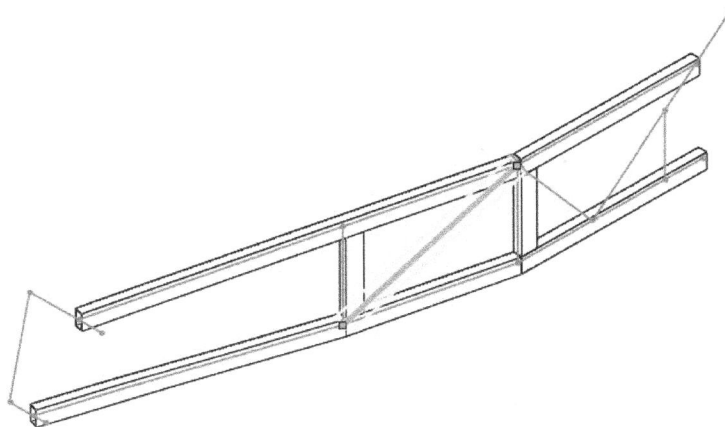

Figure 220 - Question 18 of 26 - Adding a Structural Member

ALIGNING A WELDMENT PROFILE

Click in the Alignment text area in the Structural Member PropertyManager and then in the graphics area click on the horizontal line *(Line15)* below the structural member we are currently adding then select the Align vertical axis Radio Button in the Structural Member PropertyManager under Alignment as shown in the following image .

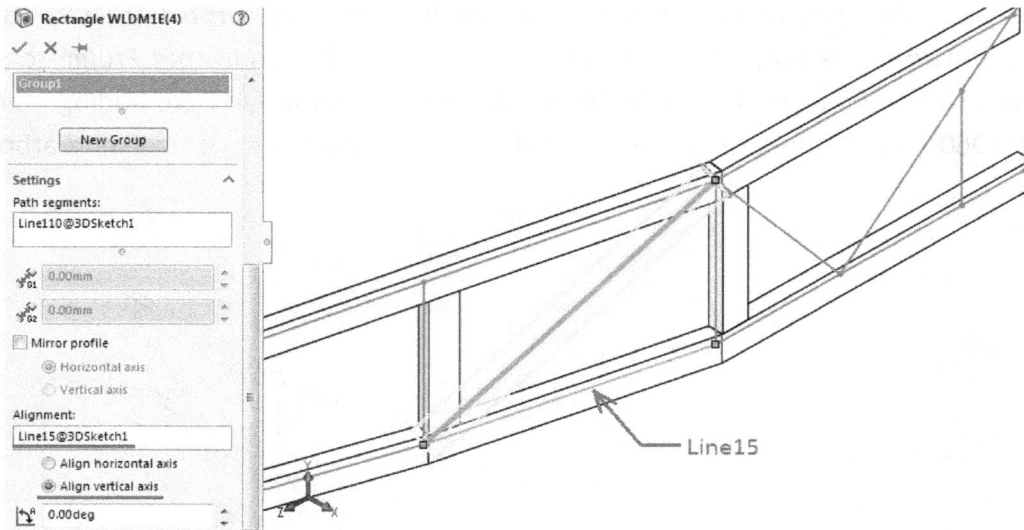

Figure 221 - Question 18 of 26 - Aligning a weldment profile

Click Ok to close the Structural Member Property Manager and have a closer look at the newly added structural member. Your part should now look as shown in the following image.

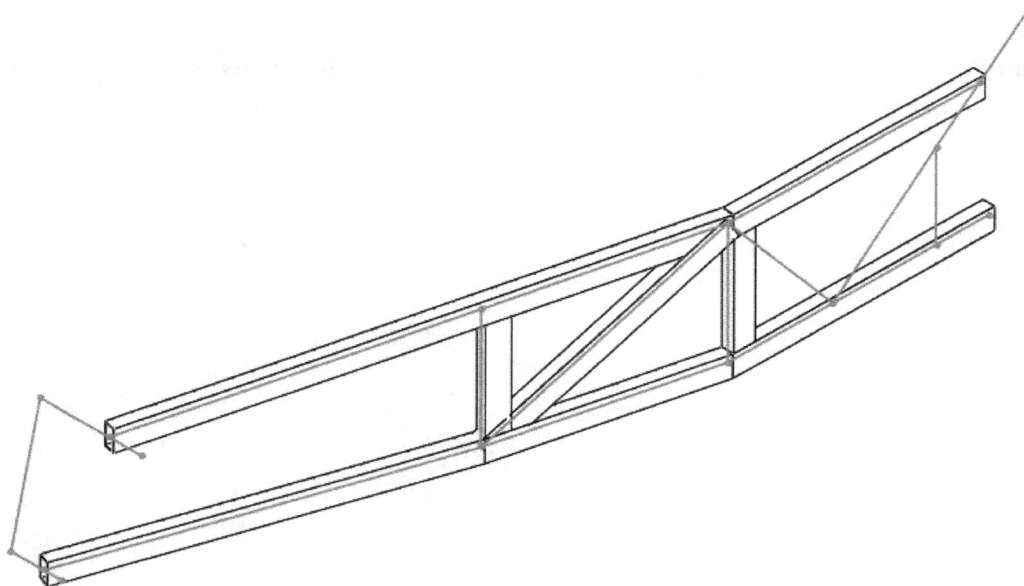

Figure 222 - Question 18 of 26 - Weldment current status

164

TRIMMING SEGMENTS

Click Trim/Extend (Weldments toolbar) or Insert > Weldments > Trim/Extend. In the PropertyManager, set options as shown in the following image.

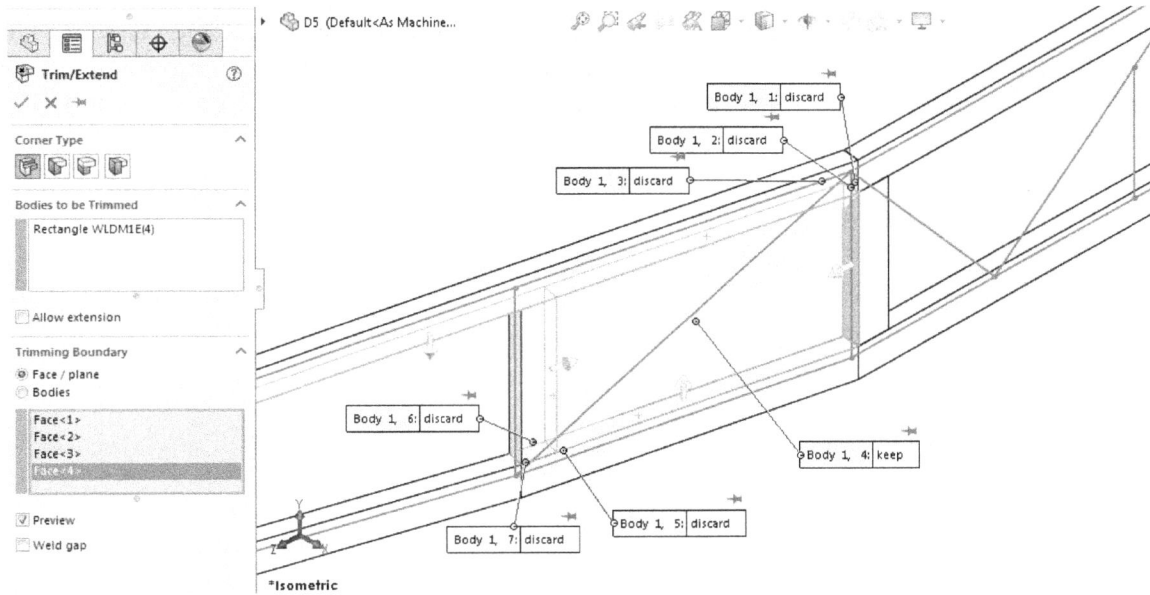

Figure 223 - Question 18 of 26 - Trimming Segments

Click Ok and save you part. Your part should now look as shown in the following image.

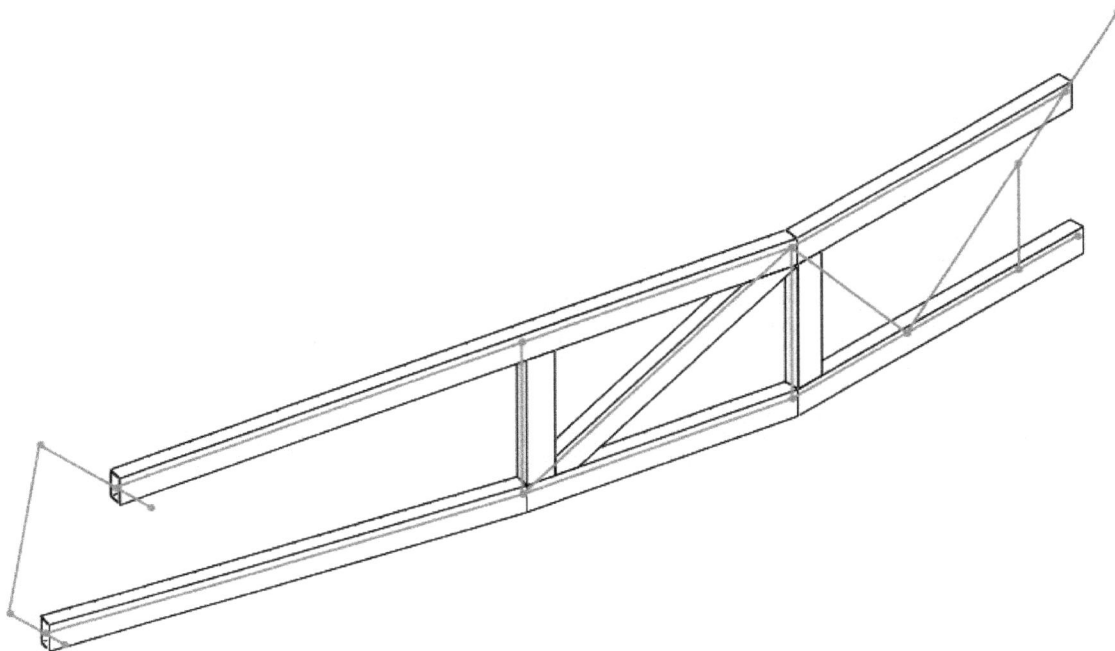

Figure 224 - Question 17 of 26 - Weldment Current Status

DISPLAYING THE CENTER OF MASS

Click Mass Properties (Tools toolbar) or Tools > Evaluate > Mass Properties. The calculated mass properties appear in the dialog box including the center of mass as shown in the following image - center of mass in underlined in red - thus X = 459.01, Y = 186.60 and Z = -1258.99.

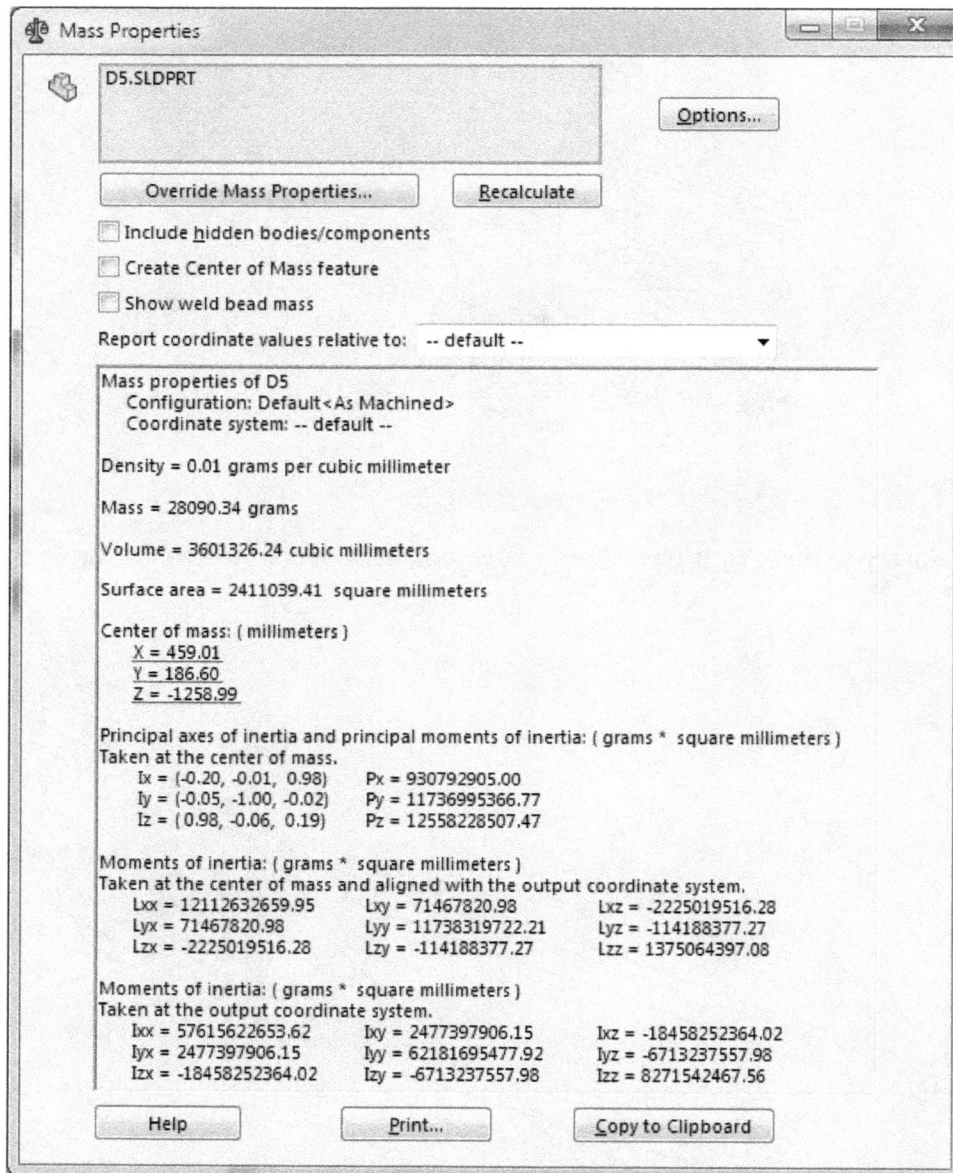

```
╔══════════════════════════════════════════════════════════════════╗
║  ⚖ Mass Properties                              ─  ☐  X            ║
╠══════════════════════════════════════════════════════════════════╣
║   🔷  D5.SLDPRT                                                     ║
║                                                    [ Options... ]  ║
║                                                                    ║
║        [ Override Mass Properties... ]   [ Recalculate ]           ║
║        ☐ Include hidden bodies/components                          ║
║        ☐ Create Center of Mass feature                             ║
║        ☐ Show weld bead mass                                       ║
║   Report coordinate values relative to:  -- default --      ▼      ║
║   ──────────────────────────────────────────────────────────      ║
║   Mass properties of D5                                            ║
║       Configuration: Default<As Machined>                          ║
║       Coordinate system: -- default --                             ║
║                                                                    ║
║   Density = 0.01 grams per cubic millimeter                        ║
║                                                                    ║
║   Mass = 28090.34 grams                                            ║
║                                                                    ║
║   Volume = 3601326.24 cubic millimeters                            ║
║                                                                    ║
║   Surface area = 2411039.41  square millimeters                    ║
║                                                                    ║
║   Center of mass: ( millimeters )                                  ║
║       X = 459.01                                                   ║
║       Y = 186.60                                                   ║
║       Z = -1258.99                                                 ║
╚══════════════════════════════════════════════════════════════════╝
```

Principal axes of inertia and principal moments of inertia: (grams * square millimeters)
Taken at the center of mass.
    Ix = (-0.20, -0.01, 0.98)   Px = 930792905.00
    Iy = (-0.05, -1.00, -0.02)   Py = 11736995366.77
    Iz = (0.98, -0.06, 0.19)   Pz = 12558228507.47

Moments of inertia: (grams * square millimeters)
Taken at the center of mass and aligned with the output coordinate system.
    Lxx = 12112632659.95   Lxy = 71467820.98   Lxz = -2225019516.28
    Lyx = 71467820.98   Lyy = 11738319722.21   Lyz = -114188377.27
    Lzx = -2225019516.28   Lzy = -114188377.27   Lzz = 1375064397.08

Moments of inertia: (grams * square millimeters)
Taken at the output coordinate system.
    Ixx = 57615622653.62   Ixy = 2477397906.15   Ixz = -18458252364.02
    Iyx = 2477397906.15   Iyy = 62181695477.92   Iyz = -6713237557.98
    Izx = -18458252364.02   Izy = -6713237557.98   Izz = 8271542467.56

[Help]   [Print...]   [Copy to Clipboard]

Figure 225 - Question 18 of 26 - Center of Mass

CHAPTER 19

QUESTION 19 EXAM SCREEN CAPTURES

QUESTION 19 EXAM SCREEN CAPTURE 1

Pro. Adv. - Advanced Weldments (CSWPA-WD)

Question 19 of 26

For 15 points:

D04005 - Create Second Diagonal Segment
Build this 3D Sketch in SolidWorks.
Unit system: MMGS (millimeter, gram, second)
Decimal places: 2
Material: Already defined in part

-Hide the existing 3D sketch.

-Create the diagonal segment indicated using weldment profile "WLDM2E".

Hint: You must create a new 3D sketch to create this new segment.

Note 1: Align the weldment profile so that it is lined up with segment 1.

Note 2: The diagonal weldment segment created must be trimmed to be flush with the segments they contact.

-Select ONLY the diagonal segment just created and measure its mass.

What is the mass of the diagonal segment just created (grams)?

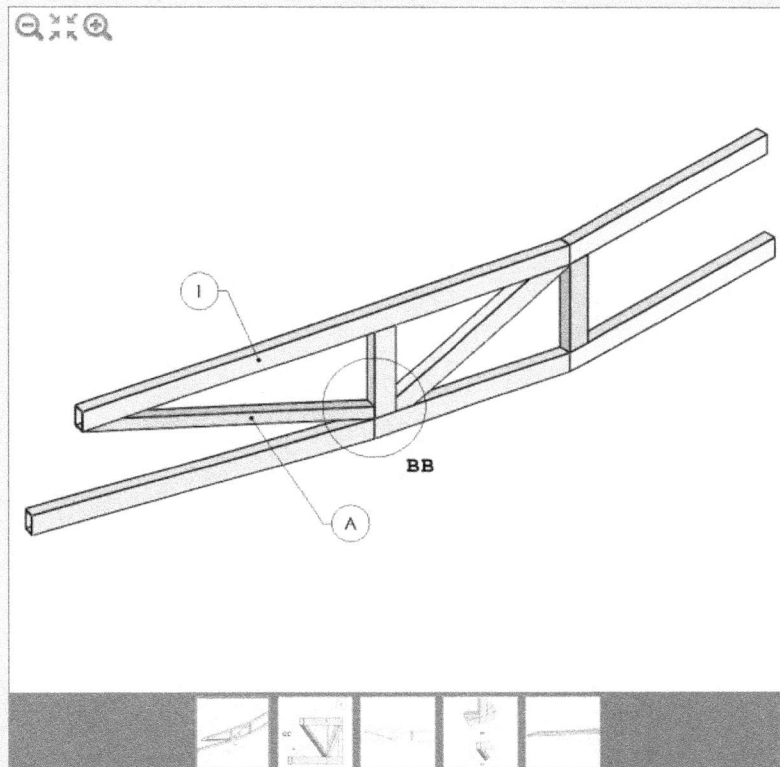

BB

A

Enter Value:

(use . (point) as decimal separator)

Figure 226 - Question 19 of 26 - Exam Screen Capture 1

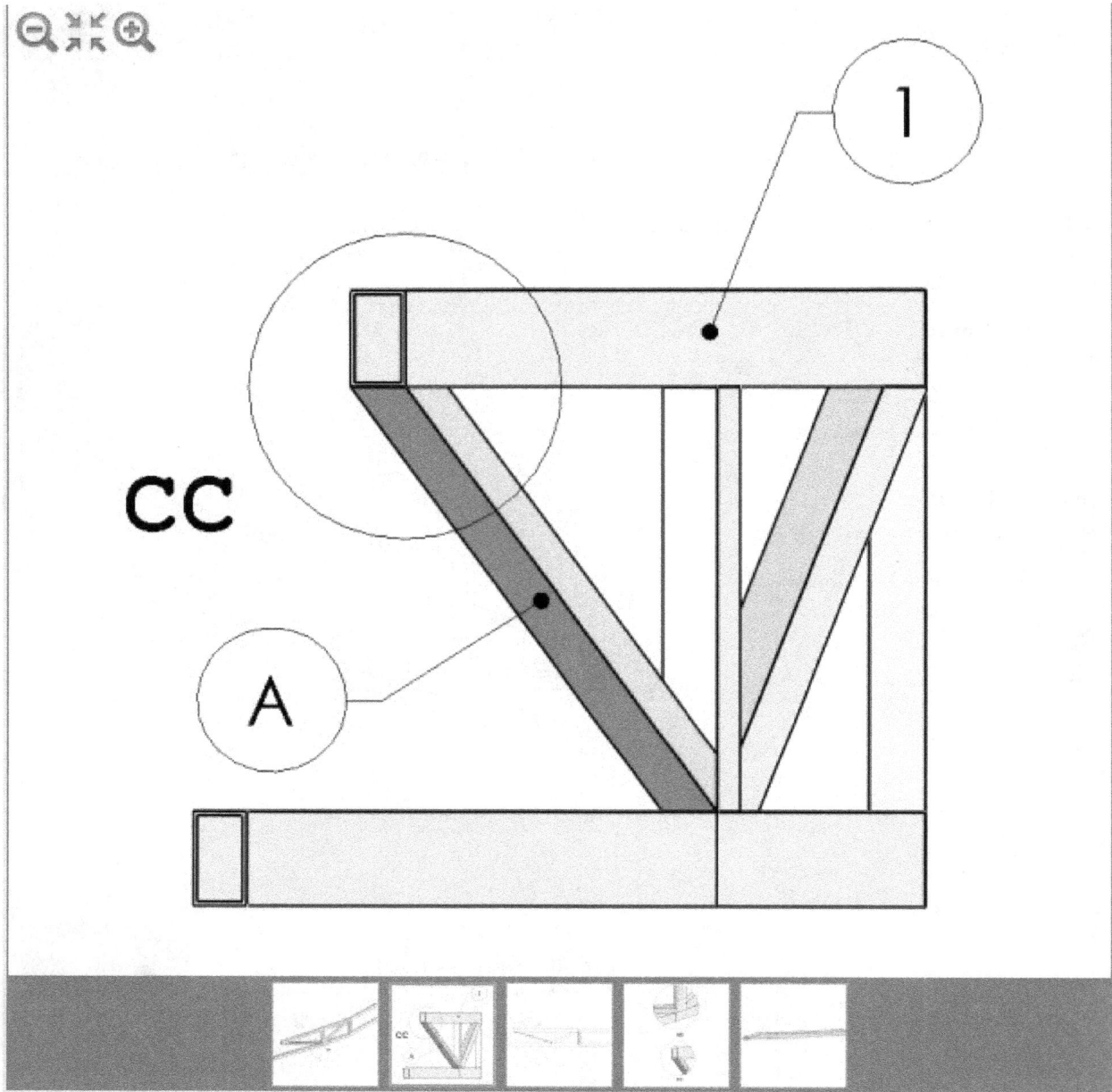

Figure 227 - Question 19 of 26 - Exam Screen Capture 2

Figure 228 - Question 19 of 26 - Exam Screen Capture 3

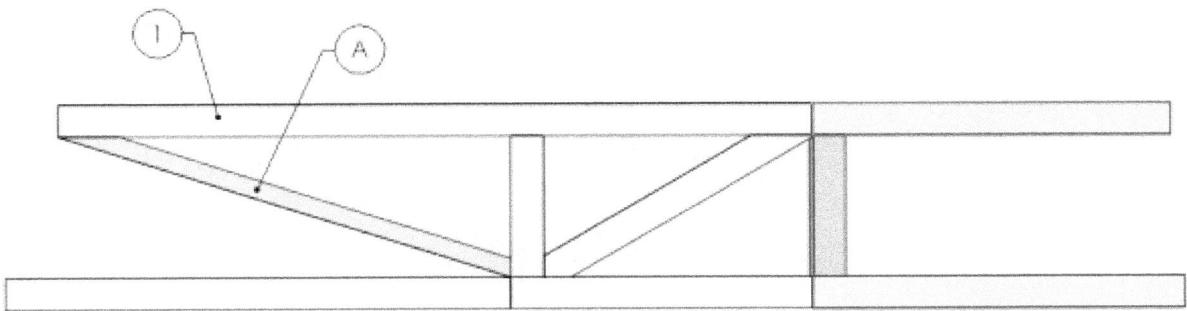

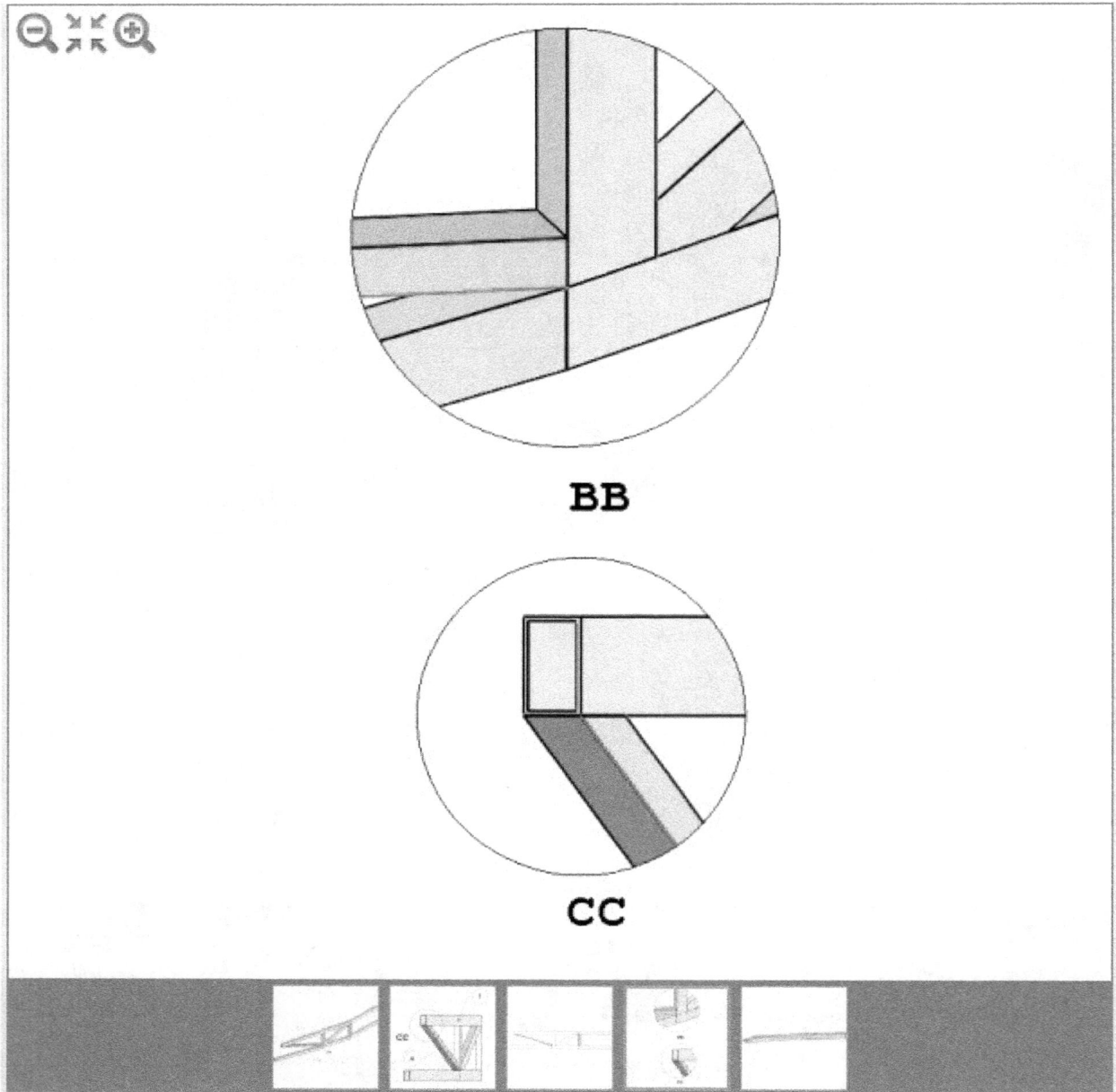

Figure 229 - Question 19 of 26 - Exam Screen Capture 4

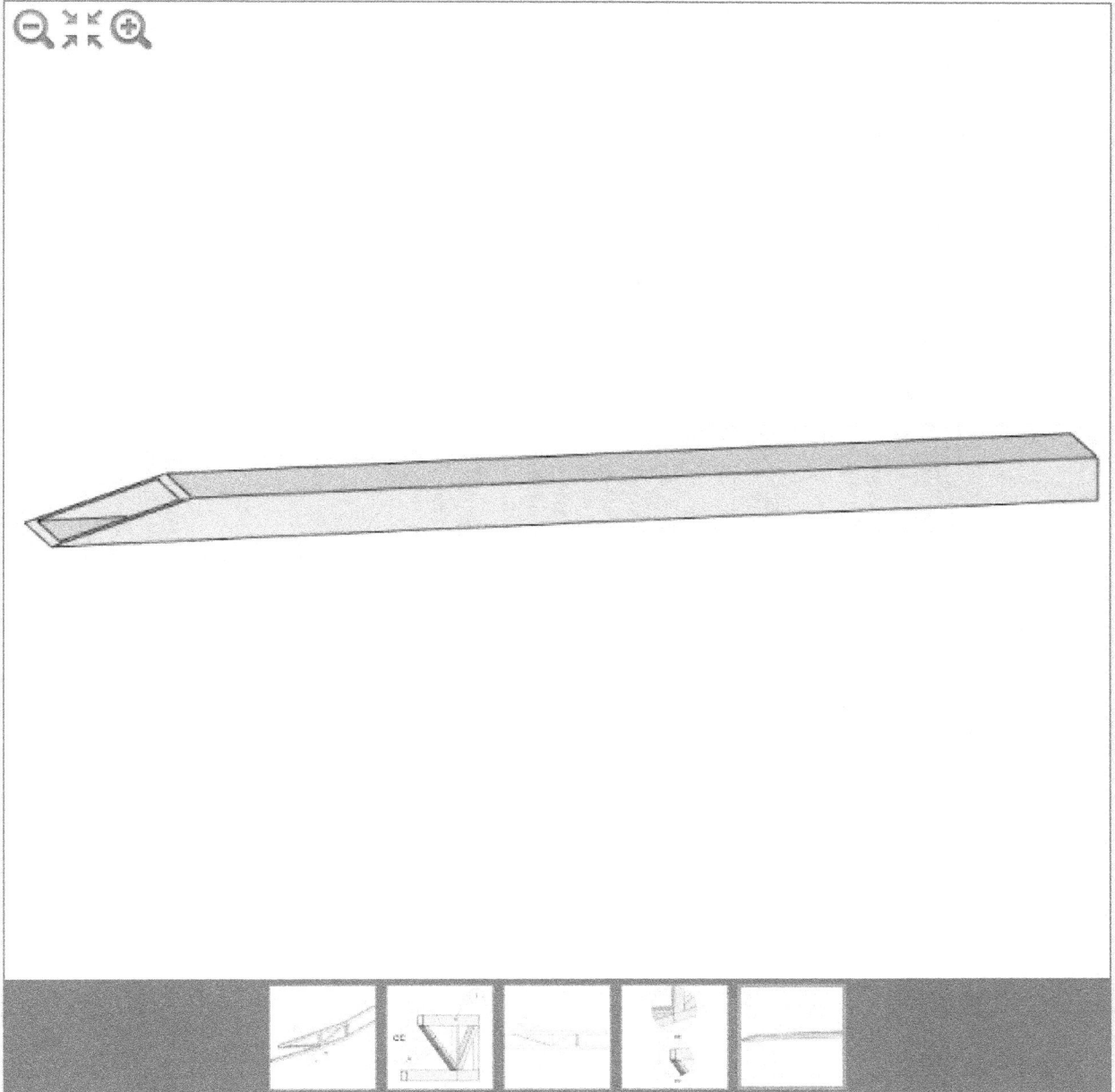

Figure 230 - Question 19 of 26 - Exam Screen Capture 5

HIDING AND SHOWING SKETCHES

To hide the 3D Sketch in the model, click on the 3D Sketch in the FeatureManager Design Tree and select Hide as shown in the following image. Follow same procedure to show the sketch.

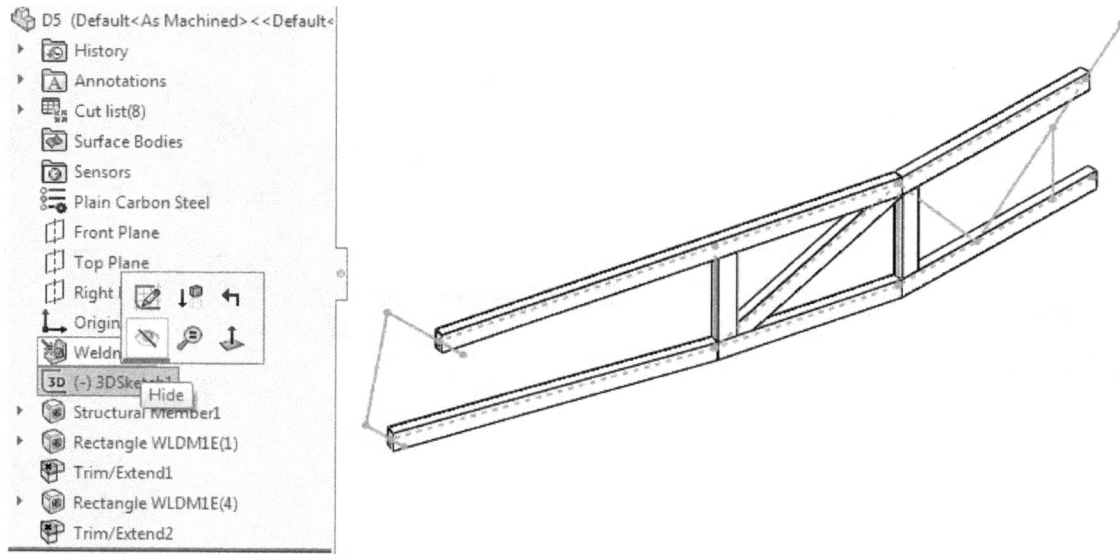

Figure 231 - Question 19 of 26 - Hiding and Showing Sketches

Your part should now look as shown in the following image with the 3D sketch hidden.

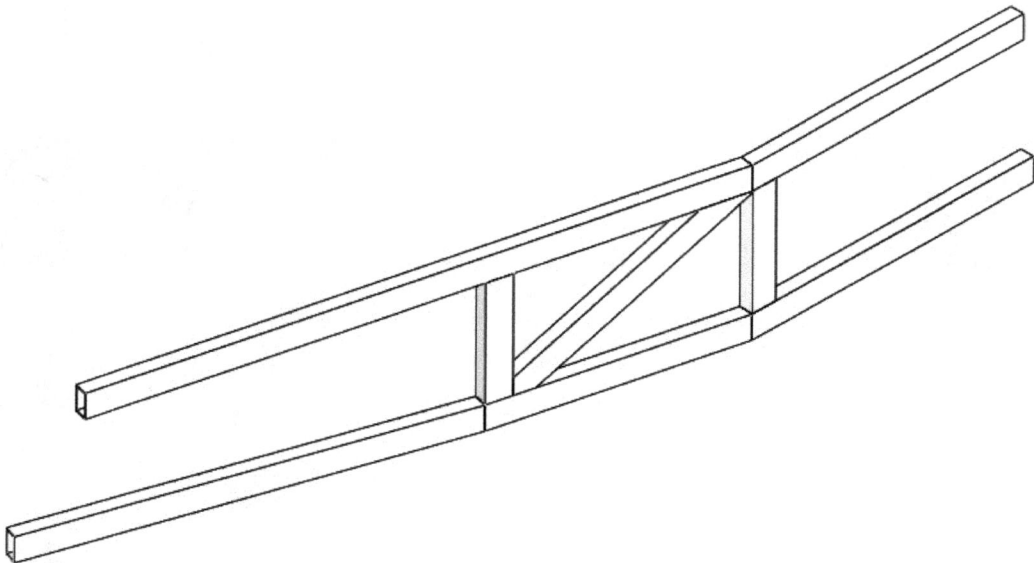

Figure 232 - Question 19 of 26 - Weldment Current Status

172

3D SKETCHING

You can create 3D sketch entities on a working plane, or at *any arbitrary point* in 3D space. Click 3D Sketch (Sketch toolbar) or Insert > 3D Sketch to open a 3D sketch in Isometric view. Click Line on the Sketch toolbar, or click Tools > Sketch Entities > Line. The 3D Line PropertyManager appears and the pointer changes to XY. Zoom to the area circled in red in the following image. Click the midpoint shown in the image to start the line - a midpoint relation is automatically added.

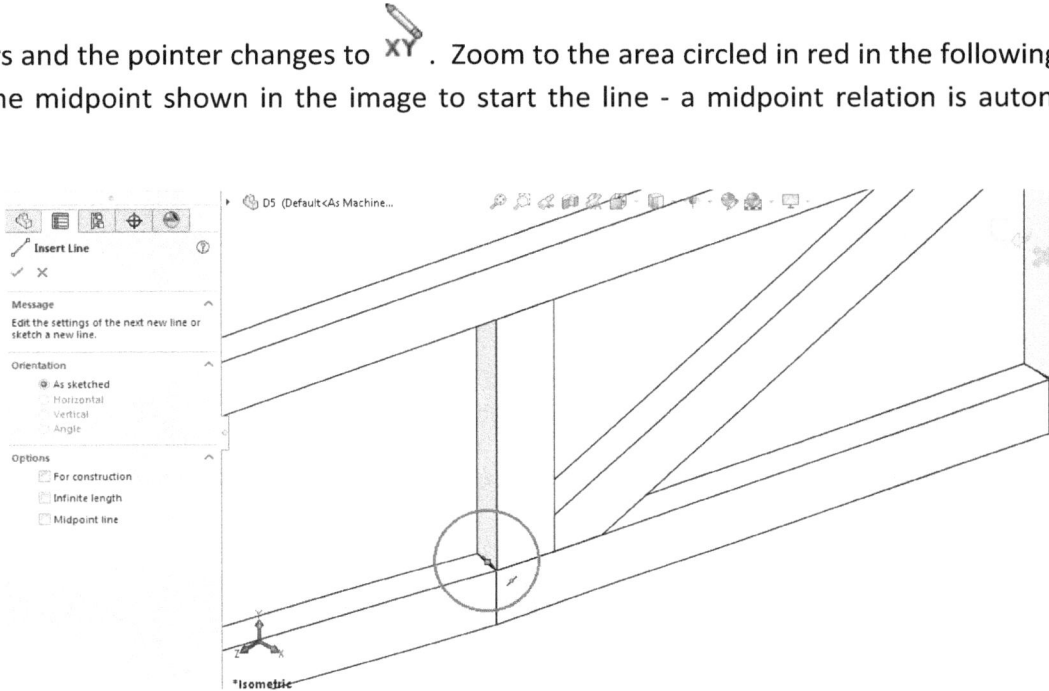

Figure 233 - Question 19 of 26 - 3D Sketching

Drag the line to the midpoint of the structural member shown in the following image and click on the midpoint of the lower horizontal outer edge then press the escape key on your keyboard to end the line segment.

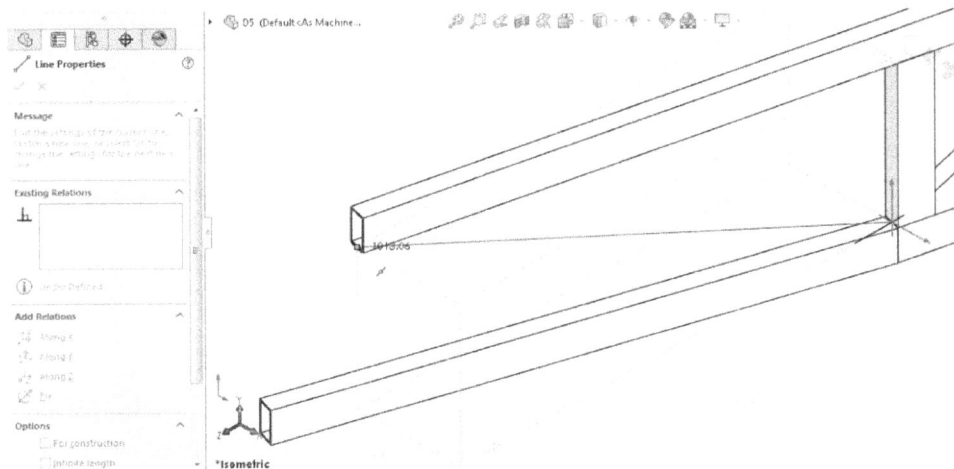

Figure 234 - Question 19 of 26 - 3D Sketching

173

Your part should now look as shown in the following image.

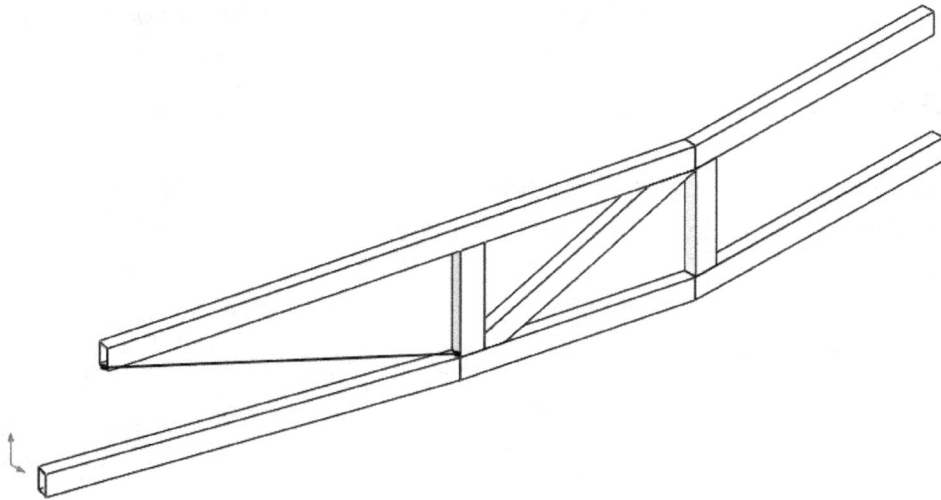

Figure 235 - Question 19 of 26 - 3D Sketch Current Status

Click the Exit Sketch icon in the Confirmation Corner to finish the sketch. Your part should now look as shown in the following image.

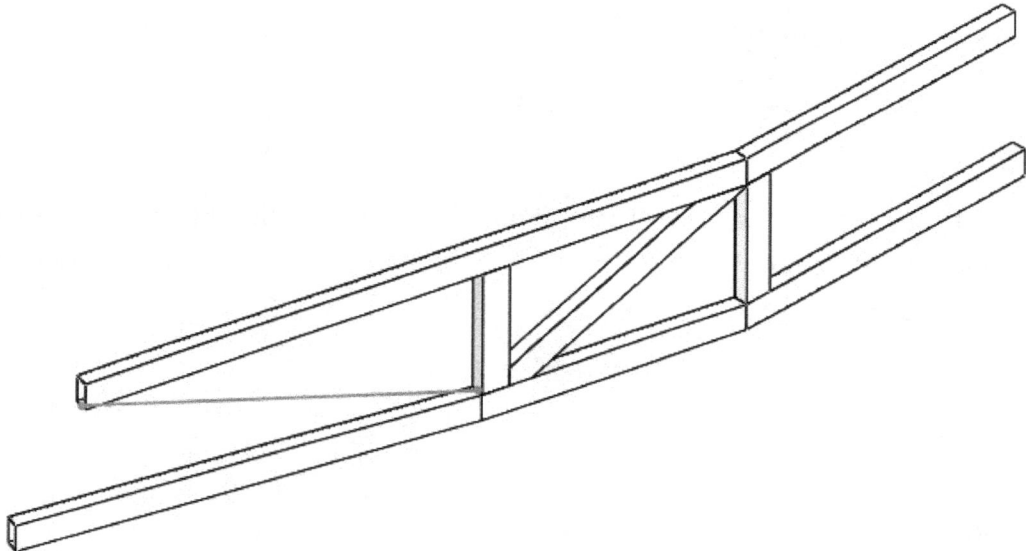

Figure 236 - Question 19 of 26 - 3D Sketch Current Status

ADDING A STRUCTURAL MEMBER

Click Structural Member under the Weldments toolbar or Insert > Weldments > Structural Member. Make selections in the PropertyManager to define the profile for the structural member as shown in the following image. **Deselect** *Transfer Material from Profile* to prevent the transfer of material from the profile to the structural member you are adding - since the

Profile has 1060 Alloy Aluminum as the material but the applied material is Plain Carbon Steel in this part.

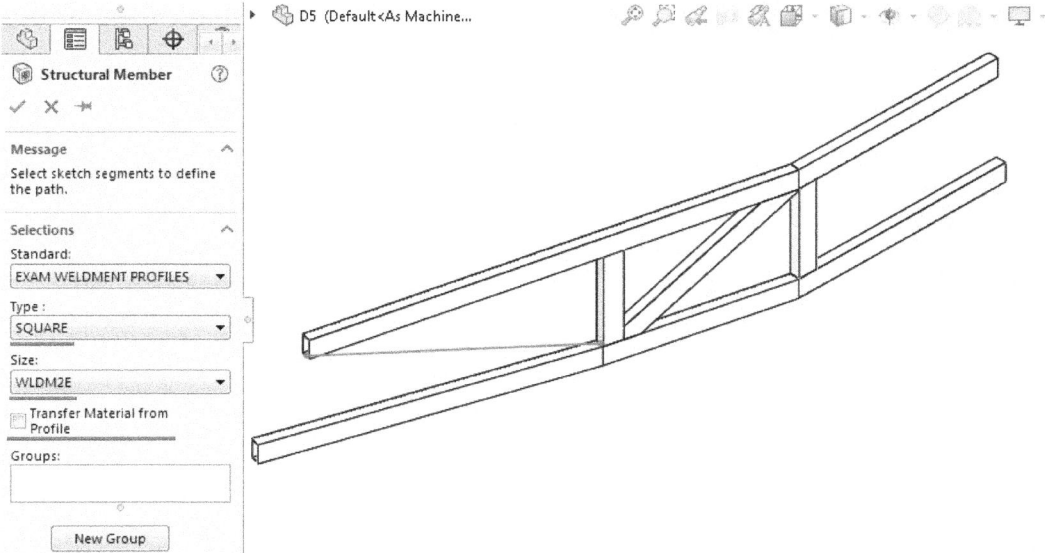

Figure 237 - Question 19 of 26 - Structural Member Property Manager

In the graphics area - select the sketch segment shown in the following image to define the path for the first structural member.

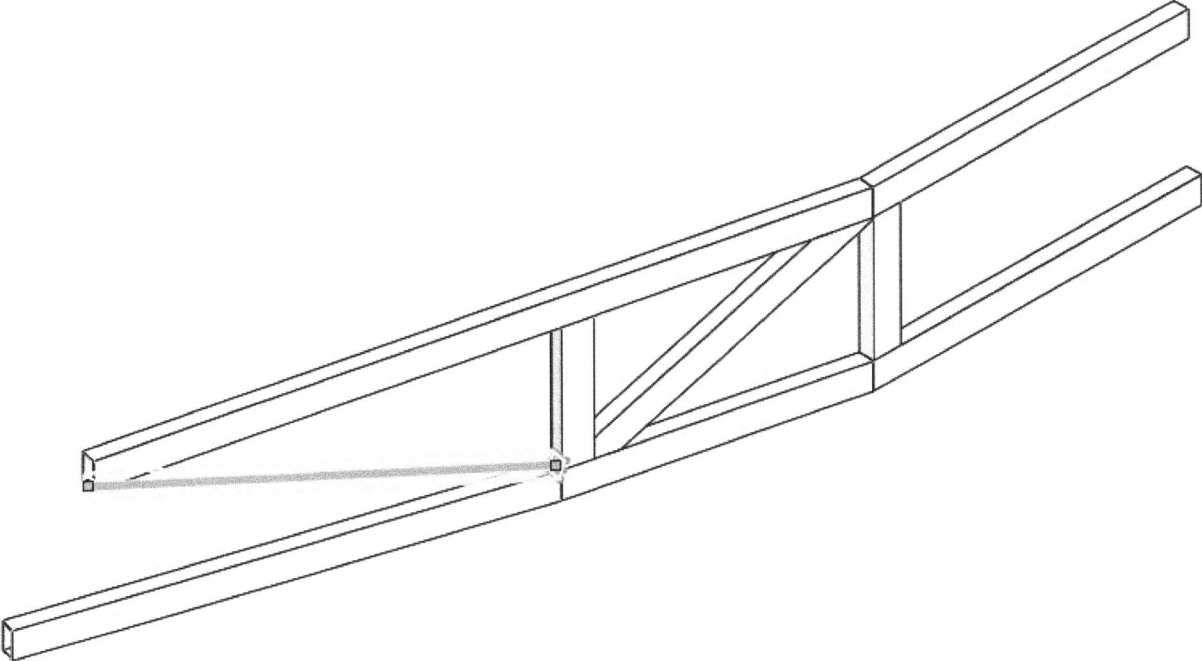

Figure 238 - Question 19 of 26 - Adding a Structural Member

ALIGNING A WELDMENT PROFILE

Click in the Alignment text area in the Structural Member PropertyManager and then in the graphics area click on the horizontal edge of the top horizontal structural member then select the Align horizontal axis Radio Button in the Structural Member PropertyManager under Alignment as shown in the following image .

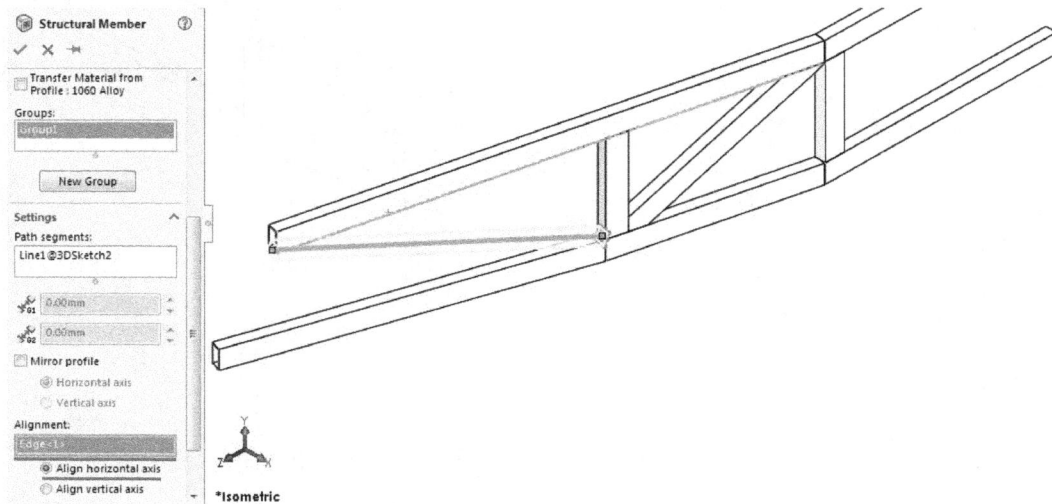

Figure 239 - Question 19 of 26 - Aligning a weldment profile

WELDMENTS - PIERCE POINTS

In the Structural Member PropertyManager, below Alignment and Rotation Angle, click the Locate Profile command button. The display zooms to the profile of the structural member as shown in the following image.

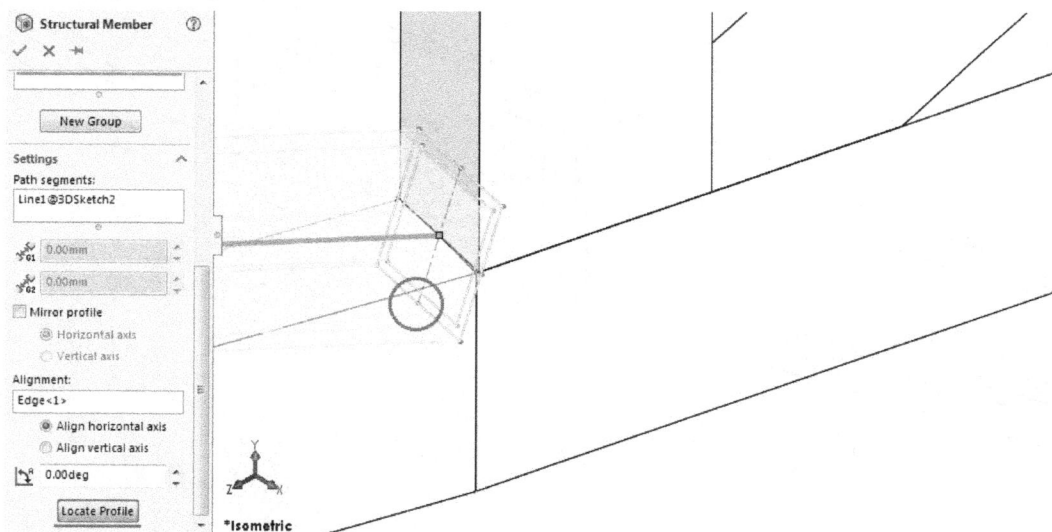

Figure 240 - Question 19 of 26 - Locate profile

176

Select a vertex point on the profile circled in the red in the previous image. The profile shifts to align the new pierce point with the structural member sketch segment as shown in the following image.

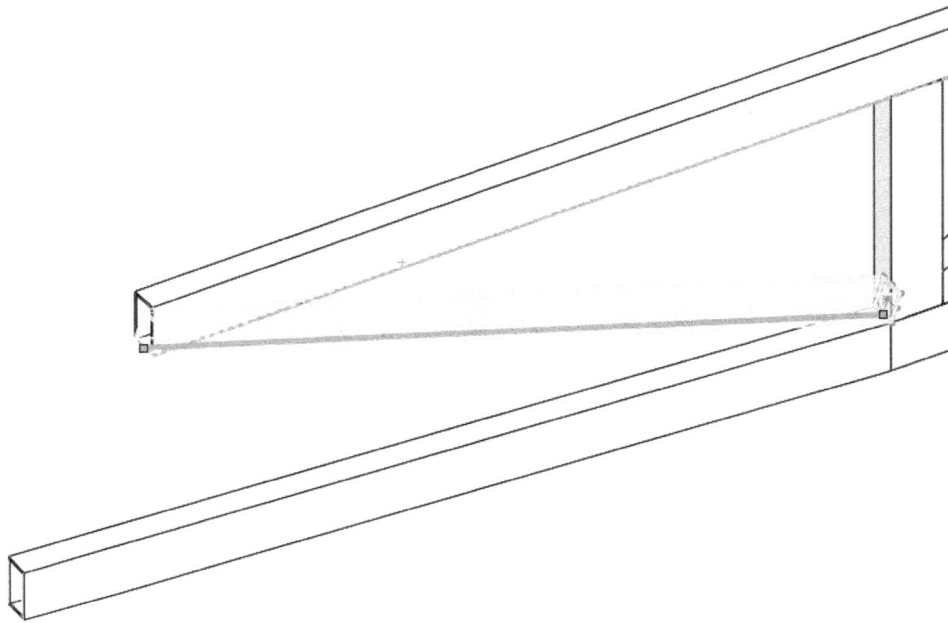

Figure 241 - Question 19 of 26 - Pierce Point

Click Ok to close the Structural Member Property Manager and have a closer look at the newly added structural member. Your part should now look as shown in the following image.

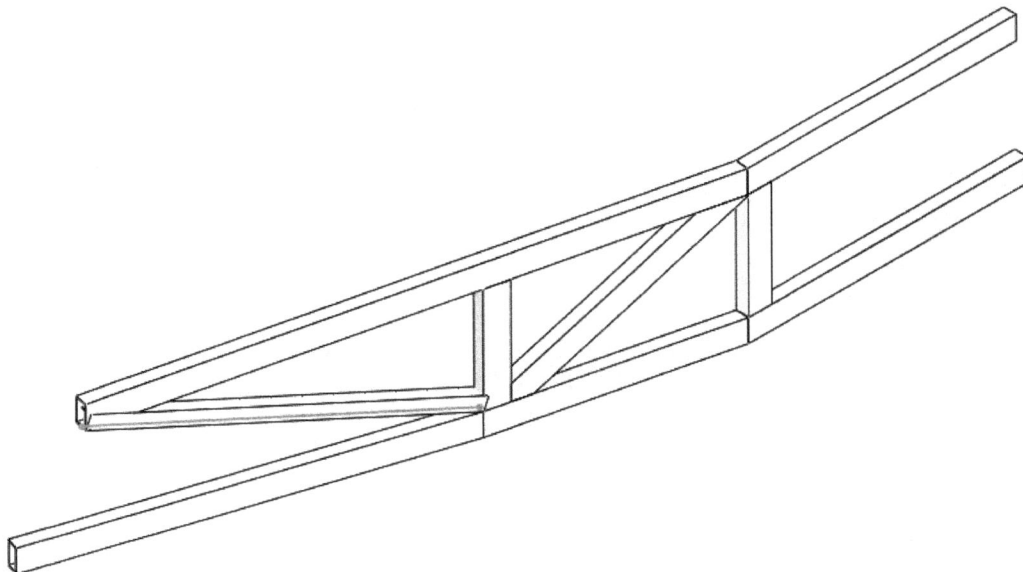

Figure 242 - Question 19 of 26 - Weldment current status

TRIMMING SEGMENTS

Click Trim/Extend (Weldments toolbar) or Insert > Weldments > Trim/Extend. In the PropertyManager, set options as shown in the following image.

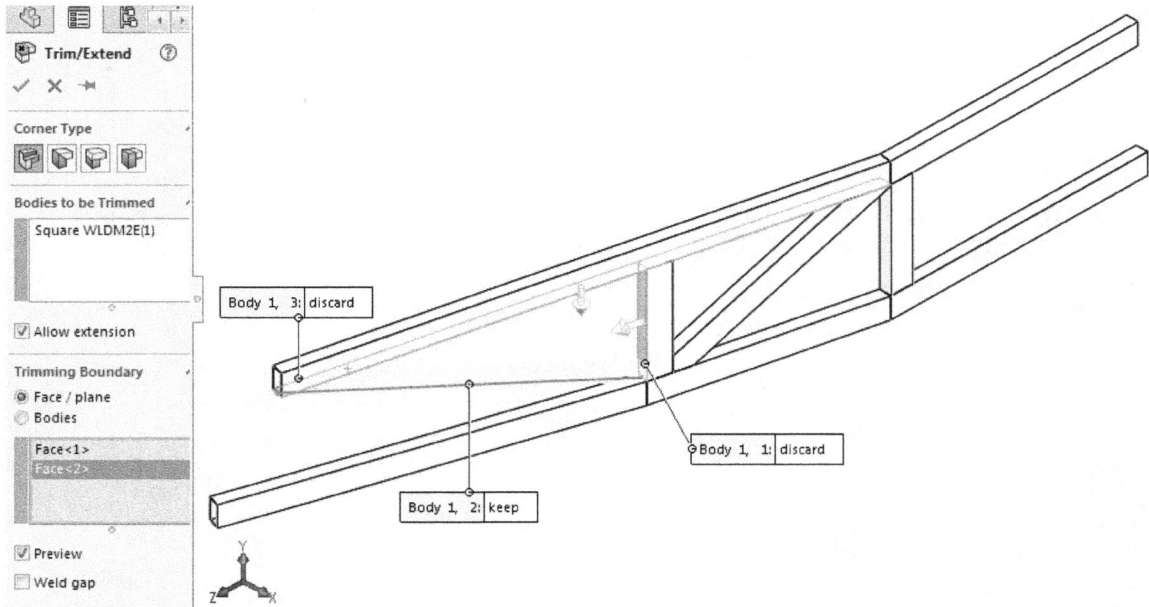

Figure 243 - Question 19 of 26 - Trimming Segments

Click Ok and save you part. Your part should now look as shown in the following image with all 3D Sketches hidden.

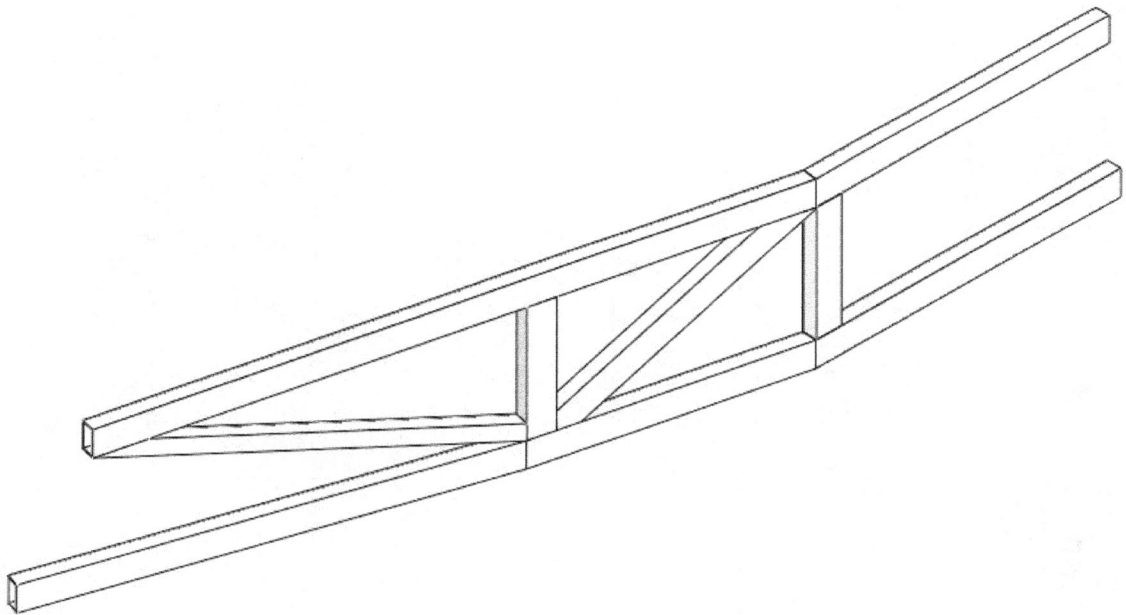

Figure 244 - Question 19 of 26 - Weldment Current Status

DISPLAYING THE MASS OF A SINGLE SEGMENT

Right click on a face on the newly added diagonal structural member and click on Isolate as shown in the following image.

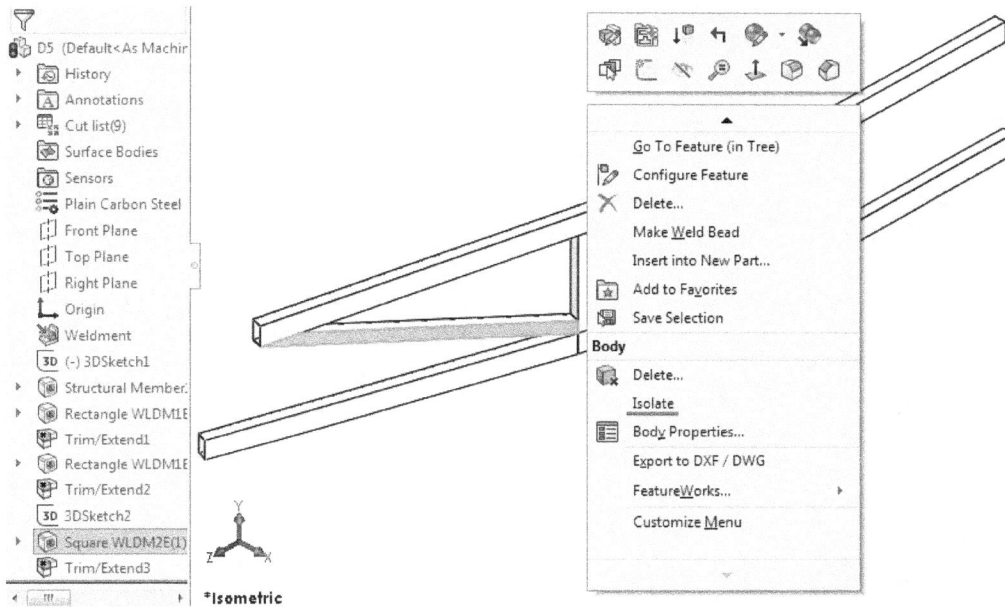

Figure 245 - Question 19 of 26 - Isolating a single segment

Your part should now look as shown in the following image.

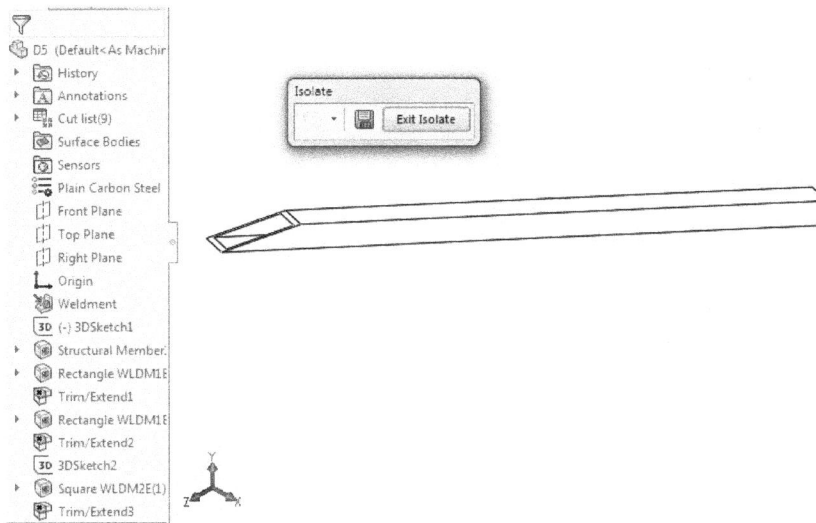

Figure 246 - Question 19 of 26 - Isolating a single segment

179

Click Mass Properties (Tools toolbar) or Tools > Evaluate > Mass Properties. The calculated mass properties appear in the dialog box as shown in the following image - make sure include hidden bodies is not checked. The mass of the diagonal segment we just created is thus 3190.62 grams.

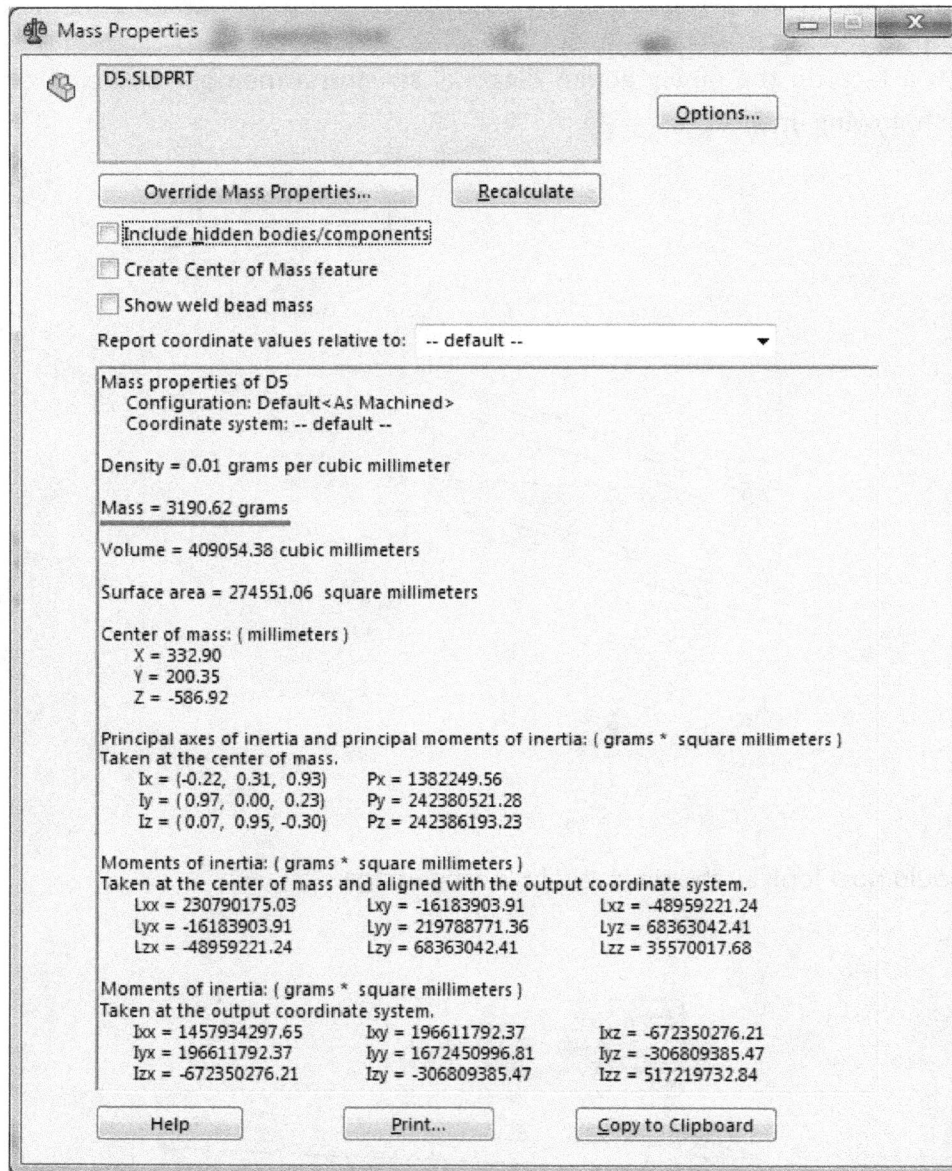

Figure 247 - Question 19 of 26 - Mass of a single segment

CHAPTER 20

QUESTION 20 EXAM SCREEN CAPTURES

QUESTION 20 EXAM SCREEN CAPTURE 1

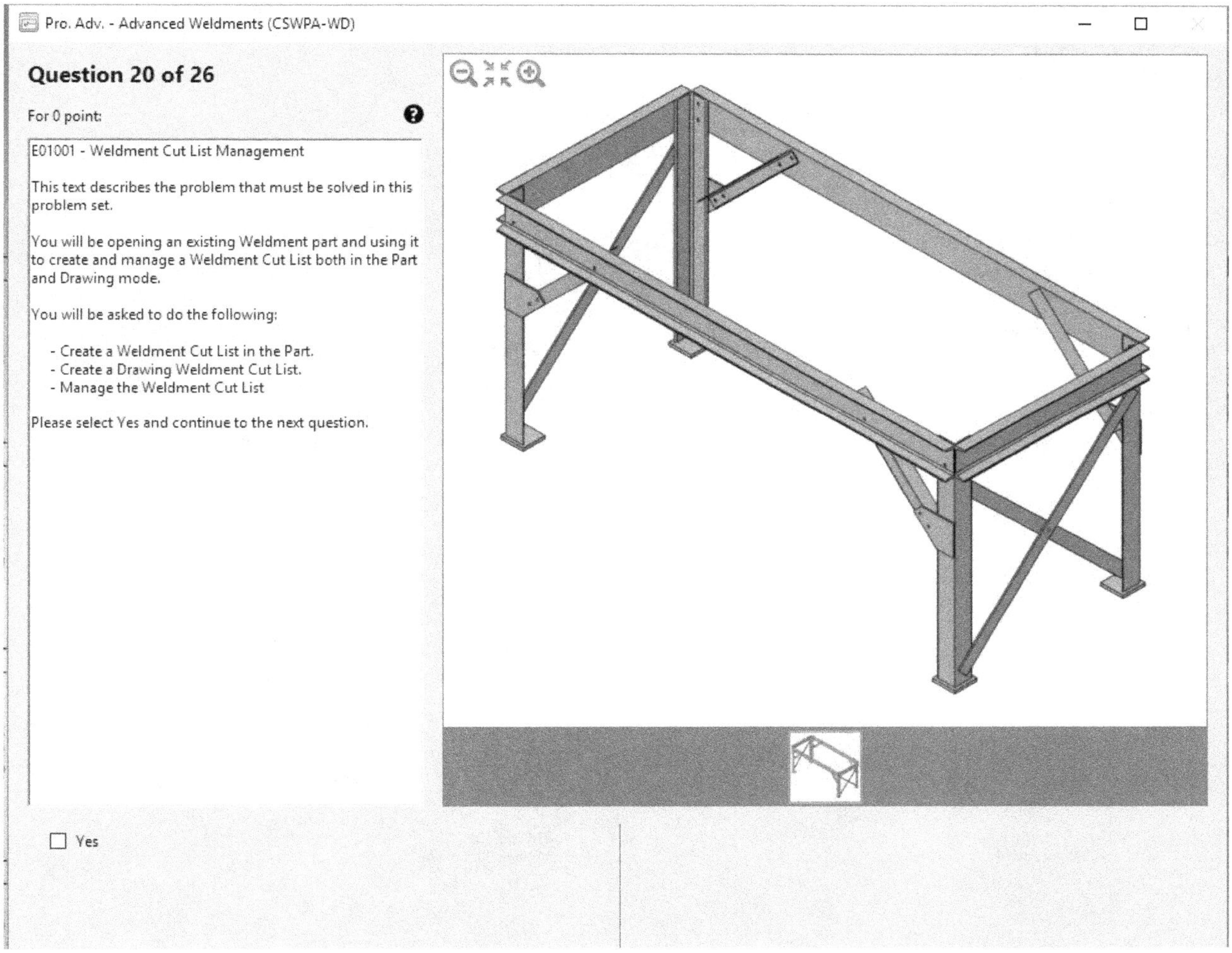

Figure 248 - Question 20 of 26 - Exam Screen Capture 1

In this question, you just read the instructions then select Yes and continue to Question 21.

CHAPTER 21

QUESTION 21 EXAM SCREEN CAPTURES

QUESTION 21 EXAM SCREEN CAPTURE 1

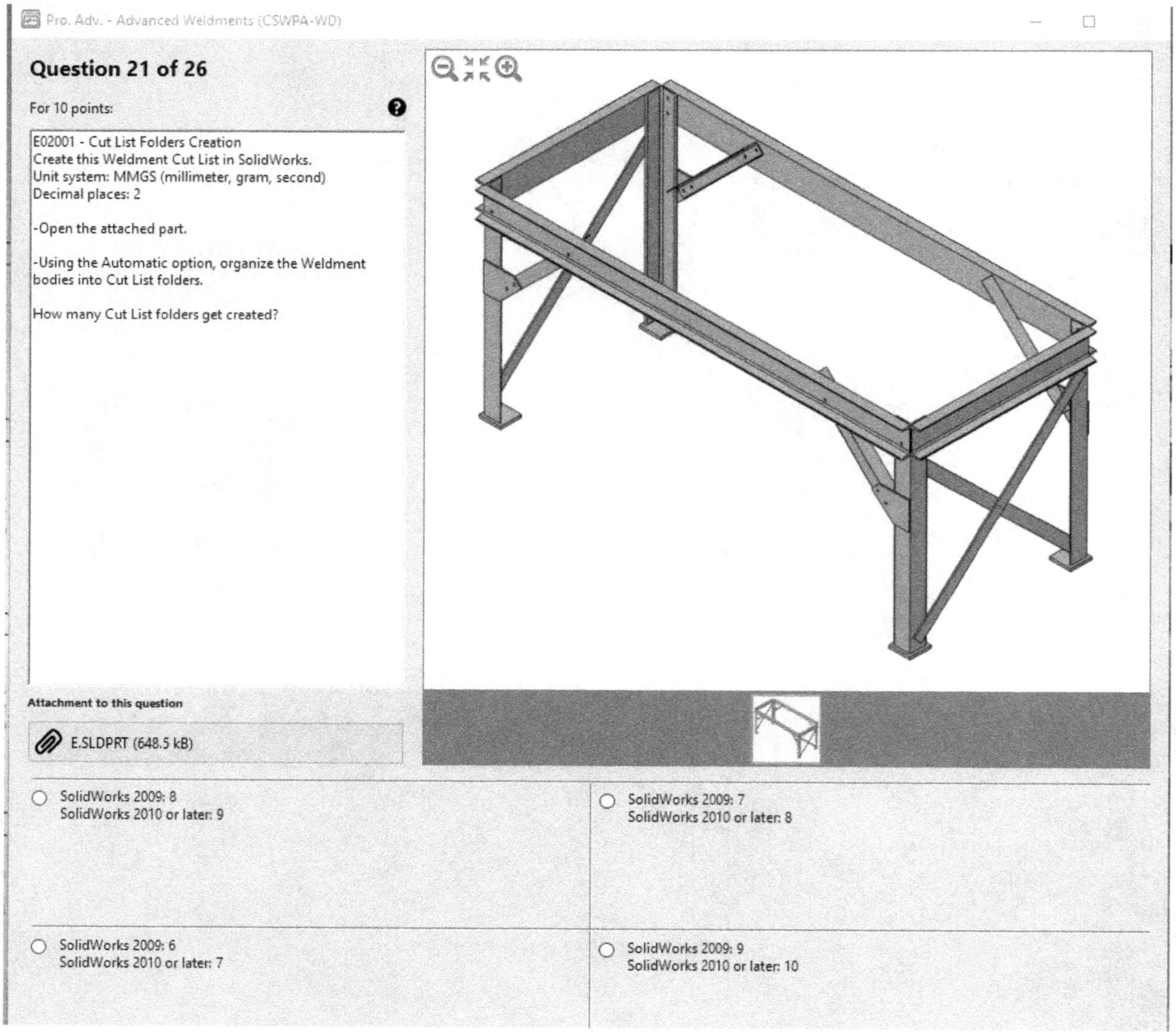

Pro. Adv. - Advanced Weldments (CSWPA-WD)

Question 21 of 26

For 10 points:

E02001 - Cut List Folders Creation
Create this Weldment Cut List in SolidWorks.
Unit system: MMGS (millimeter, gram, second)
Decimal places: 2

-Open the attached part.

-Using the Automatic option, organize the Weldment bodies into Cut List folders.

How many Cut List folders get created?

Attachment to this question

E.SLDPRT (648.5 kB)

- SolidWorks 2009: 8
  SolidWorks 2010 or later: 9

- SolidWorks 2009: 7
  SolidWorks 2010 or later: 8

- SolidWorks 2009: 6
  SolidWorks 2010 or later: 7

- SolidWorks 2009: 9
  SolidWorks 2010 or later: 10

Figure 249 - Question 21 of 26 - Exam Screen Capture 1

QUESTION 21 SOLUTION

ORGANISING WELDMENT BODIES INTO CUT LIST FOLDERS

Download part E.SLDPRT from this Google Drive Location *(http://bit.ly/CSWPA-WD)* in the Question 21 Folder. Open and save the downloaded part onto your PC. Your part should look as shown in the following image.

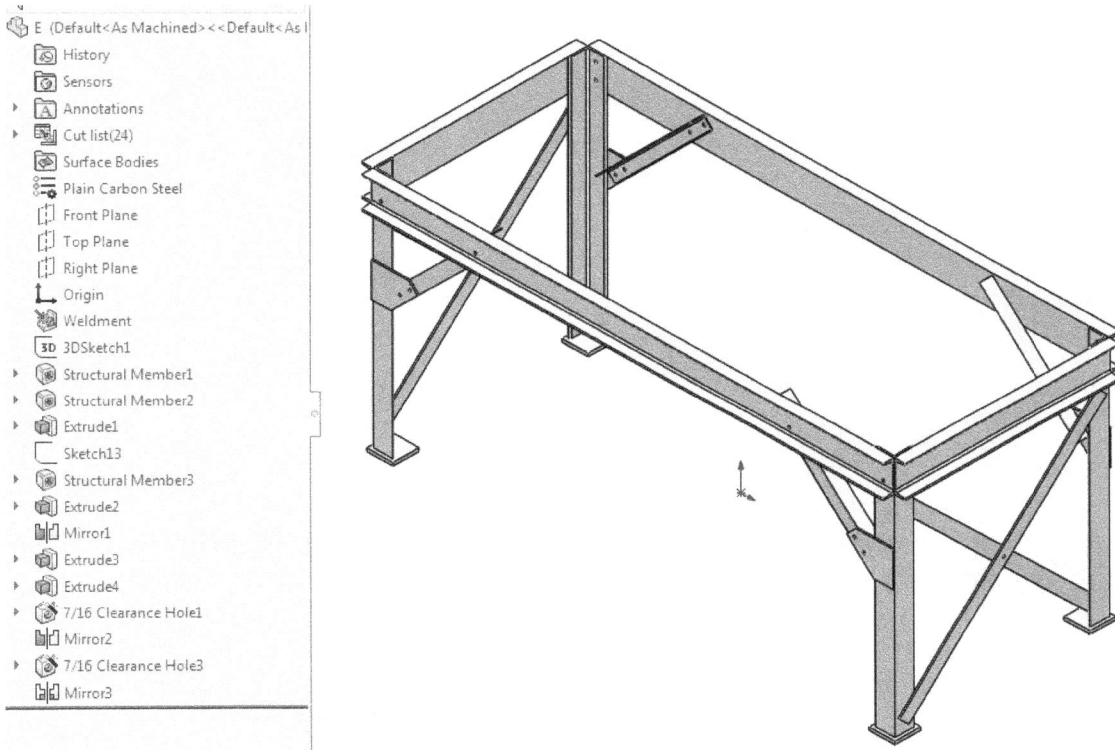

Figure 250 - Question 21 of 26 - Downloaded part E.SLDPRT

A cut list is basically an item in the FeatureManager Design Tree that groups the same entities of a part together. It is available in parts that have weldment or sheet metal features.

Click on the small black arrow next to the Cut List to expand the cut list and show what it currently looks like before we organize the weldment bodies into Cut List Folders. The following image shows what the Cut List currently looks like.

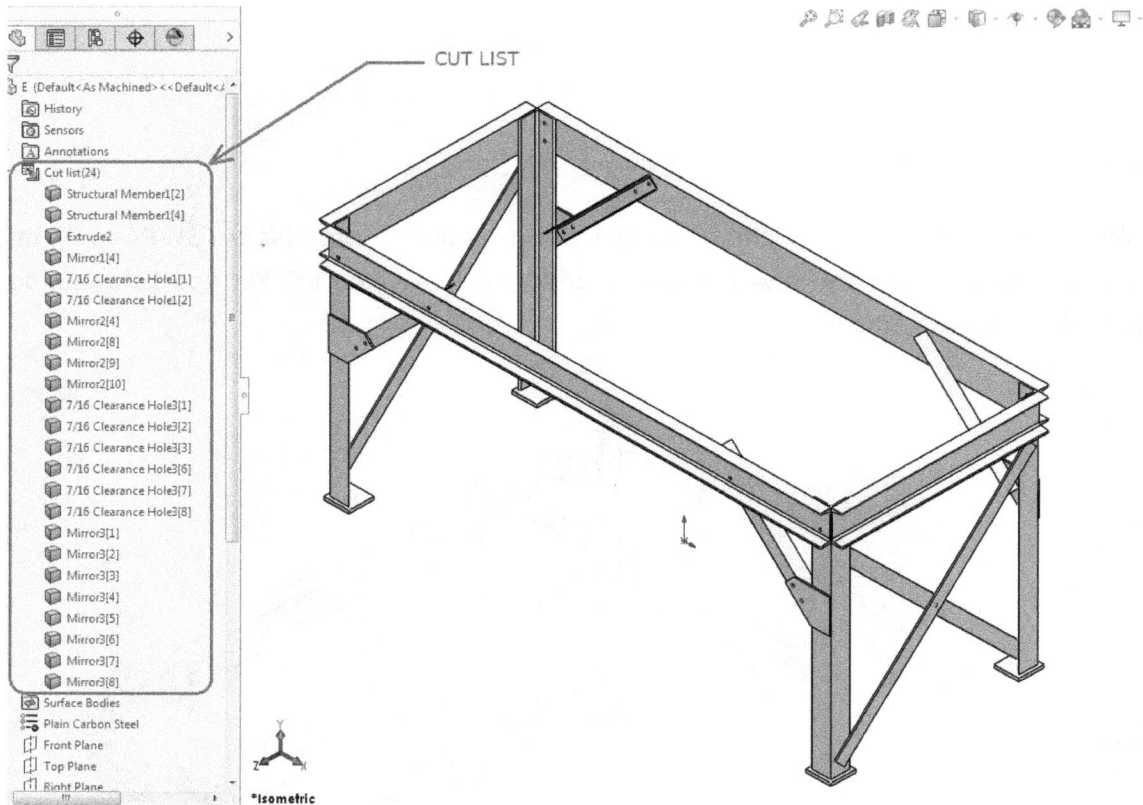

Figure 232 - Question 21 of 26 - Cut List

Right click on the Cut List item in the Feature Manager Design tree and click on *Create Cut Lists Automatically* as shown in the following image.

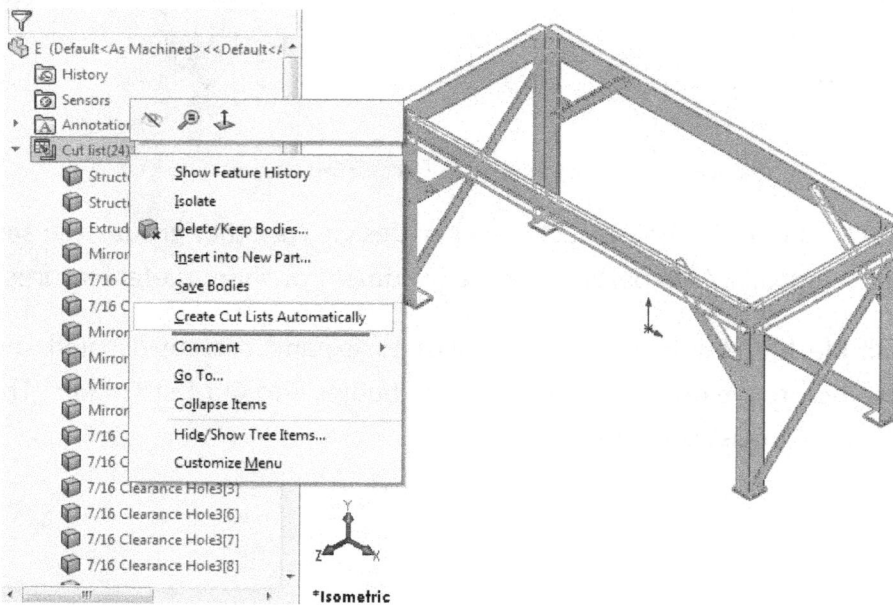

Figure 252 - Question 21 of 26 - Creating Cut Lists Automatically

The Cut List automatically organizes weldment bodies into folders as shown in the following image.

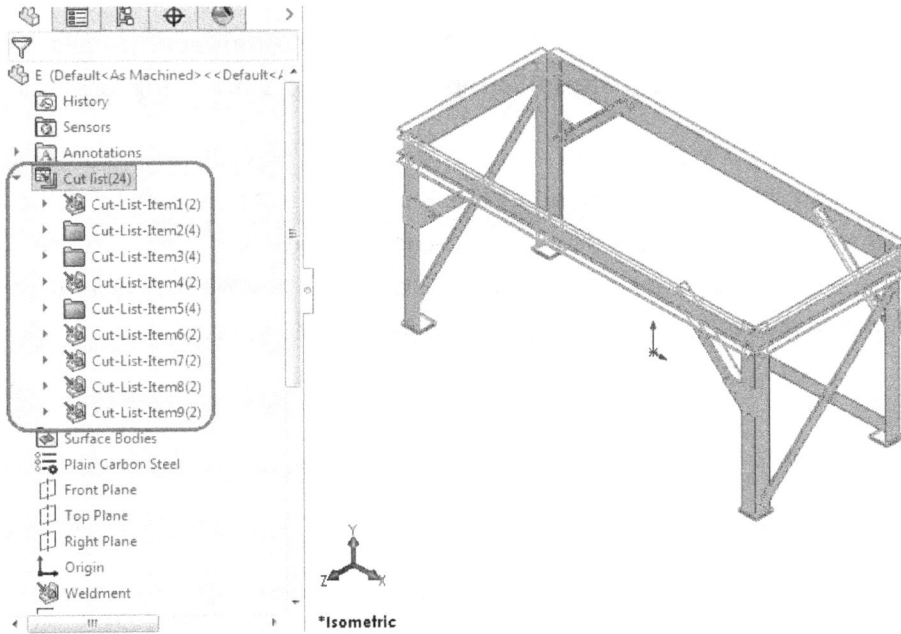

Figure 253 - Question 21 of 26 - Cut List Folders

Counting the number of folders under the Cut List item in the Design Feature Manager Tree reveals that 9 Off Folders were created. Thus 9 is the answer if you are using Solidworks 2010 or later.

TIP : The option to automatically organize all of the weldment entities in the cut list is on by default in new weldment parts. To turn it off, right-click Cut list and clear Create Cut Lists Automatically OR click Tools > Options > Document Properties > Weldments and check the Create Cut Lists Automatically checkbox as shown in the following image.

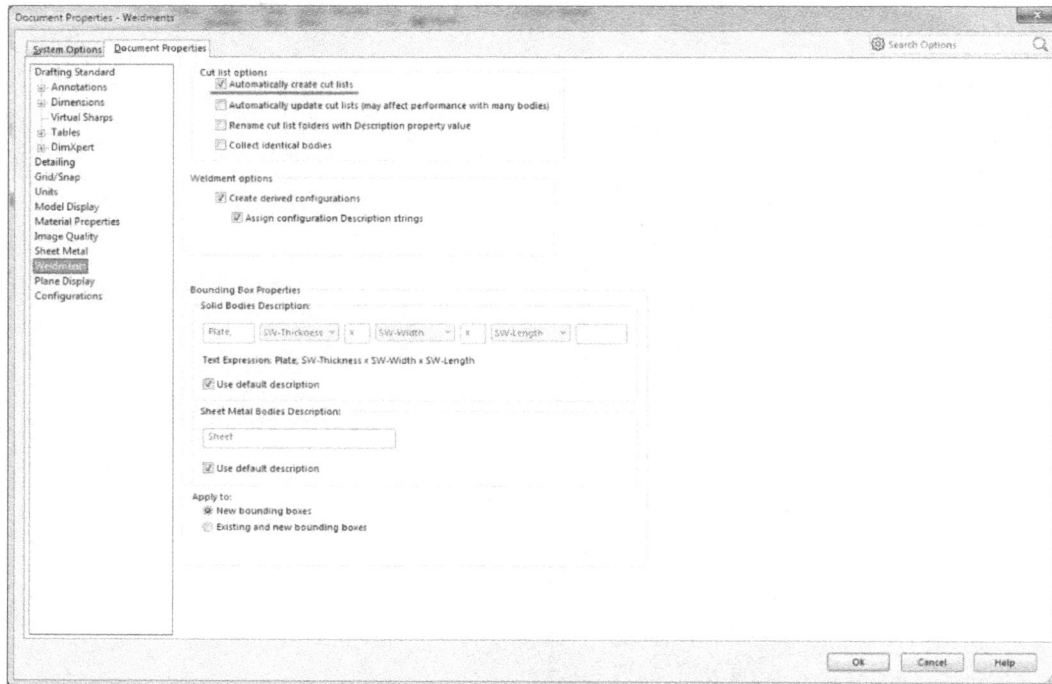

Figure 254 - Question 21 of 26 - Document Properties - Weldments

Default Behavior : -The first time you create a part document, the SOLIDWORKS software turns on the following Weldment document properties in the part template that is created: *Automatically create cut lists, Automatically update cut lists and Rename cut list folders with Description property value*. If you continue to use this part template, these options are enabled for all new part documents. To disable any of these options, clear the option, save the template, and use the saved template to create new parts.

CHAPTER 22

QUESTION 22 EXAM SCREEN CAPTURES

QUESTION 22 EXAM SCREEN CAPTURE 1

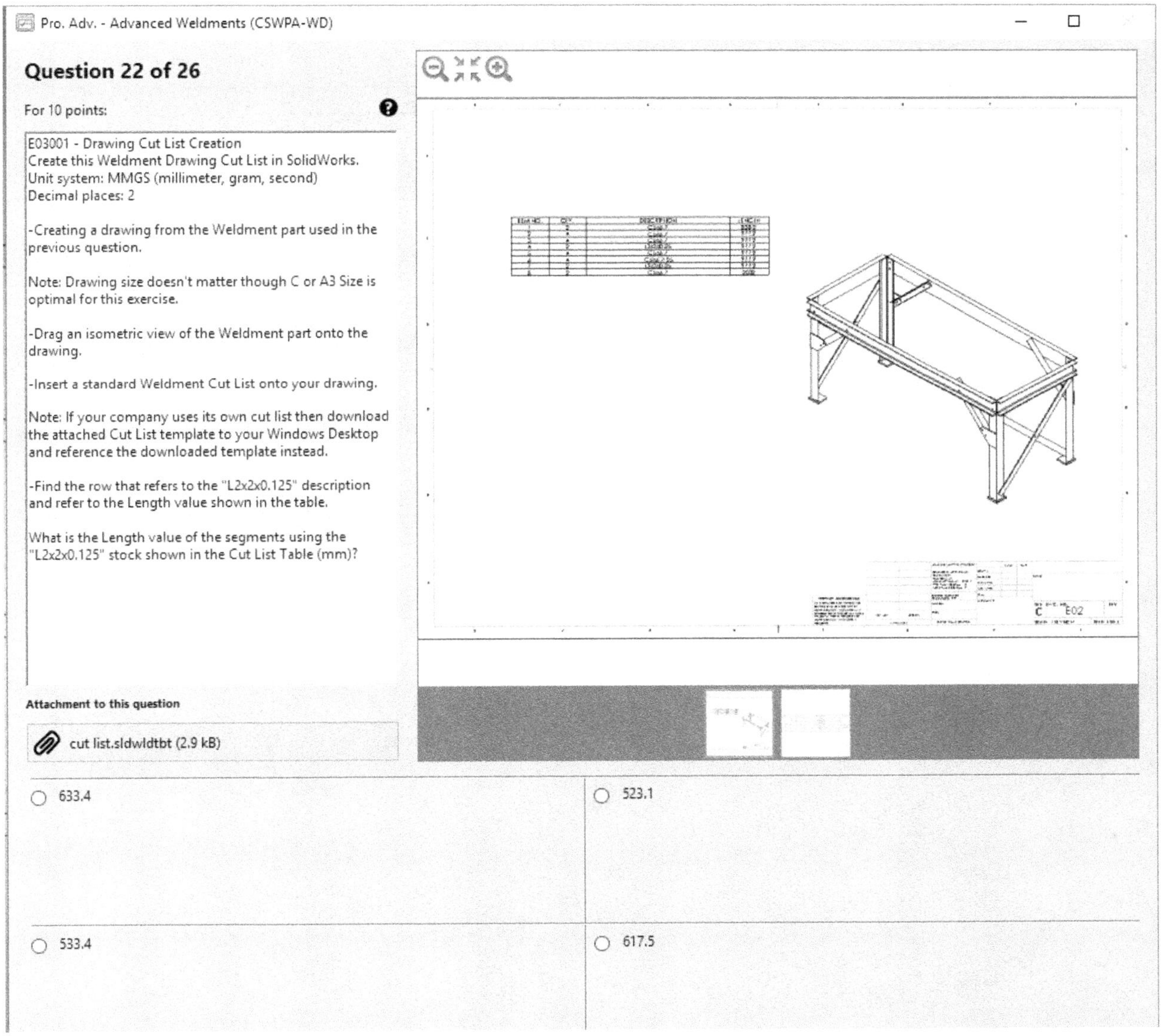

Figure 255 - Question 22 of 26 - Exam Screen Capture 1

QUESTION 22 EXAM SCREEN CAPTURE 2

ITEM NO.	QTY.	DESCRIPTION	LENGTH
1	2	C5x6.7	
2	4	Hardware	
3	4	Hardware	
4	2	L3x3x0.25	
5	4	Hardware	
6	4	L2x2x0.125	XXX.XX
7	2	L3x3x0.25	
8	2	C5x6.7	
9			

Figure 256 - Question 22 of 26 - Exam Screen Capture 2

DOWNLOAD AND SAVE CUT LIST TEMPLATE

Download the cut list template cut list.sldwldtbt from this Google Drive Location *(http://bit.ly/CSWPA-WD)* in the Question 22 Folder. Save the downloaded cut list template onto your Desktop.

CREATE A DRAWING FROM THE WELDMENT PART E.SLDPRT

Click File > Make Drawing from Part as shown in the following image.

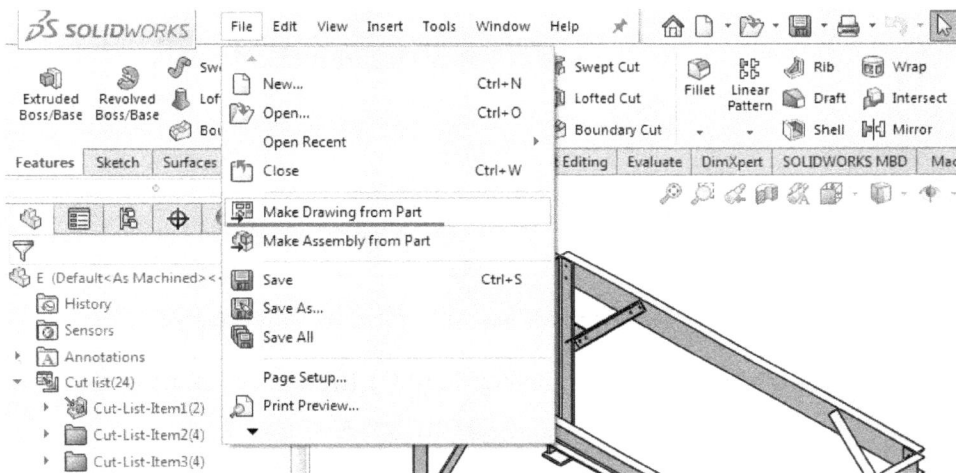

Figure 257 - Question 22 of 26 - Creating a Drawing

Another way to create a drawing is to click on the New Flyout Menu and selecting Make Drawing from Part / Assembly as shown in the following image.

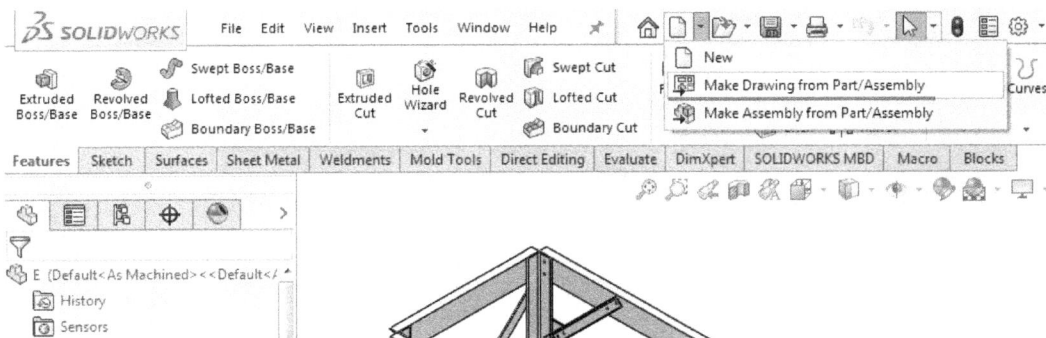

Figure 258 - Question 22 of 26 - Creating a Drawing

Once you have clicked on Make Drawing from Part / Assembly the New SOLIDWORKS Document dialog box opens and you select Drawing then click OK as shown in the following image.

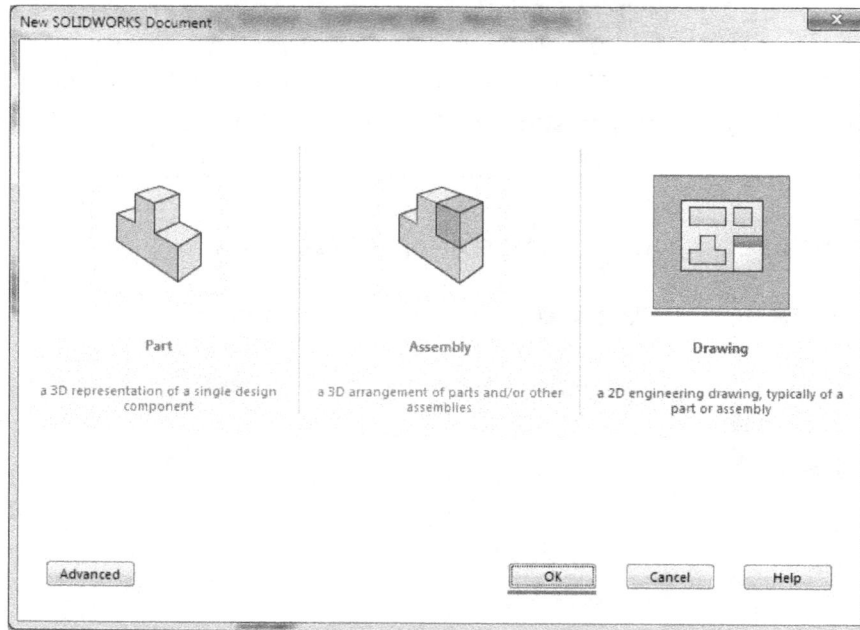

Figure 259 - Question 22 of 26 - New SOLIDWORKS Document dialog box

Under the Sheet Format/Size dialog box, select A3(ISO) then click Ok as show in the following image.

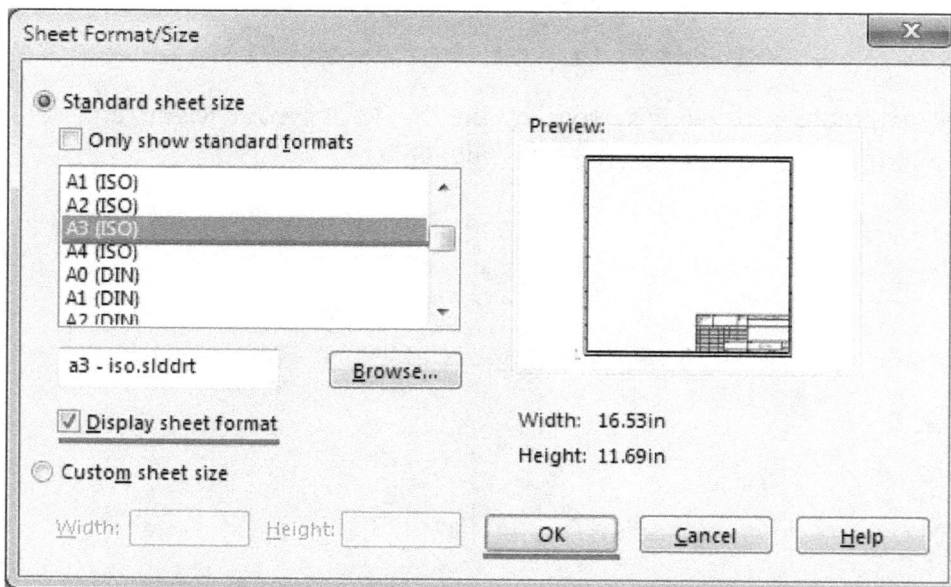

Figure 260 - Question 22 of 26 - Sheet Format/Size dialog box

In the Model View PropertyManager, drag the Isometric View onto the drawing sheet as shown in the following image.

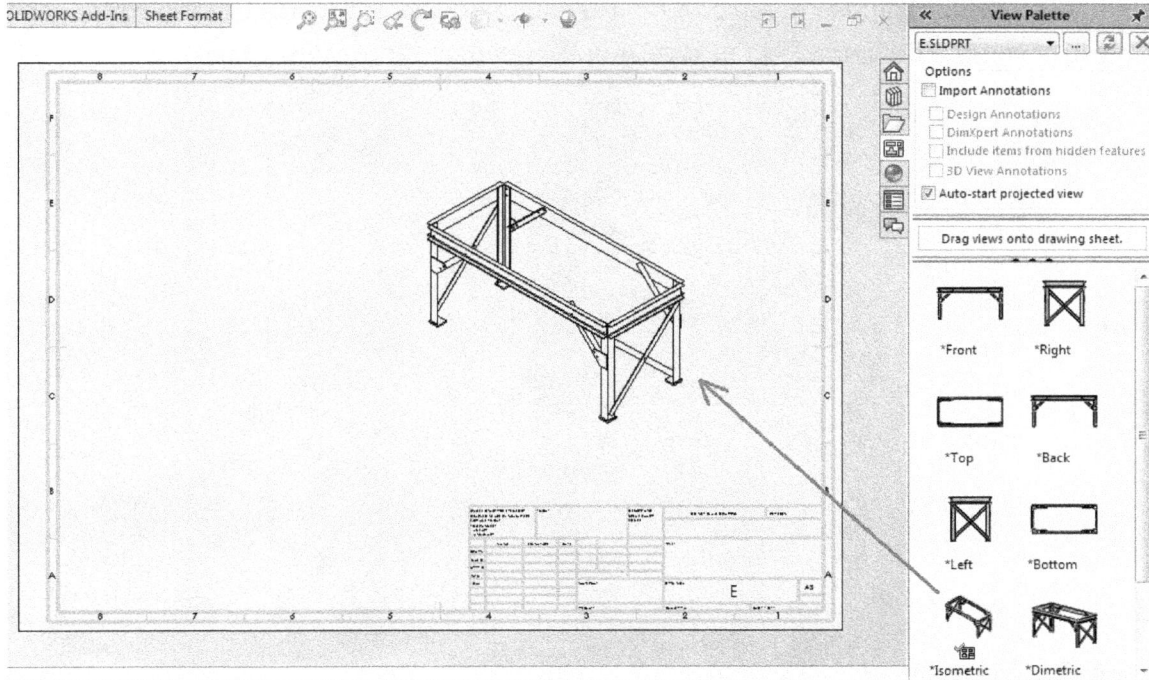

Figure 261 - Question 22 of 26 - Placing a view onto the drawing sheet

Save the drawing onto your computer.

INSERTING A WELDMENT CUT LIST TABLE INTO A DRAWING

Click Weldment Cut List (Table toolbar) or Insert > Tables > Weldment Cut List or Right Click on the Isometric View you just placed in the drawing and select Tables then Click on Weldment Cut List as shown in the following image.

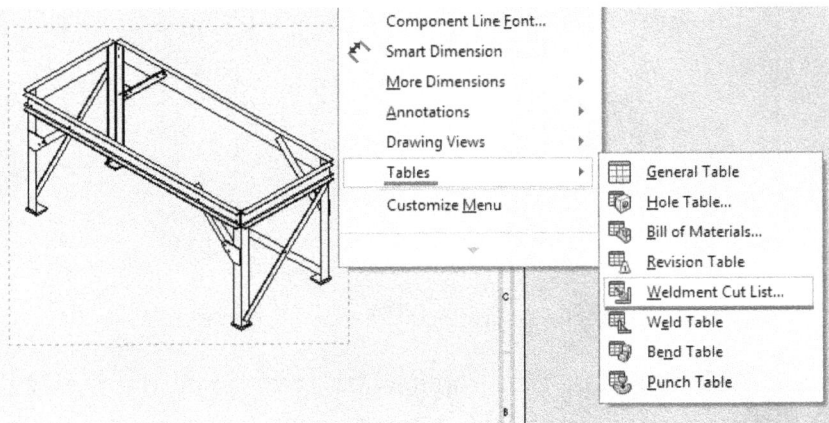

Figure 262 - Question 22 of 26 - Inserting a weldment cut list table into a drawing

In the Weldment Cut List PropertyManager, Under Table Template - Click Browse for template to choose the custom template we downloaded at the beginning of this chapter as shown in the following image if you saved it on your Desktop otherwise browse to the location where you saved the downloaded Cut List Template - select it then click Open.

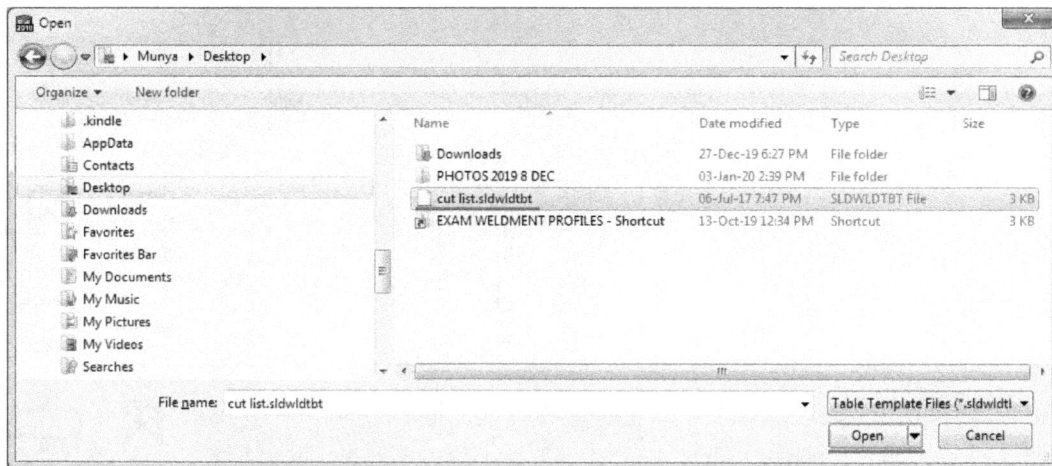

Figure 263 - Question 22 of 26 - Choosing a Table Template

Your drawing should appear as shown below:

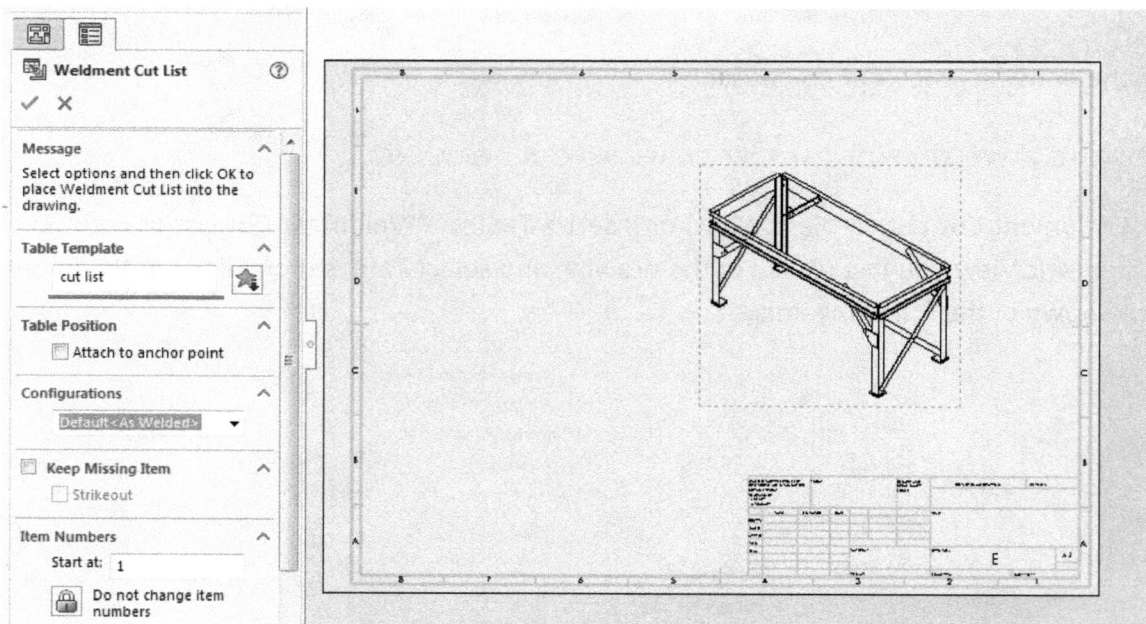

Figure 264 - Question 22 of 26 - Choosing a Table Template

Click ✓ in the Weldment Cut List PropertyManager. If you did not select Attach to anchor point in the PropertyManager, click in the graphics area to place the cut list.

Your drawing should now look as shown in the following image.

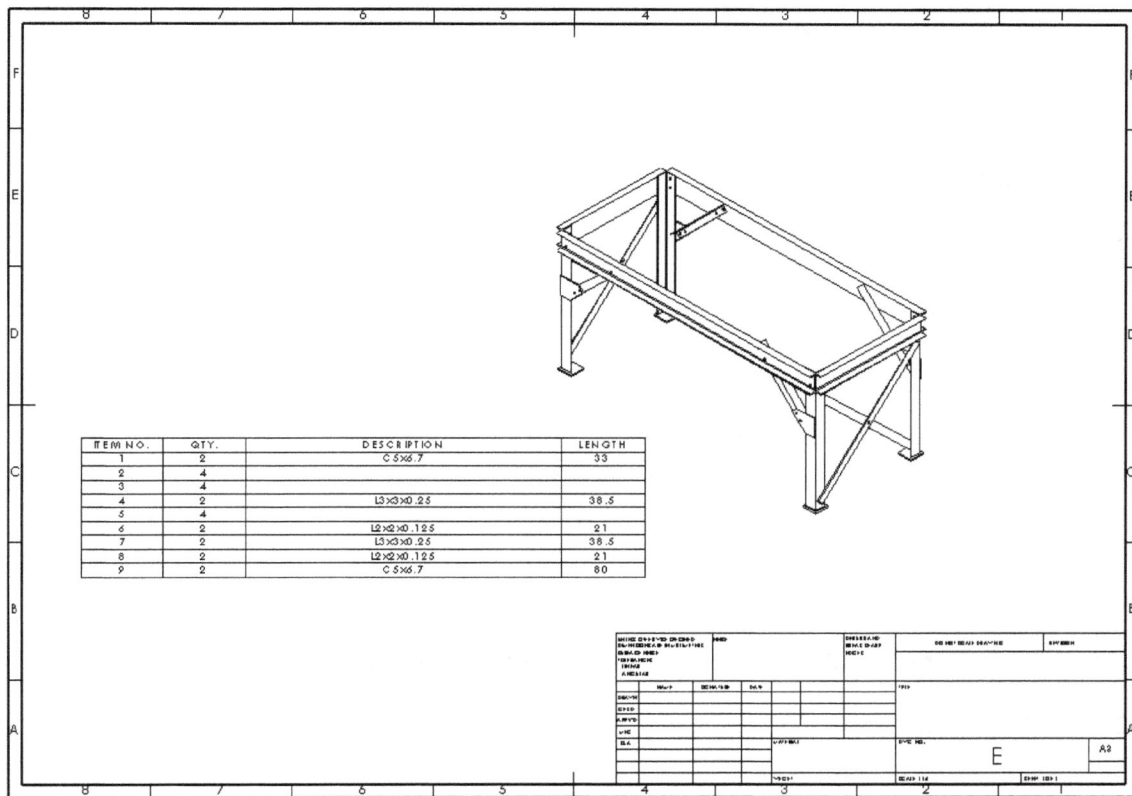

However, you will notice that the Length dimension on the cut list table is in inches instead of millimeters as requested in this question under Question 22 Exam Screen Capture 1. Hence, we need to change the unit system.

CHANGING THE UNIT SYSTEM

Go to Tools > Options > Document Properties > Units to change the Unit System to MMGS (millimeter, gram, second) and to also set the number of decimal places to two decimal places as shown in the following image.

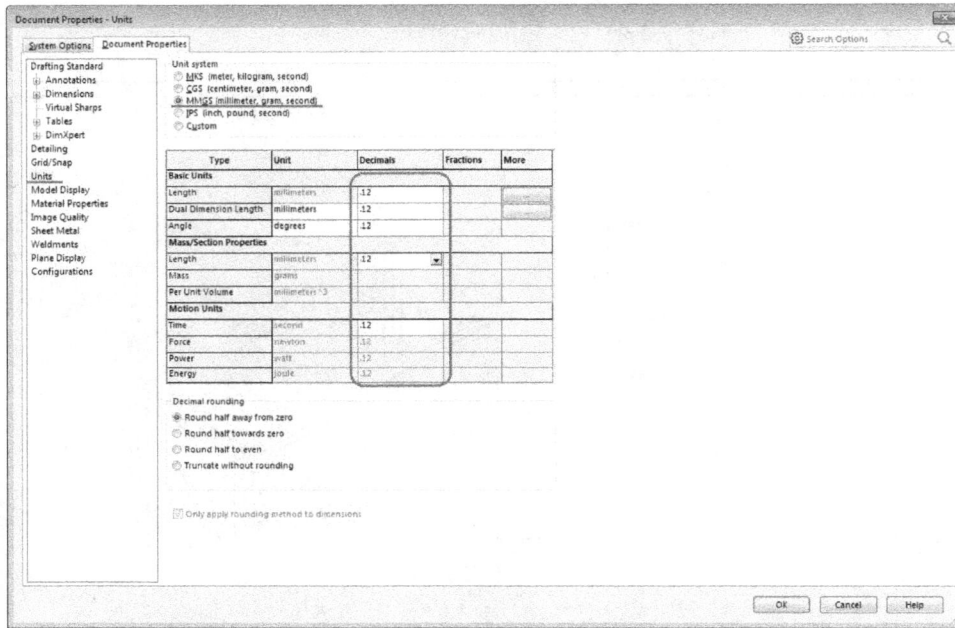

Figure 265 - Question 22 of 26 - Document Properties - Unit System

Click OK. Your drawing should now look as shown in the following image.

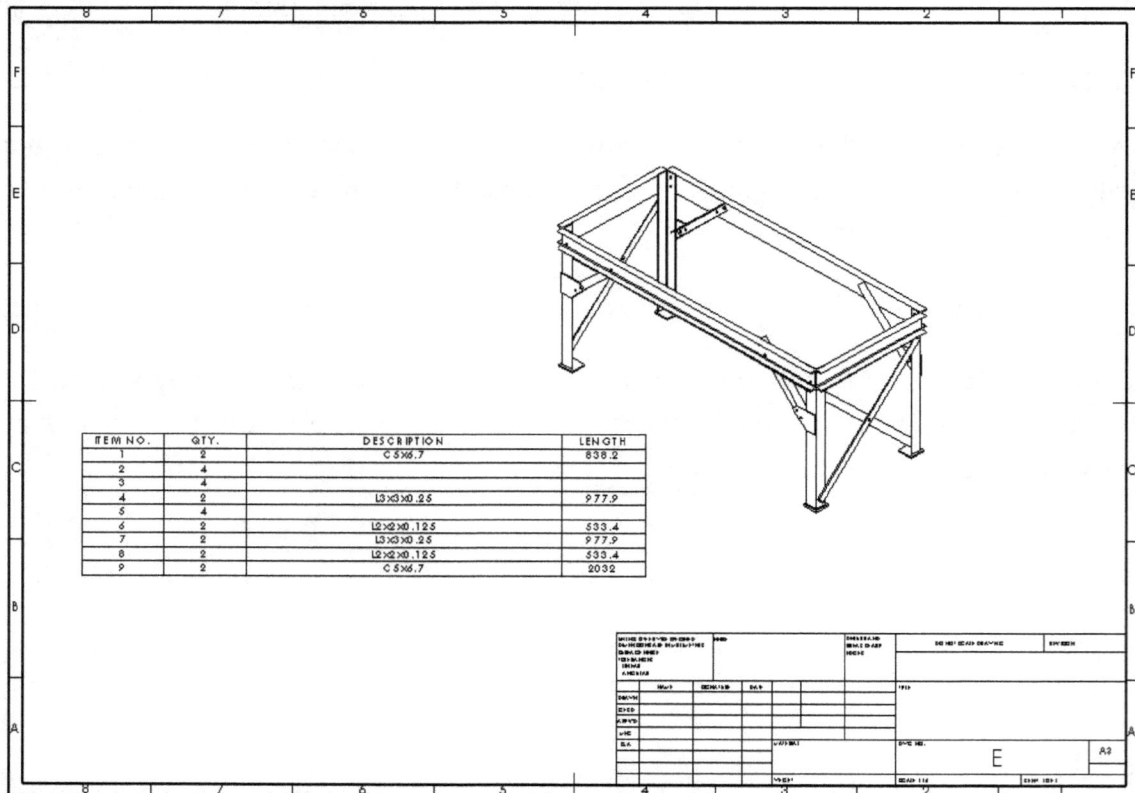

ITEM NO.	QTY.	DESCRIPTION	LENGTH
1	2	C 5×6.7	838.2
2	4		
3	4		
4	2	L3×3×0.25	977.9
5	4		
6	2	L2×2×0.125	533.4
7	2	L3×3×0.25	977.9
8	2	L2×2×0.125	533.4
9	2	C 5×6.7	2032

Figure 266 - Question 22 of 26 - Cut List Table

194

QUESTION 22 ANSWER

The Length Value of the segments using the "L2x2x0.125" stock as shown in the Cut List Table below on our drawing is 533.4mm.

ITEM NO.	QTY.	DESCRIPTION	LENGTH
1	2	C5x6.7	838.2
2	4		
3	4		
4	2	L3x3x0.25	977.9
5	4		
6	2	L2x2x0.125	533.4
7	2	L3x3x0.25	977.9
8	2	L2x2x0.125	533.4
9	2	C5x6.7	2032

Figure 267 - Question 22 of 26 - Cut List Table

Save and close the drawing as well as part E to move onto Question 23.

CHAPTER 23

QUESTION 23 EXAM SCREEN CAPTURES

QUESTION 23 EXAM SCREEN CAPTURE 1

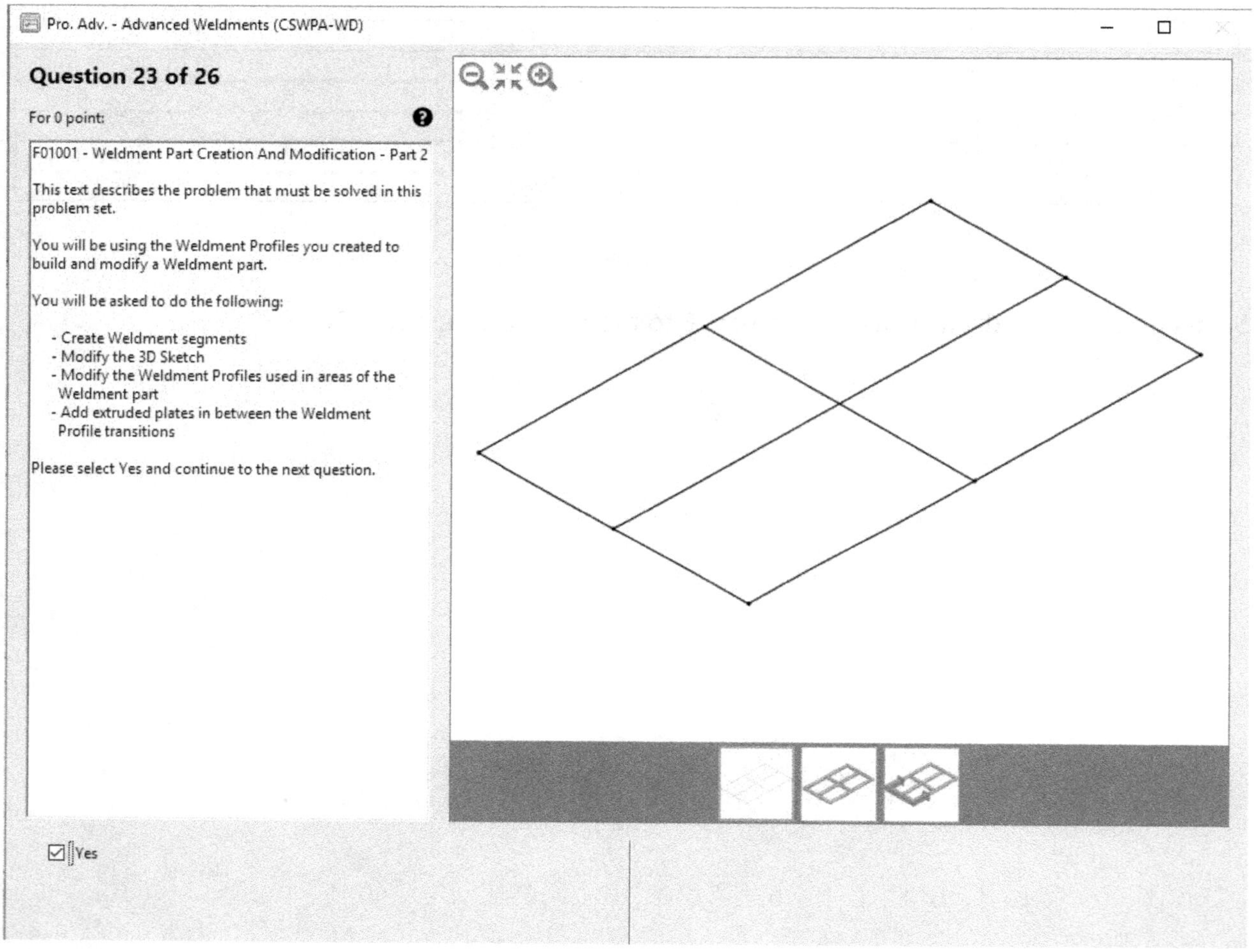

Figure 268 - Question 23 of 26 - Exam Screen Capture 1

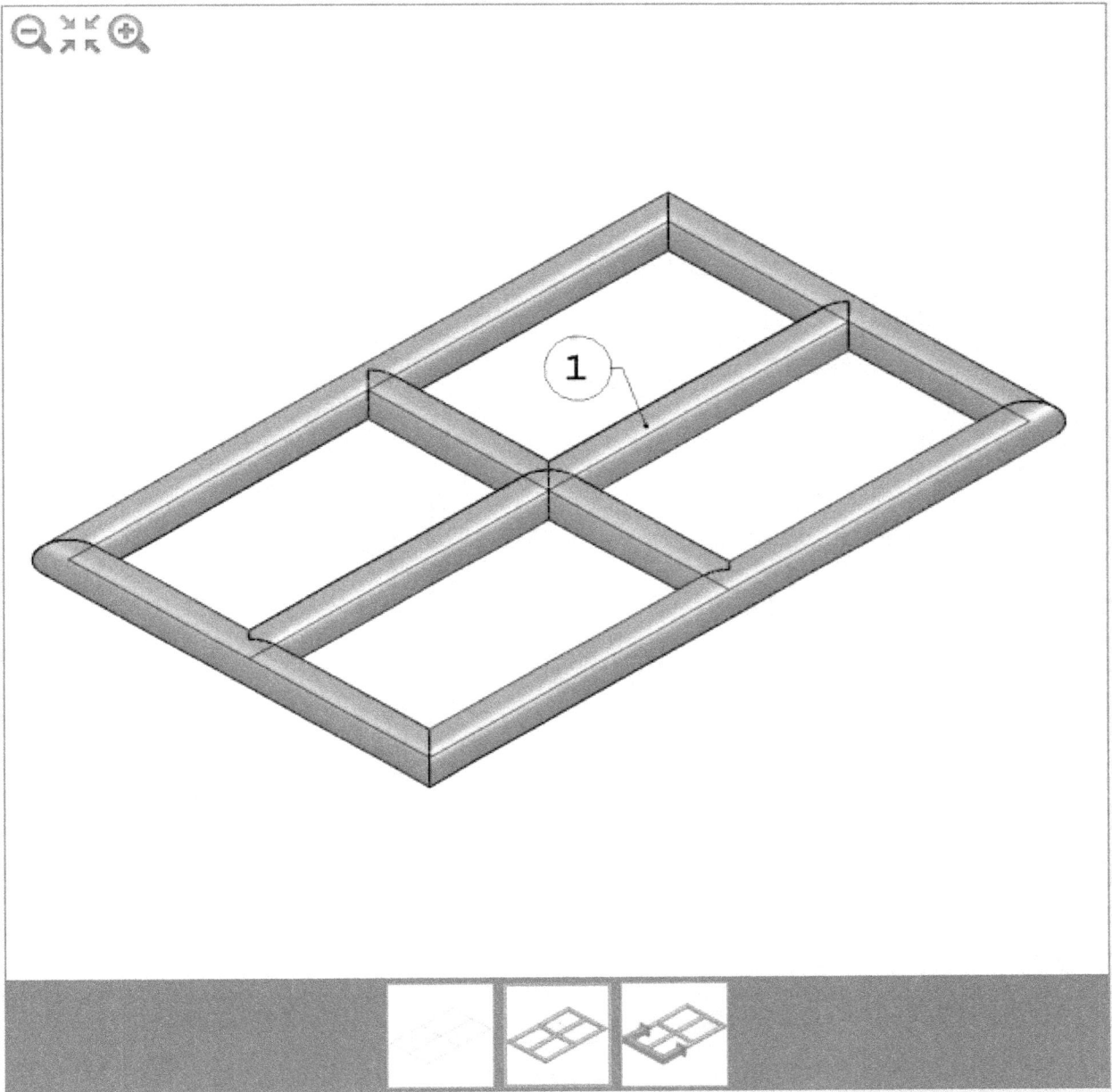

Figure 269 - Question 23 of 26 - Exam Screen Capture 2

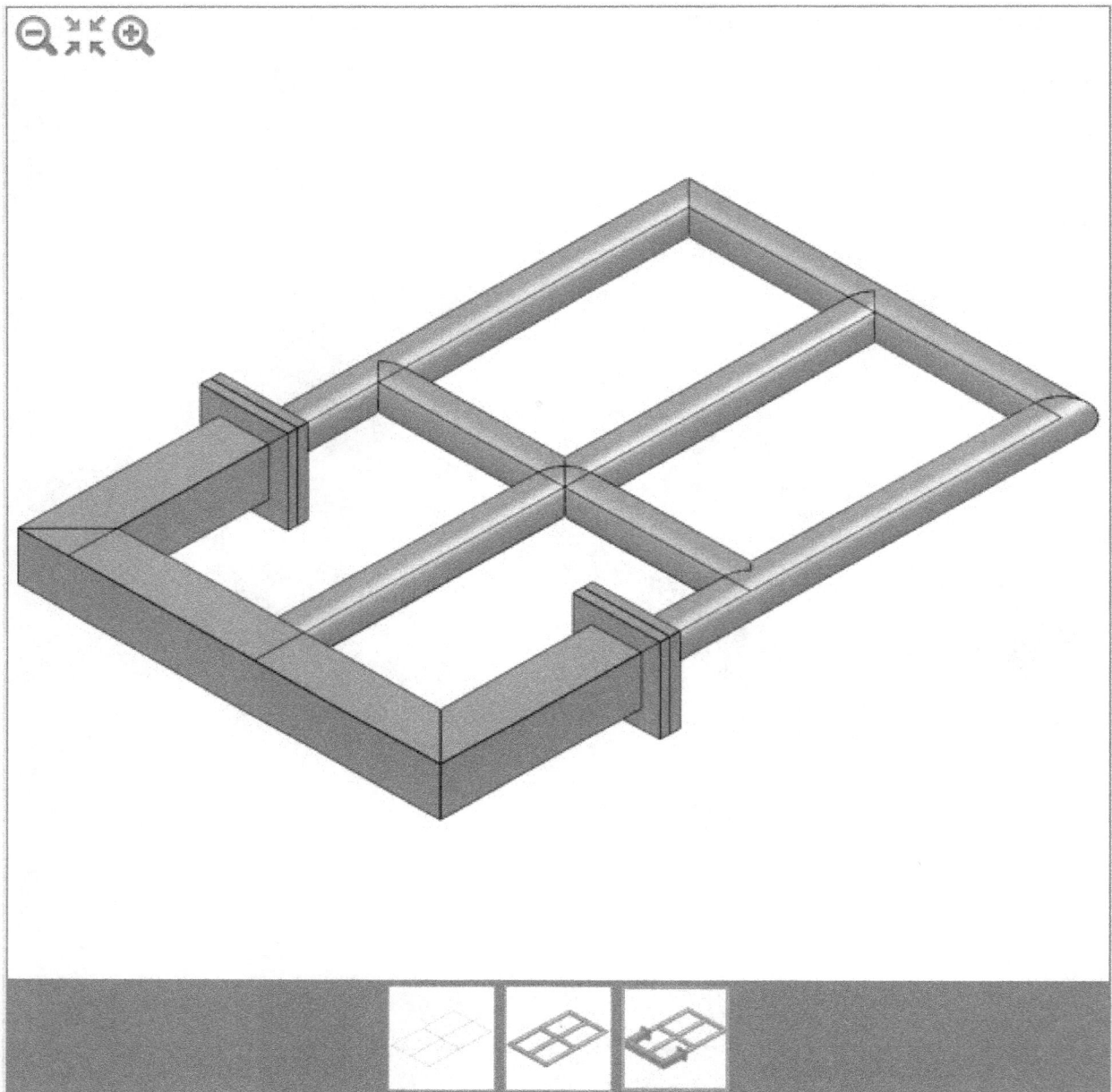

Figure 270 - Question 23 of 26 - Exam Screen Capture 2

In this Question you just read the instructions and then click Yes and continue to Question 24.

CHAPTER 24

QUESTION 24 EXAM SCREEN CAPTURES

QUESTION 24 EXAM SCREEN CAPTURE 1

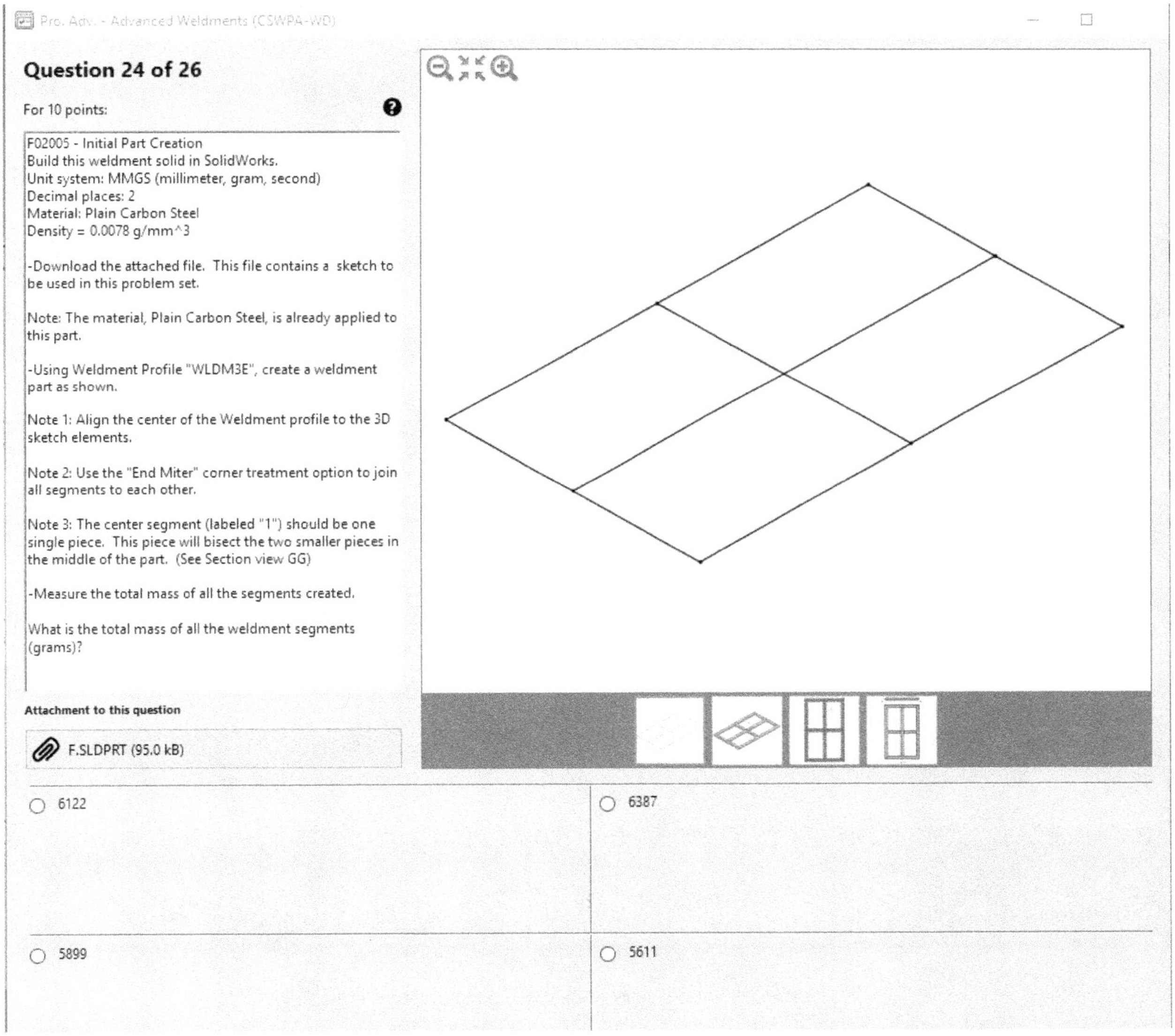

Pro. Adv. - Advanced Weldments (CSWPA-WD)

Question 24 of 26

For 10 points:

F02005 - Initial Part Creation
Build this weldment solid in SolidWorks.
Unit system: MMGS (millimeter, gram, second)
Decimal places: 2
Material: Plain Carbon Steel
Density = 0.0078 g/mm^3

-Download the attached file. This file contains a sketch to be used in this problem set.

Note: The material, Plain Carbon Steel, is already applied to this part.

-Using Weldment Profile "WLDM3E", create a weldment part as shown.

Note 1: Align the center of the Weldment profile to the 3D sketch elements.

Note 2: Use the "End Miter" corner treatment option to join all segments to each other.

Note 3: The center segment (labeled "1") should be one single piece. This piece will bisect the two smaller pieces in the middle of the part. (See Section view GG)

-Measure the total mass of all the segments created.

What is the total mass of all the weldment segments (grams)?

Attachment to this question

F.SLDPRT (95.0 kB)

○ 6122

○ 6387

○ 5899

○ 5611

Figure 271 - Question 24 of 26 - Exam Screen Capture 1

QUESTION 24 EXAM SCREEN CAPTURE 2

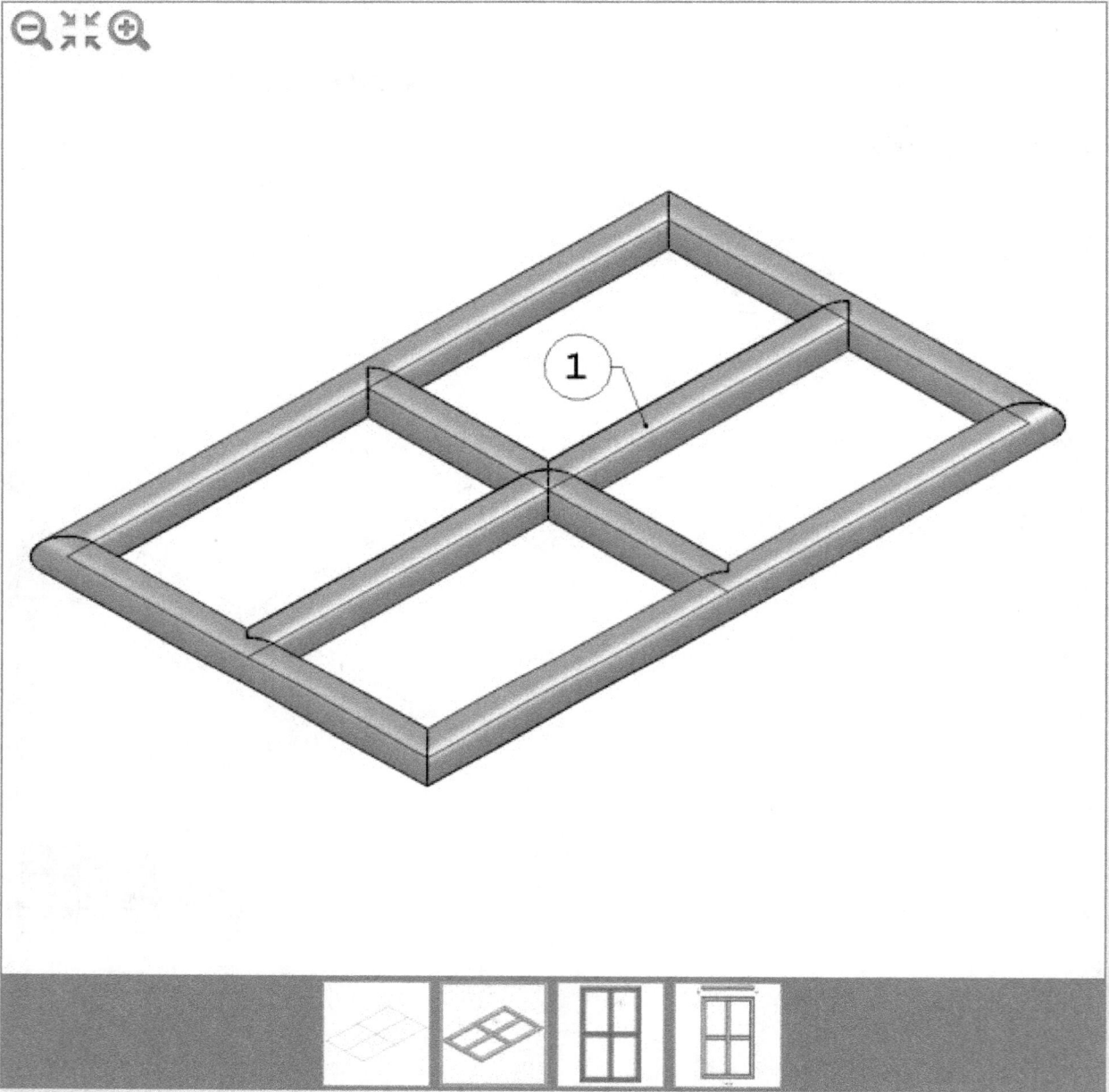

Figure 272 - Question 24 of 26 - Exam Screen Capture 2

Figure 273 - Question 24 of 26 - Exam Screen Capture 3

Figure 274 - Question 24 of 26 - Exam Screen Capture 4

DOWNLOAD PART F.SLDPRT

Download the part F.SLDPRT from this Google Drive Location *(http://bit.ly/CSWPA-WD)* in the Question 24 Folder. Save the downloaded part onto your PC.

Open the downloaded part F.SLDPRT and it should look as shown in the following image.

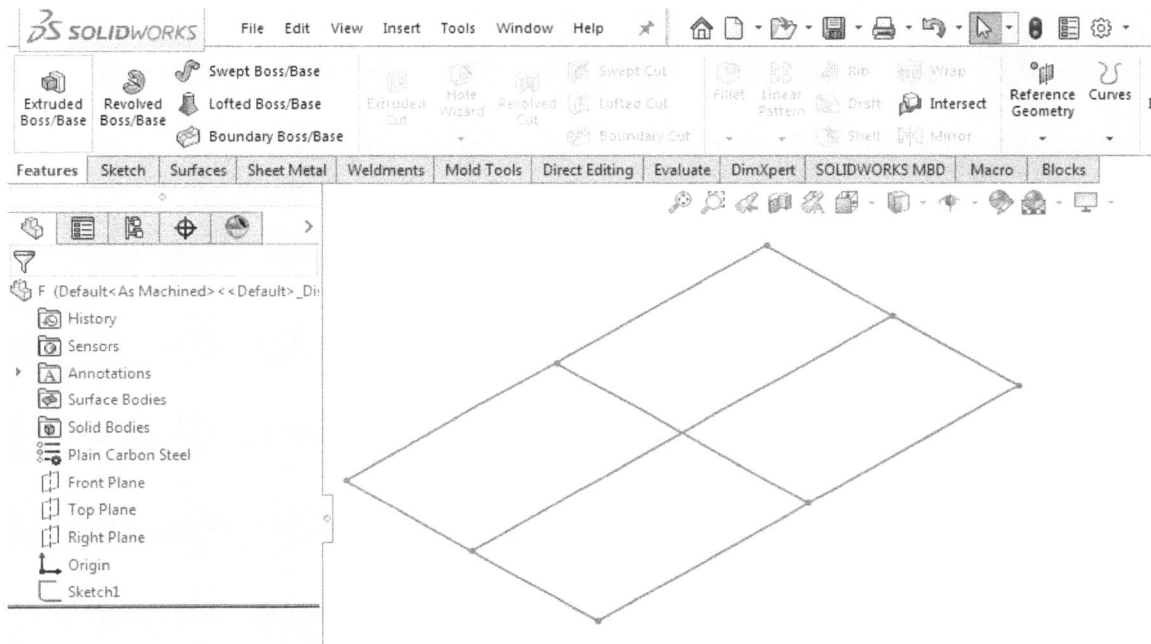

Figure 275 - Question 24 of 26 - Downloaded part F.SLDPRT

ADDING A STRUCTURAL MEMBER

Click Structural Member under the Weldments toolbar or Insert > Weldments > Structural Member. Make selections in the PropertyManager to define the profile for the structural member as shown in the following image. **Deselect** Transfer Material from Profile to prevent the transfer of material from the profile to the weldment you are creating - since the Profile has 1060 Alloy Aluminum as the material but the material is specified as Plain Carbon Steel in this question.

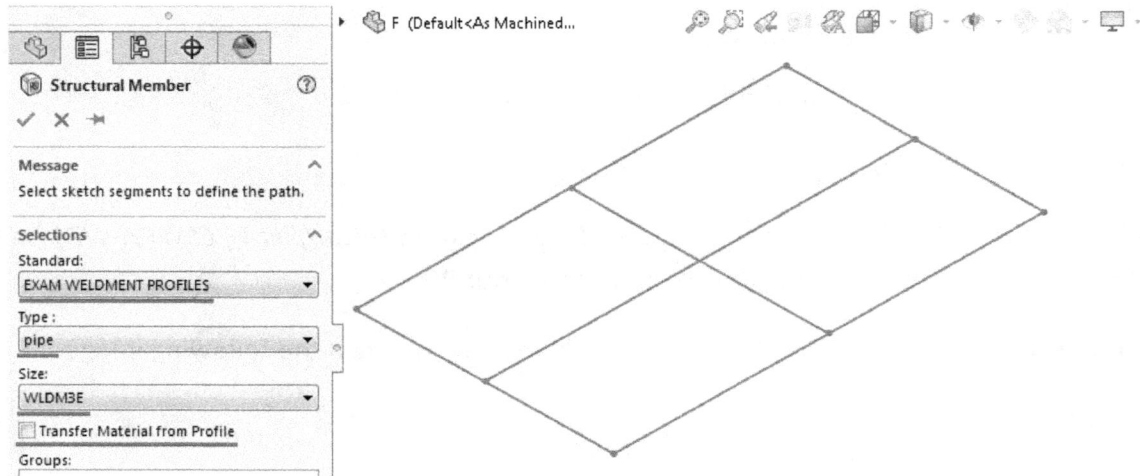

Figure 276 - Question 24 of 26 - Structural Member Property Manager

In the graphics area - select sketch segments shown in the following image to define the path for the structural member. **Uncheck** Transfer Material from Profile: 1060 Alloy.

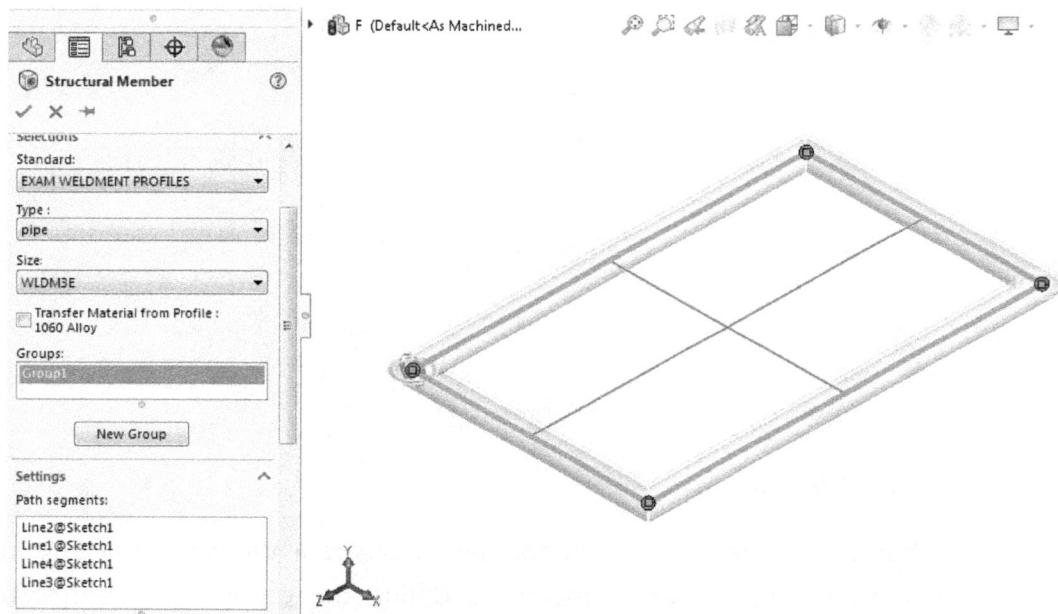

Figure 277 - Question 24 of 26 - Adding a Structural Member

Select Apply Corner Treatment and click on the End Miter option as shown in the following image.

Figure 278 - Question 24 of 26 - Corner Treatment

Click the New Group Command Button on the Structural Member Property Manager and select the sketch segment shown in the following image to define the path for the structural member.

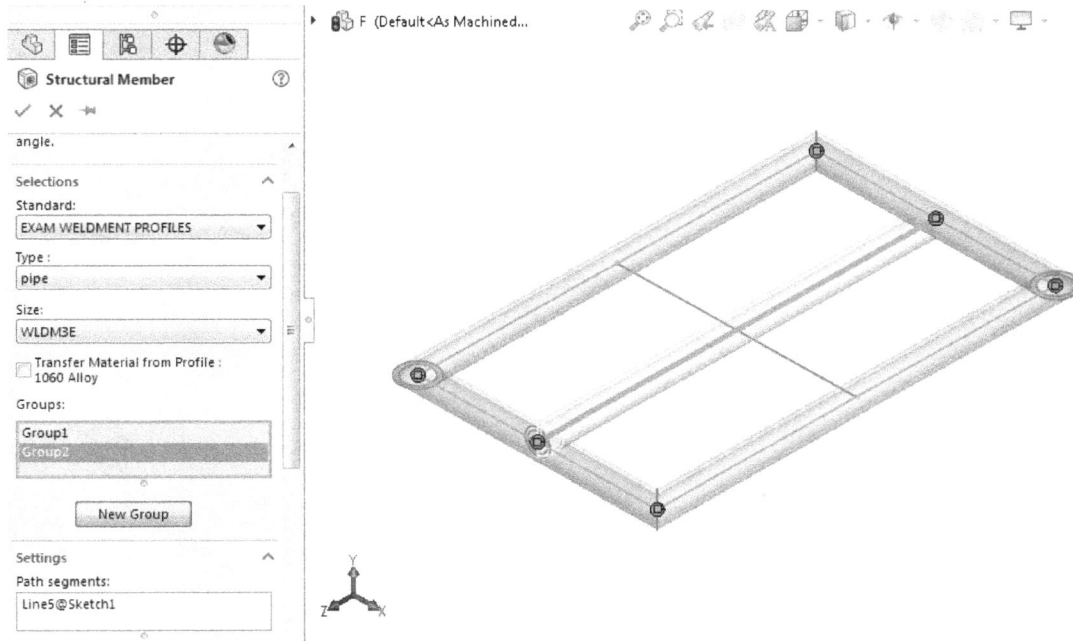

Figure 279 - Question 24 of 26 - Adding a Structural Member

Click the New Group Command Button on the Structural Member Property Manager and select the sketch segment shown in the following image to define the path for the structural member.

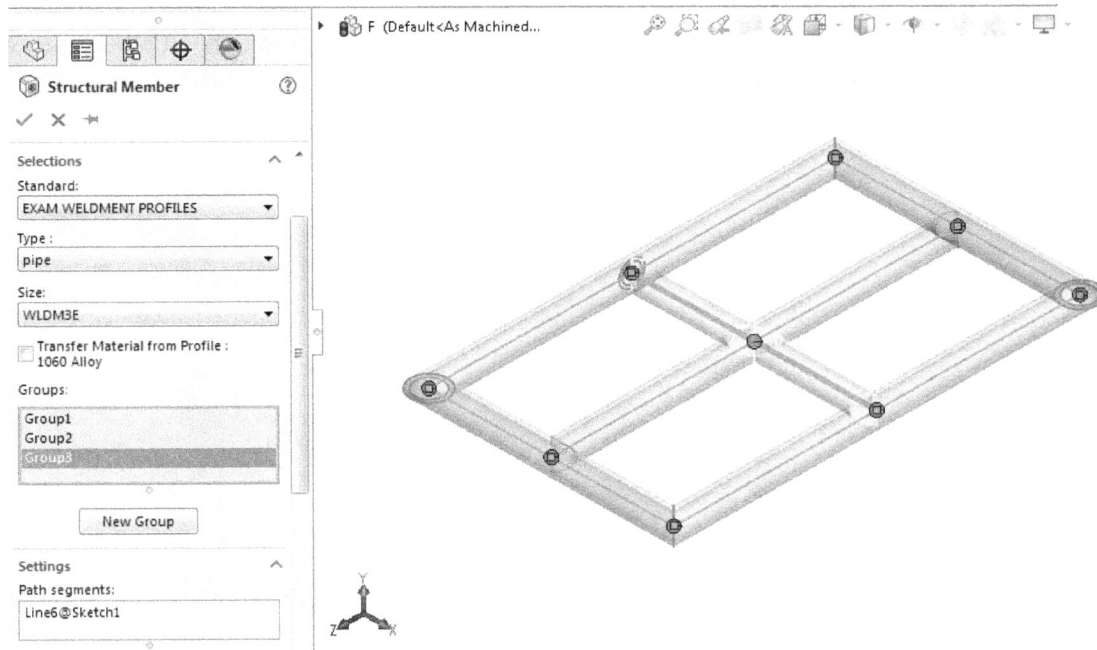

Figure 280 - Question 24 of 26 - Adding a Structural Member

Click OK to close the Structural Member PropertyManager. Save your part. Your part should now look as shown in the following image.

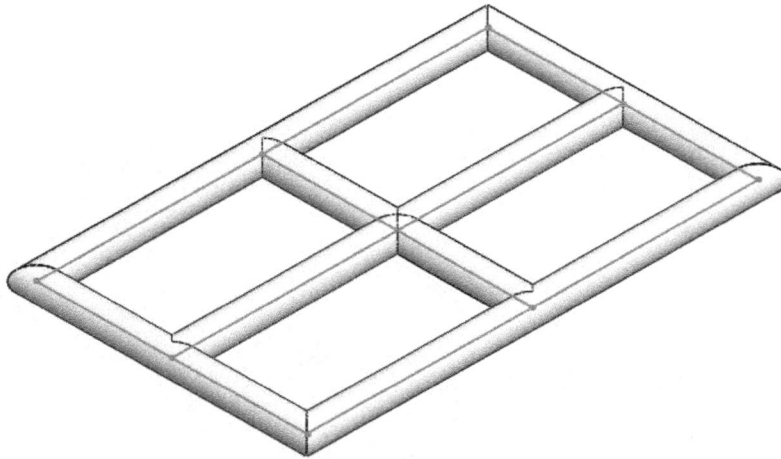

Figure 281 - Question 24 of 26 - Weldment Current Status

CREATING A SECTION VIEW IN A MODEL

Now, one of the requirements in this question is that the center segment (labeled "1") in Question 24 Exam Screen Capture 2 and Exam Screen Capture 3 should be one single piece. We can check if this requirement has been met by creating a cross section in the model view on the Top Plane.

Click Section View (View toolbar) or View > Display > Section View. In the Section View PropertyManager, select options as shown in the following image.

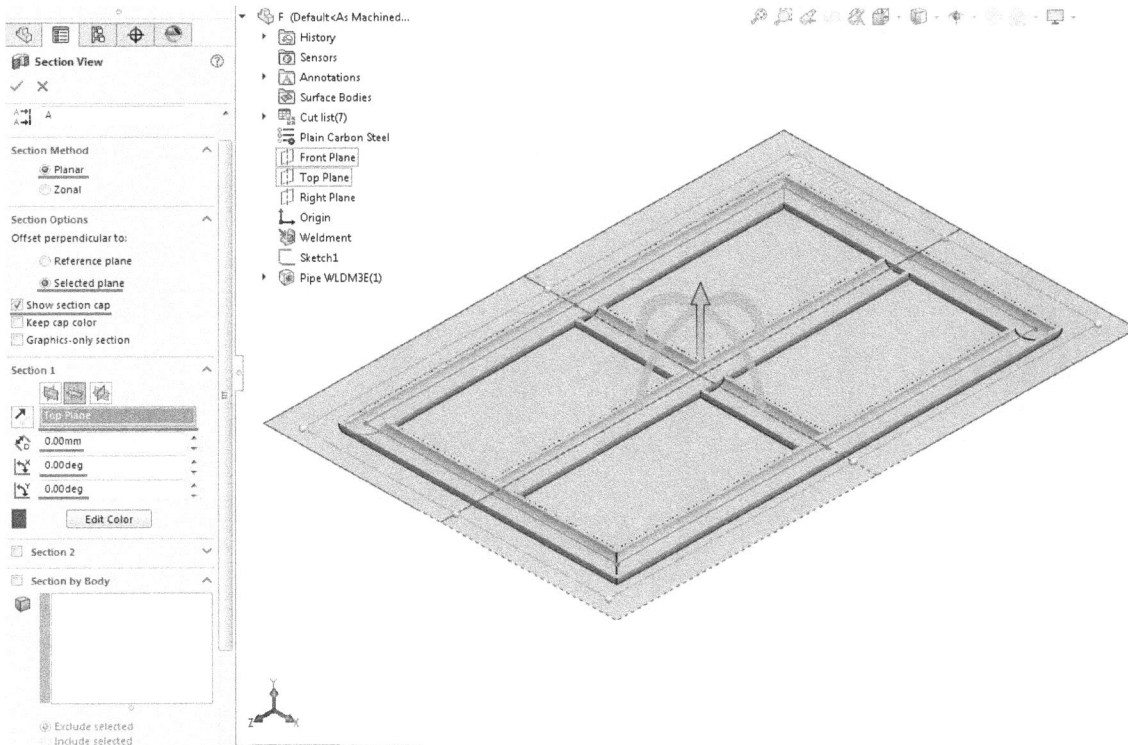

Figure 282 - Question 24 of 26 - Section View in Model

Click OK. Your part should now look as shown in the following image which is the same as on Question 24 Exam Screen Capture 4 (Section View G-G).

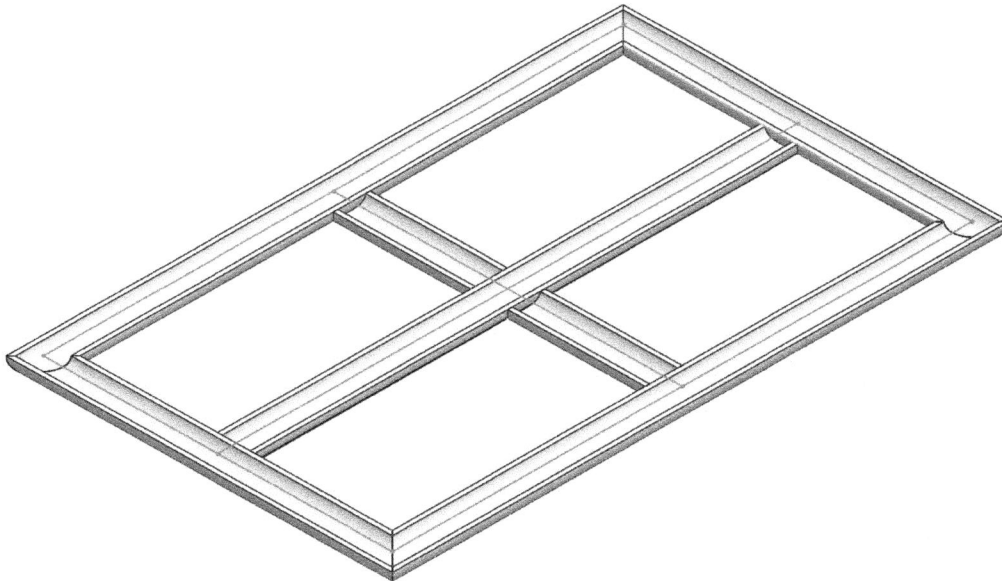

Figure 283 - Question 24 of 26 - Section View in Model

Click Section View (View toolbar) or View > Display > Section View again to return the model to full view. Rebuild (Click Edit > Rebuild) and Save your part.

MEASURING THE MASS OF ALL SEGMENTS IN A WELDMENT PART

Under the Evaluate Tab, click on the Mass Properties feature and the Mass Properties Dialog Box appears showing that the total mass of the part is 6387.20 grams. So the answer is 6387 to the nearest whole number.

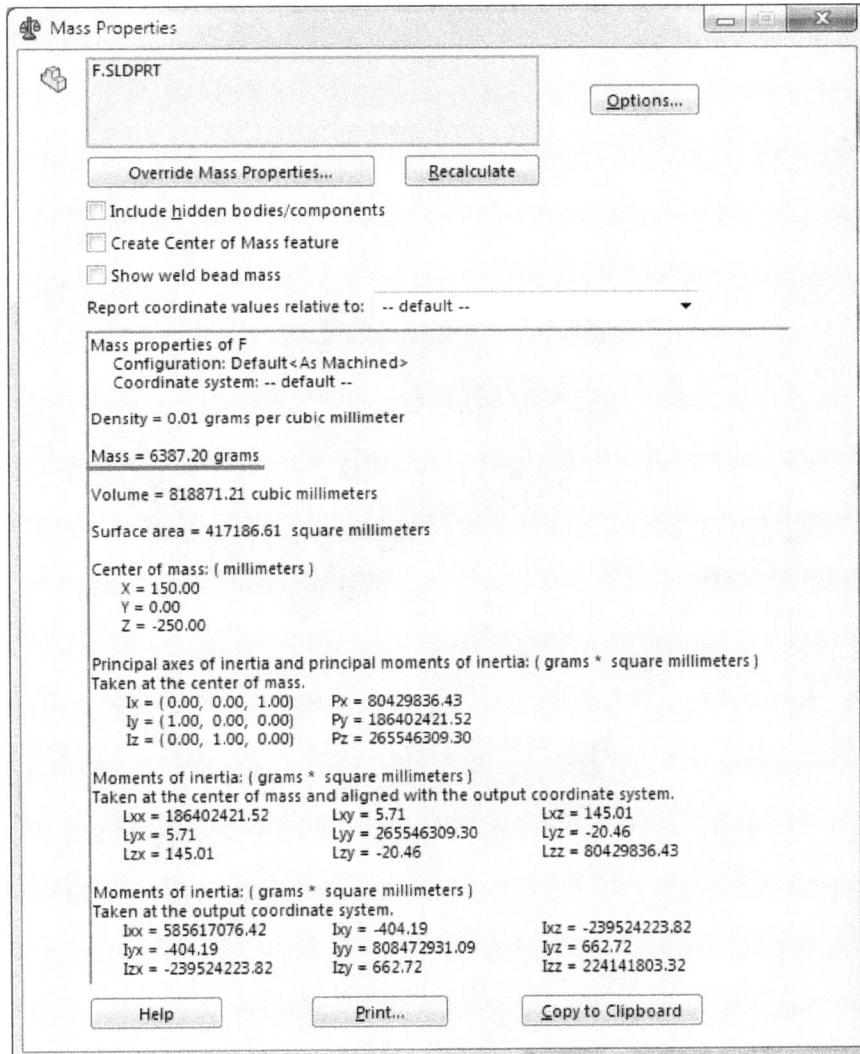

Figure 284 - Question 24 of 26 - Mass Properties

Click save to save the Part.

CHAPTER 25

QUESTION 25 EXAM SCREEN CAPTURES

QUESTION 25 EXAM SCREEN CAPTURE 1

Pro. Adv. - Advanced Weldments (CSWPA-WD)

Question 25 of 26

For 10 points:

F03005 - Weldment Part Modification
Build this weldment solid in SolidWorks.
Unit system: MMGS (millimeter, gram, second)
Decimal places: 2
Material: Plain Carbon Steel
Density = 0.0078 g/mm^3

-Modify the sketch and Weldment part to replace the bottom area segments of the part using the square Weldment Profile "WLDM2E".

-Add a total of 4 square solid plates to transition from the round Weldment Profile to the square Weldment Profile.

Note 1: The newly modified Weldment segments and square plates should be centered on the underlying sketch.

Note 2: Use the "End Miter" corner treatment option to join all segments to each other.

Note 3: All Weldment segments should be trimmed to the faces they intersect with no weldment gap.

-Measure the center of mass of the entire weldment part.

What is the center of mass of the entire weldment part (mm)?

Enter Coordinates: X:
                     Y:
                     Z:
(use . (point) as decimal separator)

Figure 285 - Question 25 of 26 - Exam Screen Capture 1

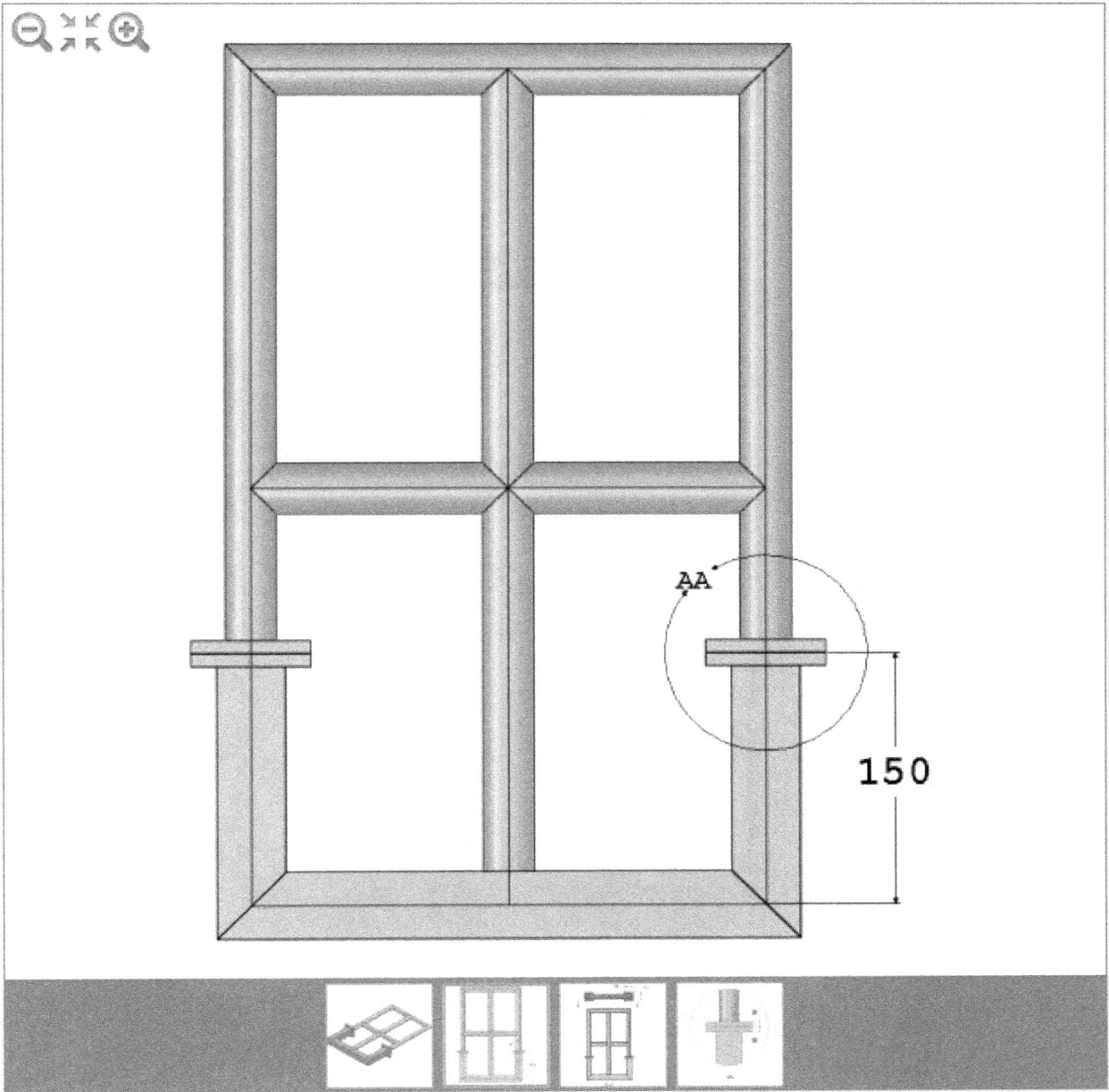

Figure 286 - Question 25 of 26 - Exam Screen Capture 2

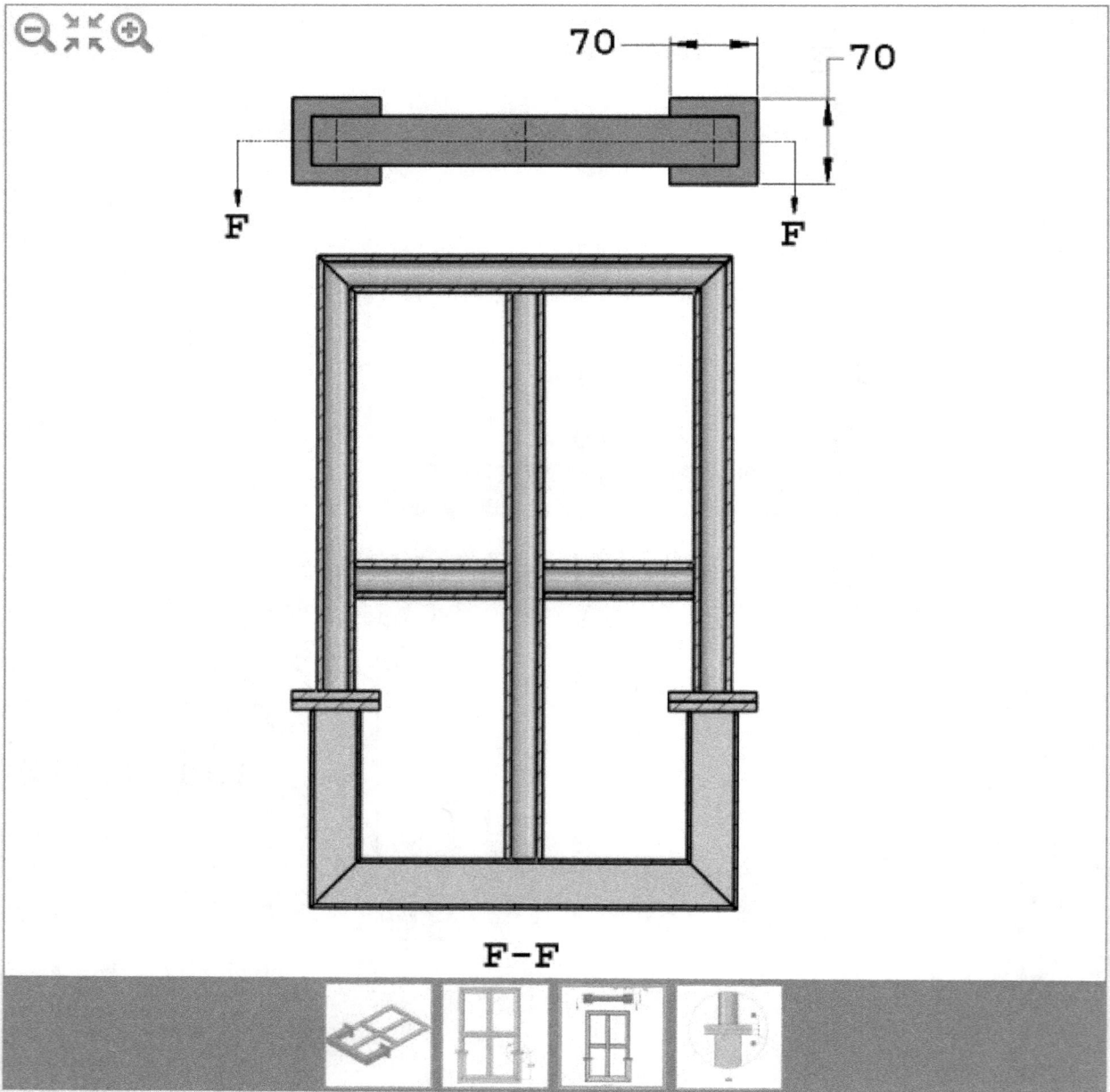

Figure 287 - Question 25 of 26 - Exam Screen Capture 3

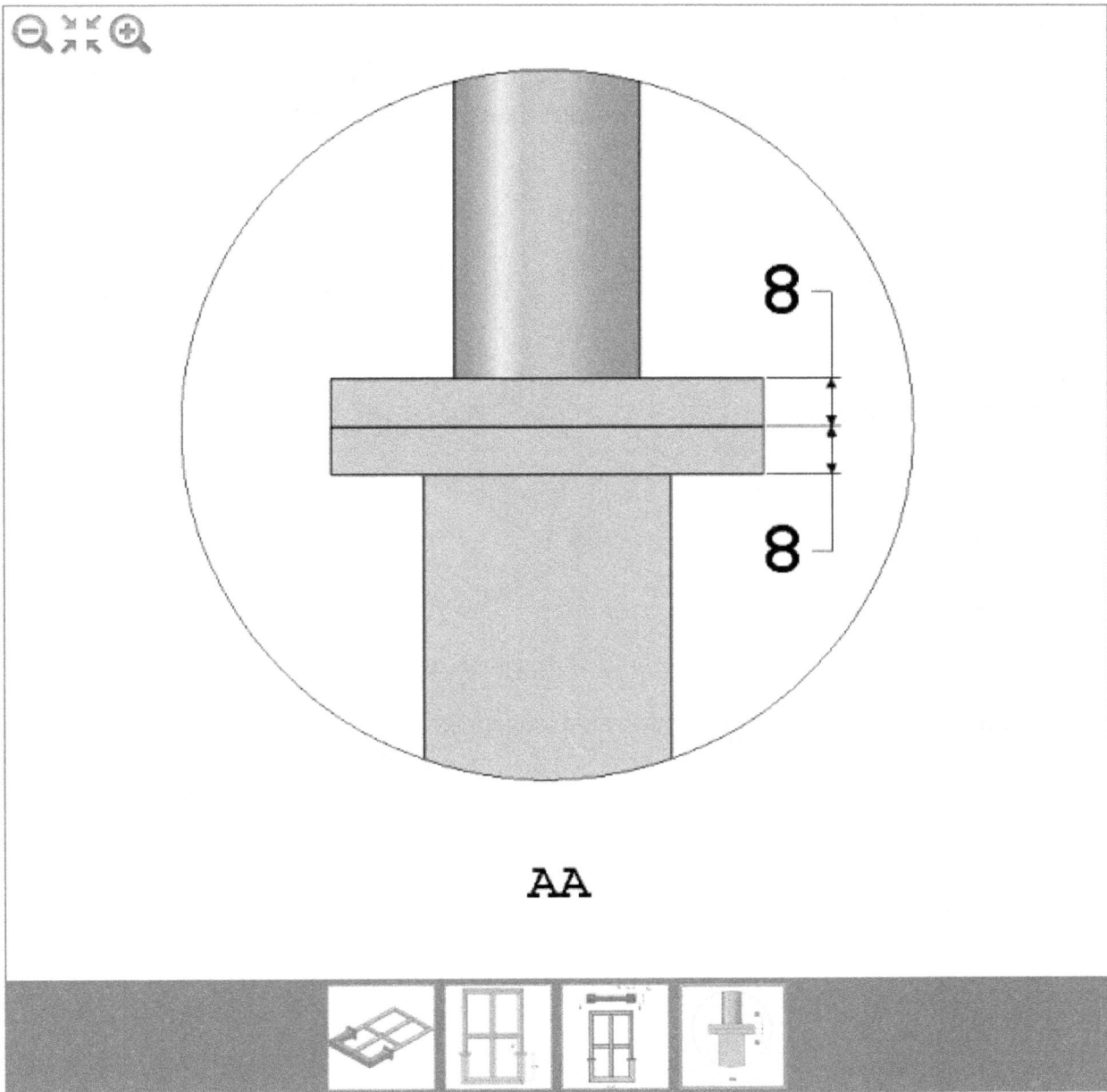

Figure 288 - Question 25 of 26 - Exam Screen Capture 4

EDITING AN EXISTING SKETCH

In the Feature Manager Design Tree, click or right click on Sketch1 and select Edit Sketch as shown in the following image. You may also click Sketch on the Sketch toolbar, or click Insert > Sketch then select an existing sketch to edit.

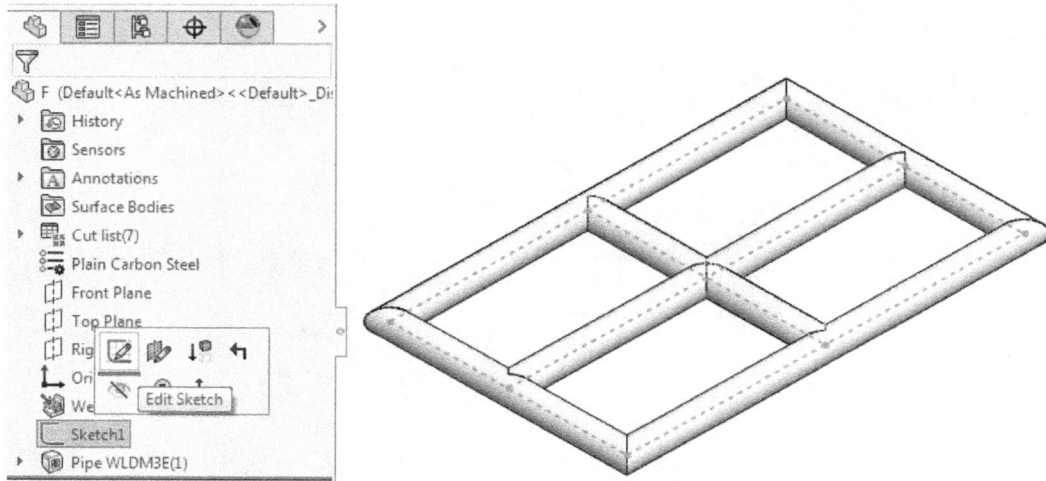

Figure 289 - Question 25 of 26 - Editing an existing sketch

Your part should now look as shown in the following image.

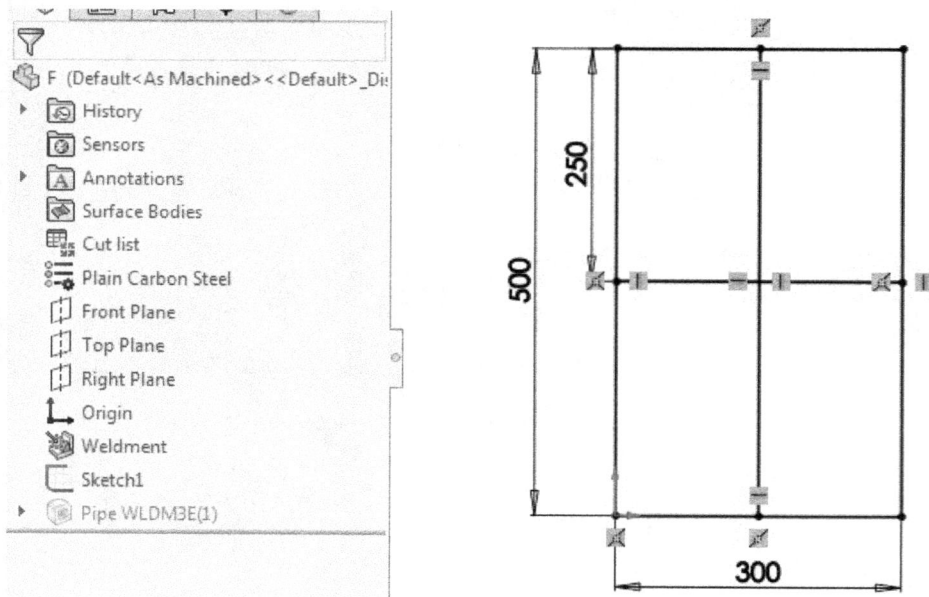

Figure 290 - Question 25 of 26 - Editing an existing sketch

Click Split Entities (Sketch toolbar - *underlined in blue in the following image*) or Tools > Sketch Tools > Split Entities *(underlined in red in the following image)* as shown in the following image.

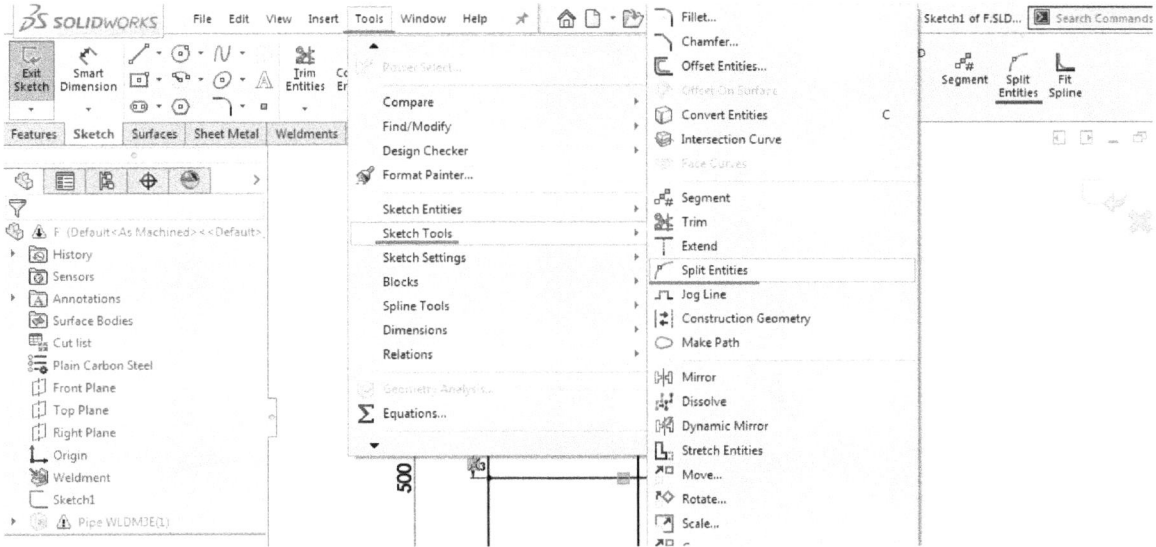

Figure 291 - Question 25 of 26 - Using Split Entities

The pointer changes to . Click the two outer vertical lines at the two locations where we want the splits to occur as shown in the following image. Each line splits into two entities, and a split point is added between the two sketch entities - see the following image.

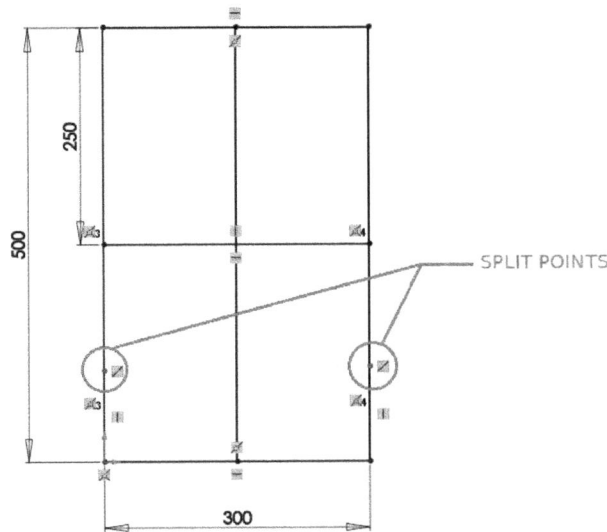

Figure 292 - Question 25 of 26 - Using Split Entities

215

DIMENSIONING A SKETCH ENTITY AND ADDING RELATIONS

Click Smart Dimension on the Dimensions/Relations toolbar, or click Tools > Dimensions > Smart. Click on one of the Sketch Entities as shown in the following image and enter a dimension of 150mm then click OK.

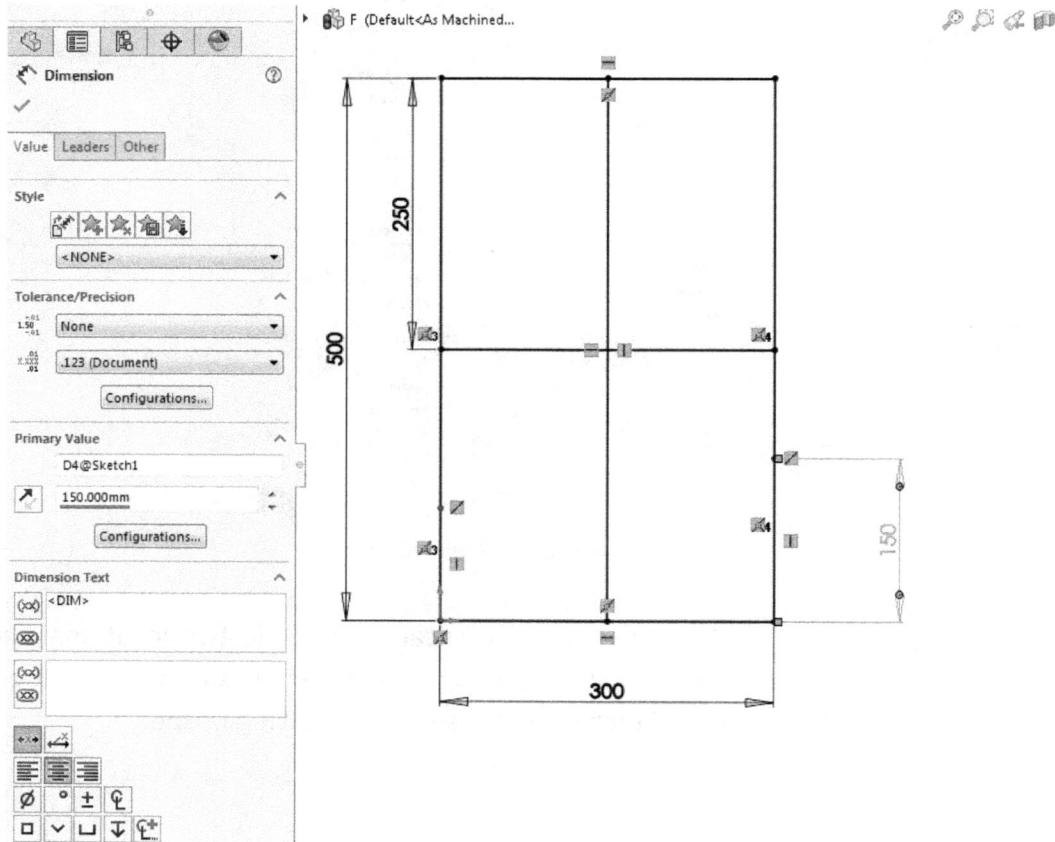

Figure 293 - Question 25 of 26 - Dimensioning a Sketch

Click Add Relations (Dimensions/Relations toolbar) as shown in the following image.

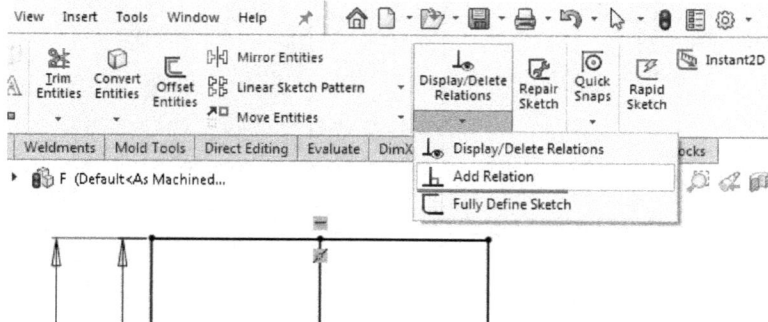

Figure 294 - Question 25 of 26 - Corner Treatment

Under Selected Entities in the Add Relations PropertyManager, select the sketch segments shown in the following image and click on the Equal Relation under Add Relations then click OK.

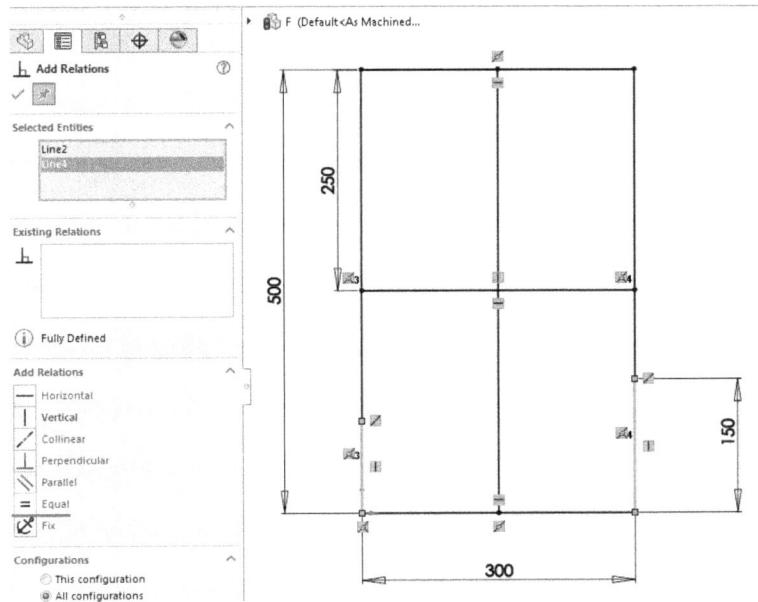

Figure 295 - Question 25 of 26 - Adding Relations

Your Fully Defined Sketch should now look as shown in the following image - take note of the added equal relation.

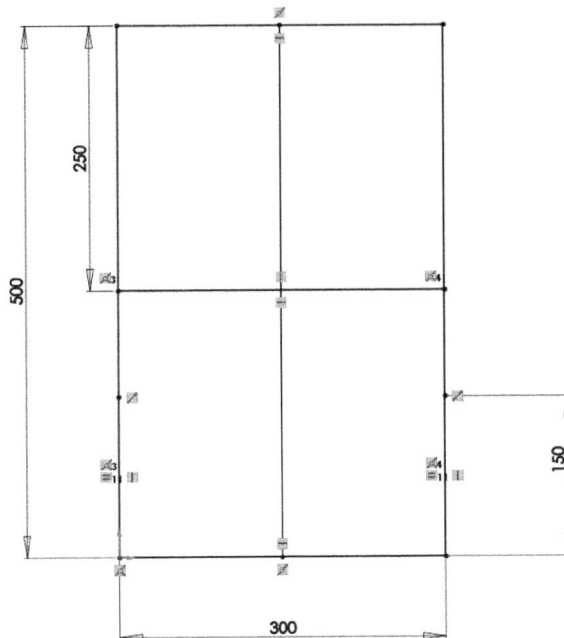

Figure 296 - Question 25 of 26 - Adding a Structural Member

Exit the Edit Sketch Mode. Your part will now look as shown in the following image or slightly different but with a warning appearing in the Feature Manager Design Tree on the structural member.

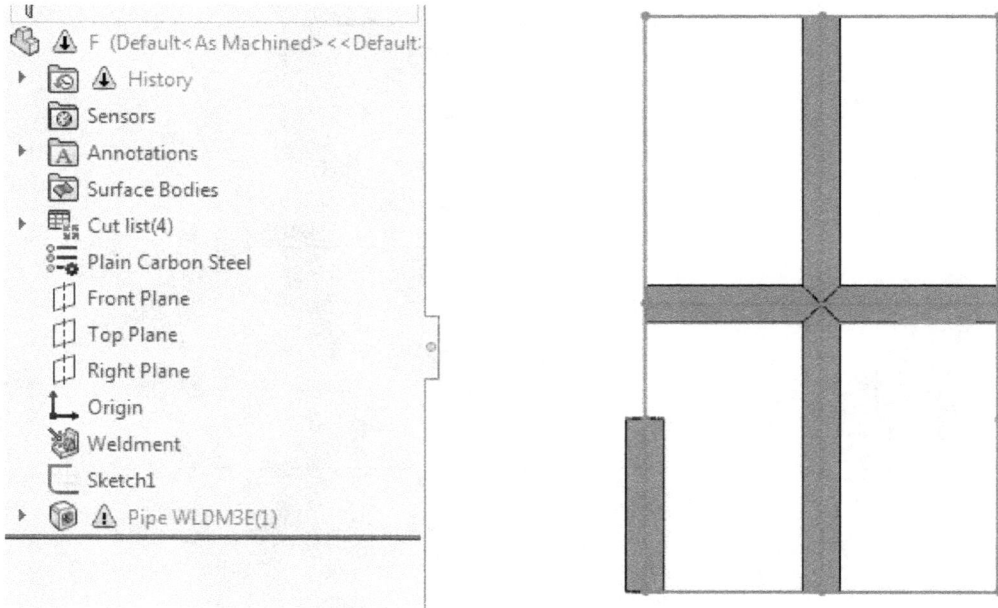

Figure 297 - Question 25 of 26 - Weldment Current Status

If you hover the mouse pointer above the structural member feature in the Feature Manager Design Tree you will notice a warning message - *Warning: Some of the selected path segments are no longer valid* - as shown in the following image.

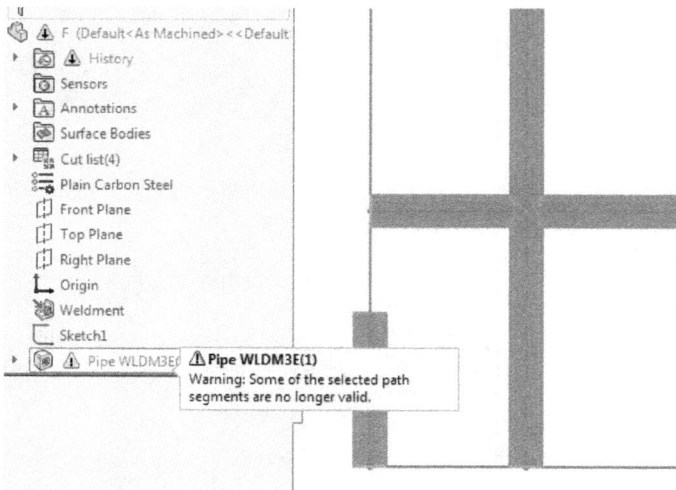

Figure 298 - Question 25 of 26 - Weldment Current Status

Right click or Click on the Pipe WLDM3E(1) structural member in the Feature Manager Design Tree and select Edit Feature as shown in the following image.

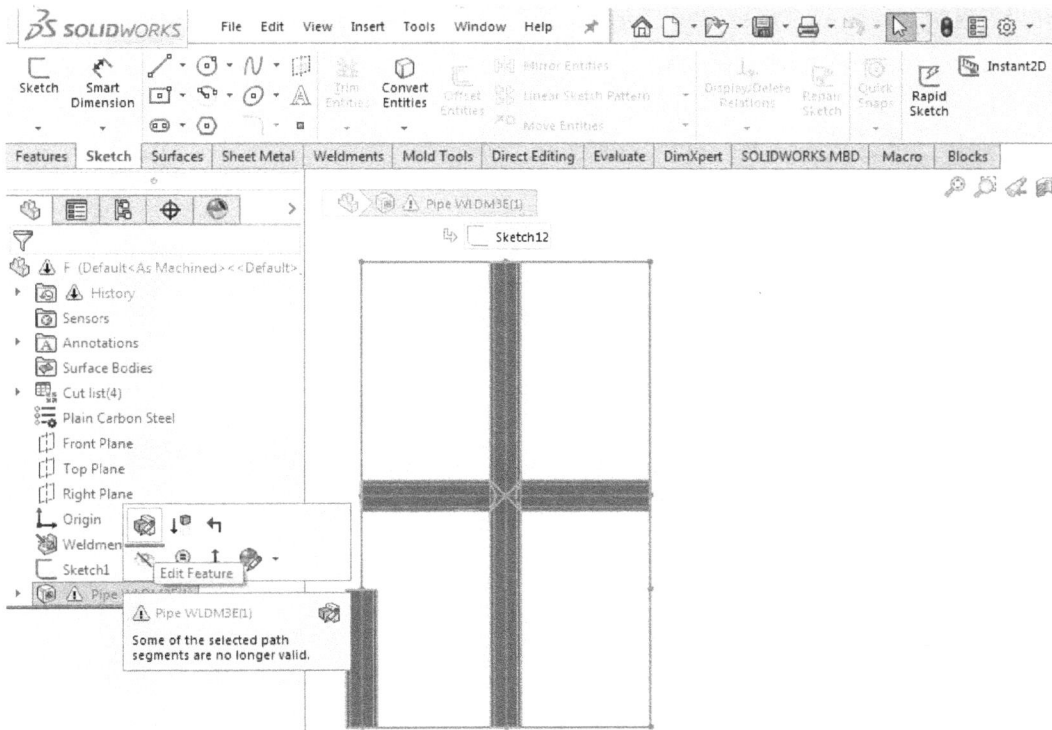

Figure 299 - Question 25 of 26 - Editing a structural member

Change the model view to Isometric View (View Toolbar) - see the following image below.

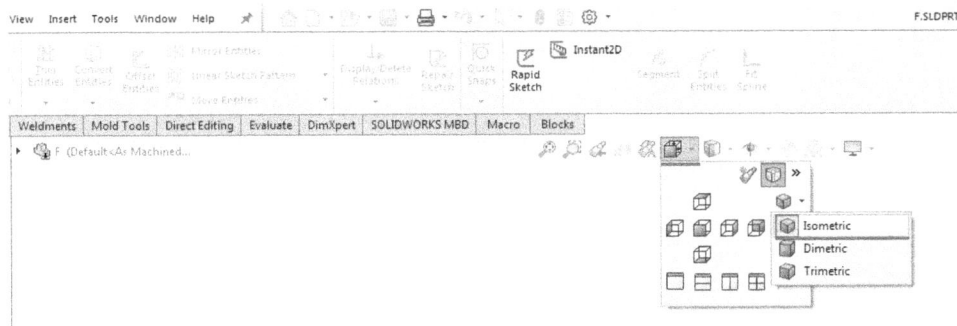

Figure 300 - Question 25 of 26 - Editing a structural member

TIP : You can also use the View Selector to see and select model views in context. To do so, Press Ctrl + Spacebar or click View Selector in the Orientation dialog box. The View Selector helps you see what the right, left, front, back, top, and axonometric views of your model will look like when selected. Click a face of the View Selector to select a view. Press Alt to select views on the back of the View Selector cube. Use the flyout button in the Orientation dialog box to set which type of axonometric view (isometric, dimetric, or trimetric) is displayed when you select a View Selector corner. You can see a preview by moving the pointer over a View Selector plane. As you move the pointer over different planes, the preview window updates. If a view is a standard view, the view name appears in the upper left corner of the preview window - see the following image.

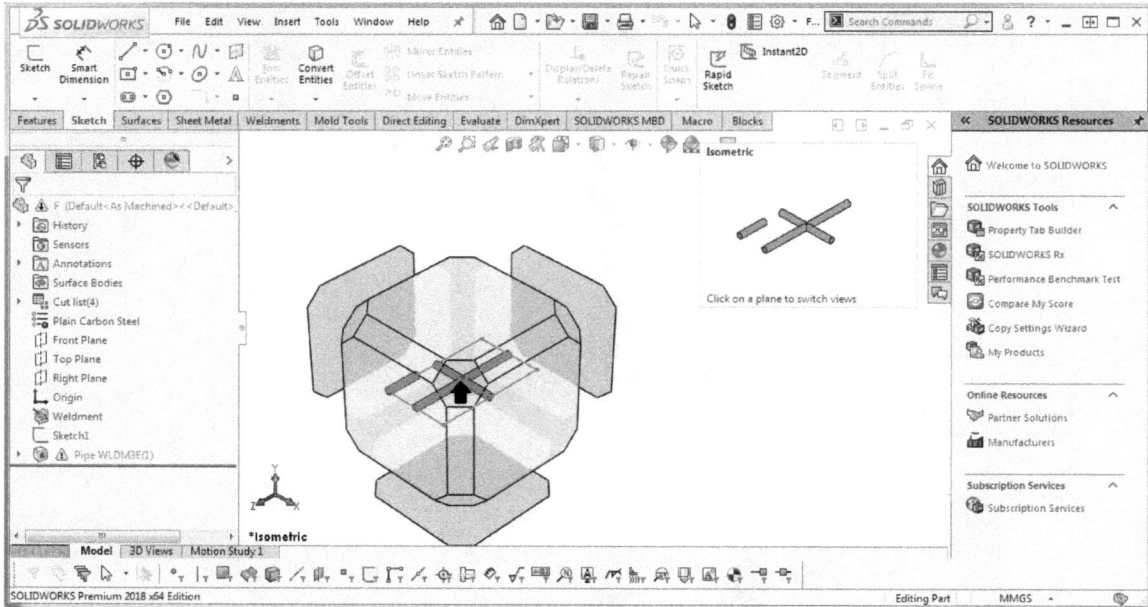

Figure 301 - Question 25 of 26 - Using the View Selector

Your part should now look as shown in the following image.

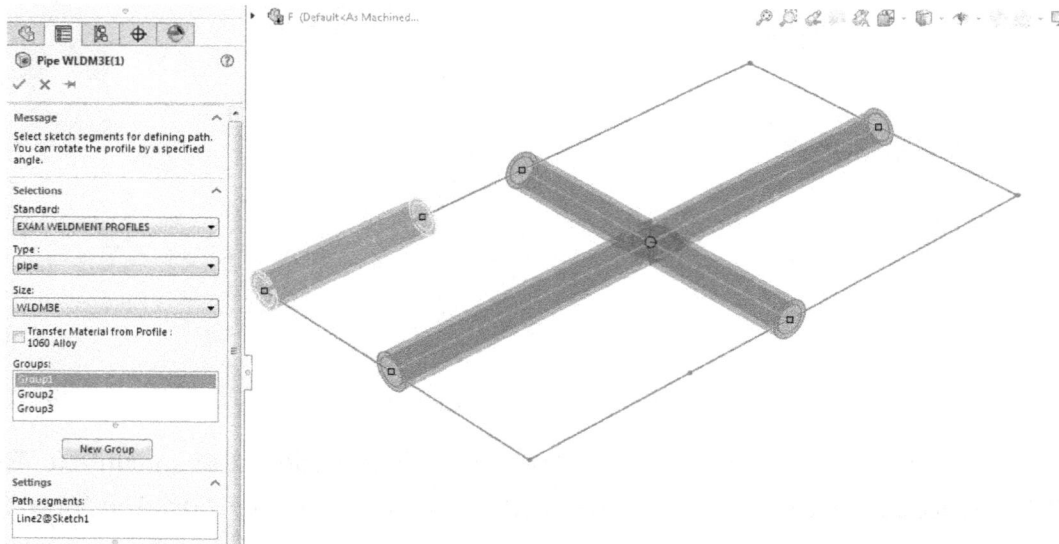

Figure 302 - Question 25 of 26 - Editing a structural member

220

In the Structural Member Property Manager, select Group 1 under Groups and under Path Segments Select 3 Line Segments as shown in the following image in the order 1,2 and 3 in red.

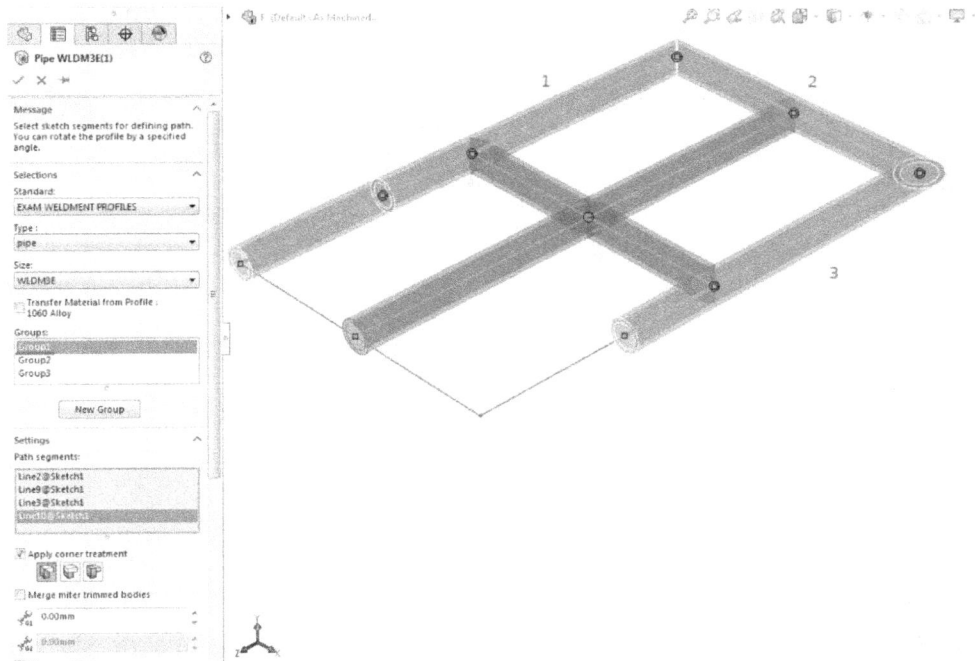

Figure 303 - Question 25 of 26 - Editing a structural member

Right Click on Line2 under Path Segments and select Delete as shown in the following image. Click OK to close the Structural Member Property Manager.

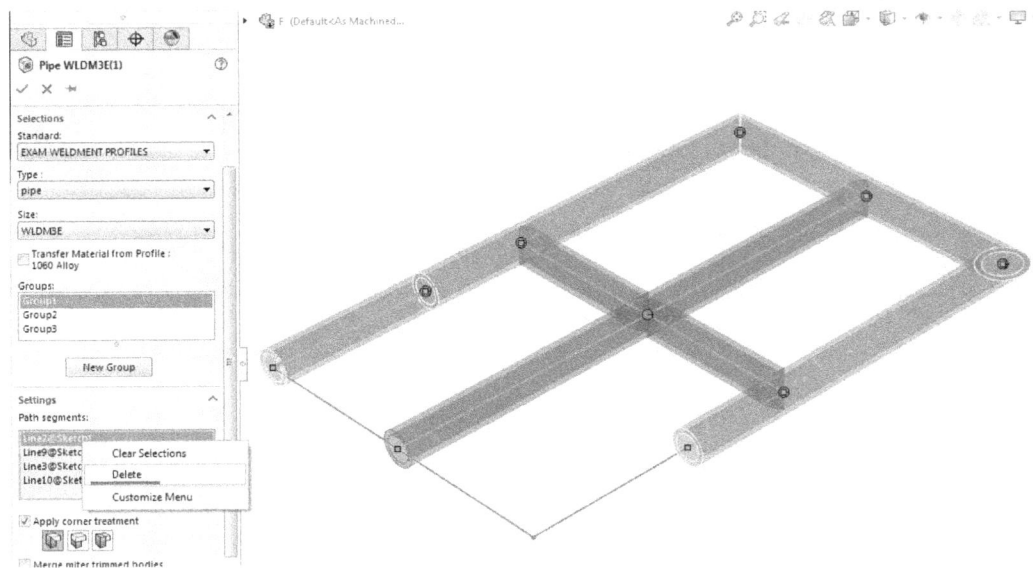

Figure 304 - Question 25 of 26 - Editing a structural member

Your Part should now look as shown in the following image.

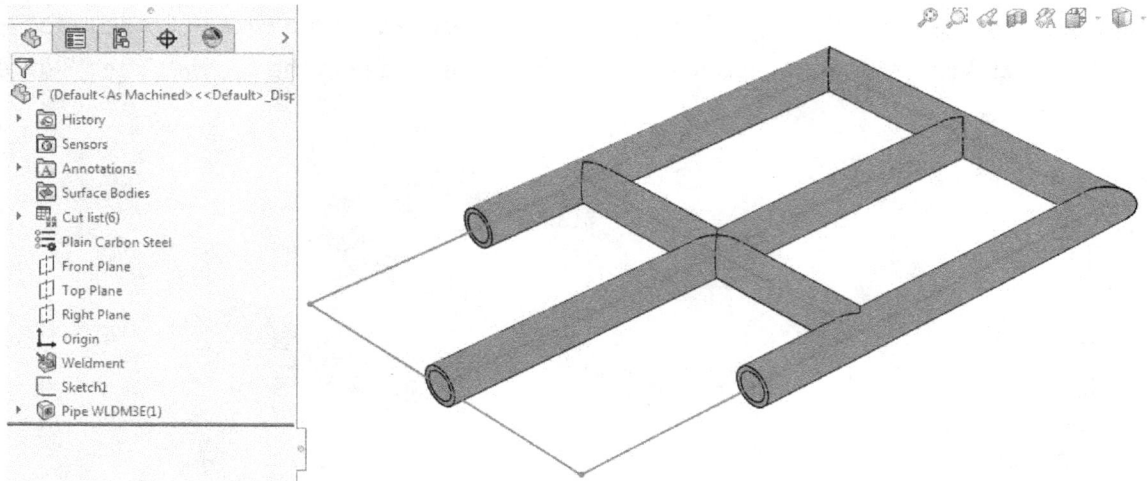

Figure 305 - Question 25 of 26 - Weldment Current Status

ADDING A STRUCTURAL MEMBER

Click Structural Member (Weldments toolbar) or Insert > Weldments > Structural Member. Make selections in the Structural Member PropertyManager to define the profile for the structural member and also select the three line segments and options as shown in the following image.

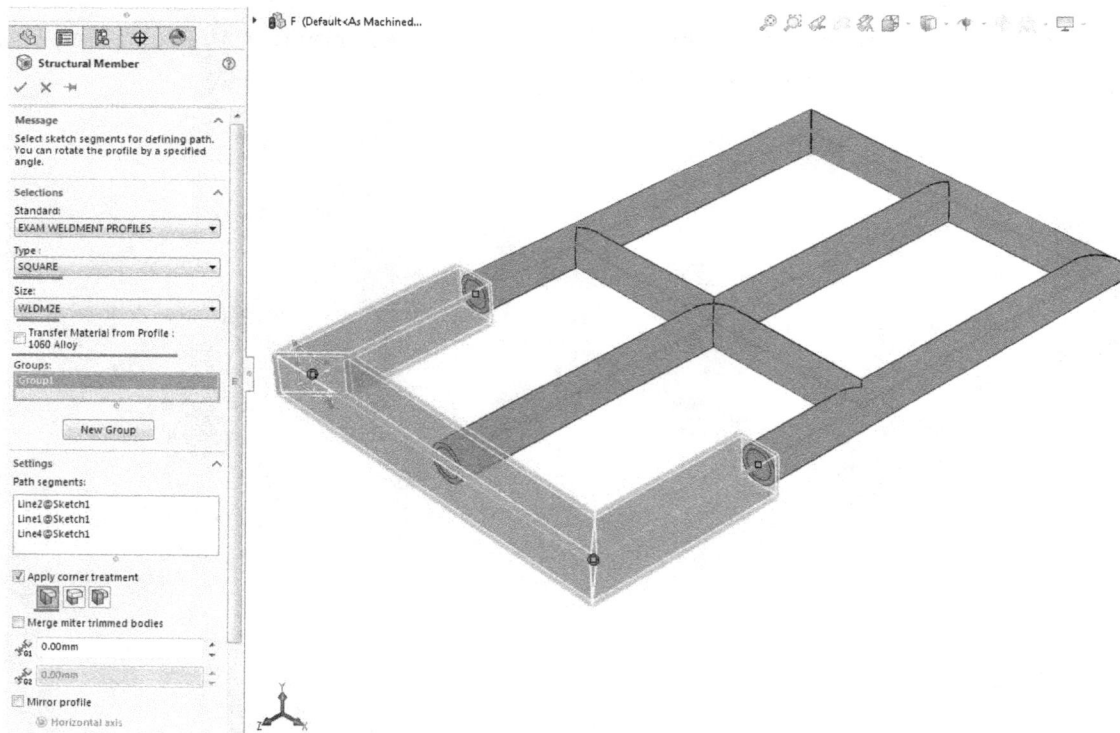

Figure 306 - Question 25 of 26 - Adding a structural member

Click OK to close the Structural Member Property Manager. Your part should now look as shown in the following image.

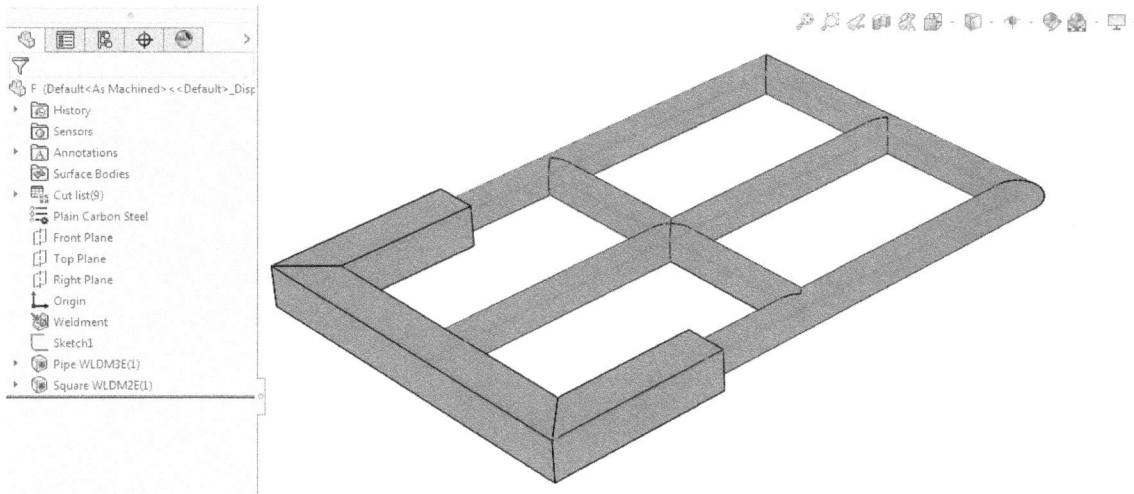

Figure 307 - Question 25 of 26 - Weldment current status

ADDING A BOSS EXTRUDE

Click on the face shown in the following image and select Sketch to start a sketch on the selected face.

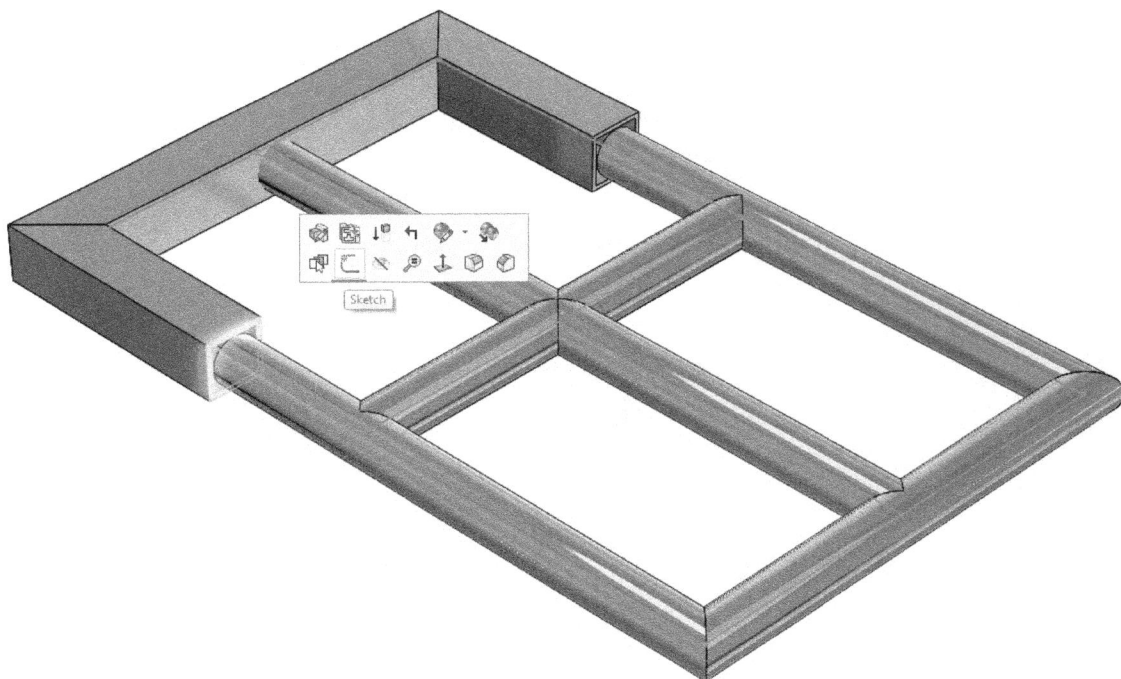

Figure 308 - Question 25 of 26 - Creating a sketch on a selected face

Sketch a Center Rectangle with its center coincident to the origin and dimensioned as shown in the following image.

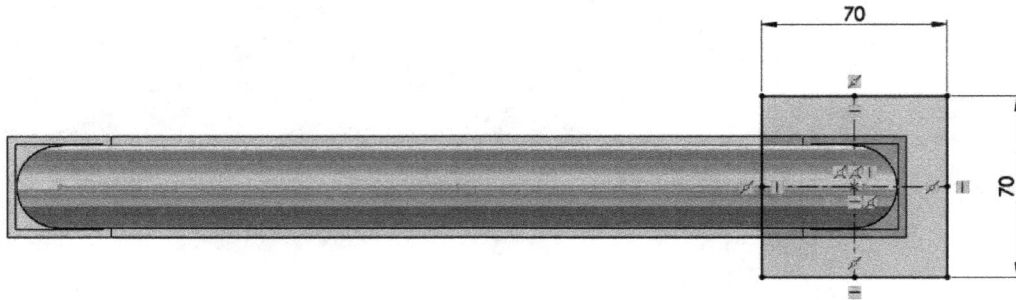

Figure 309 - Question 25 of 26 - Center rectangle

While still in Sketch Mode, under the Feature Tab, click the Extruded Boss/Base feature and enter options and parameters as shown in the following image then click OK to close the Extruded Boss/Base feature Property Manager.

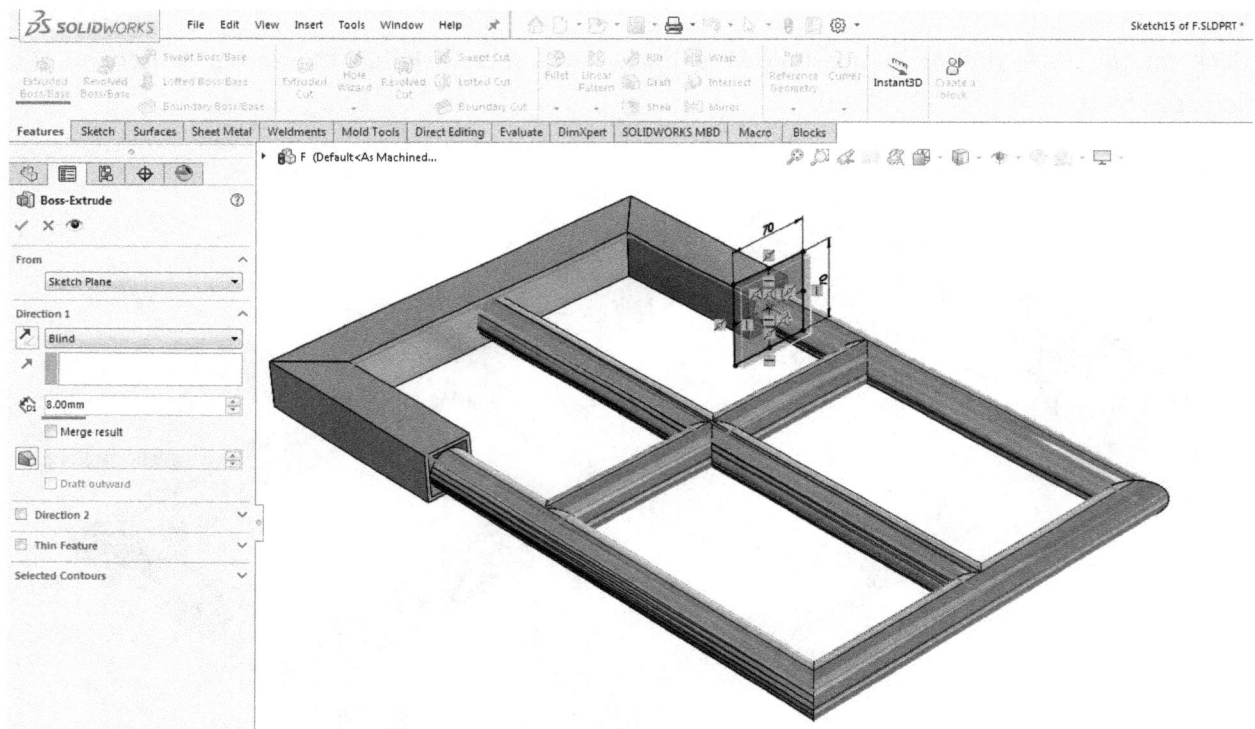

Figure 310 - Question 25 of 26 - Extruded Boss/Base feature

Your parts should now look as shown in the following image.

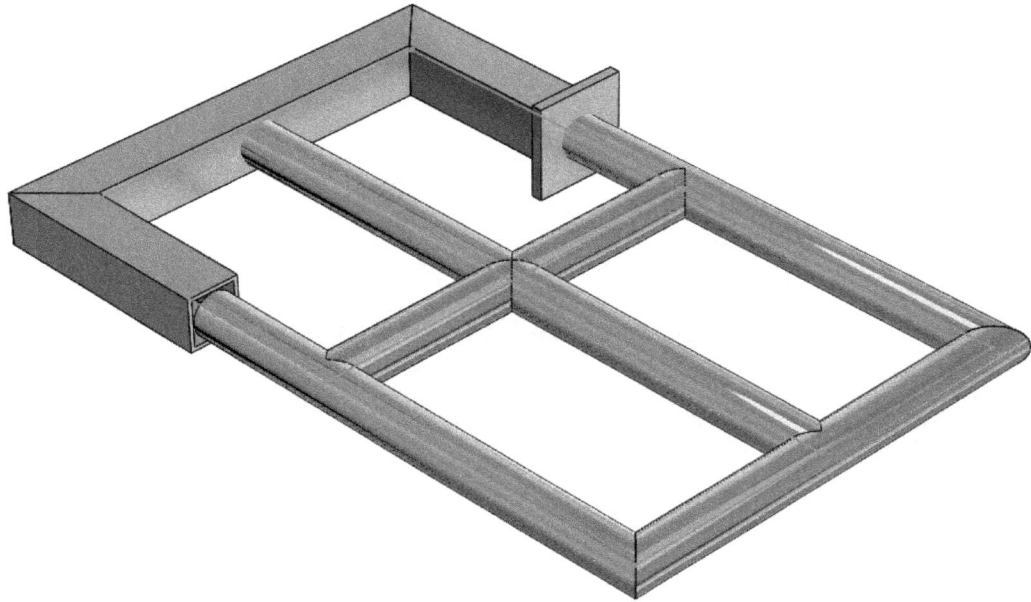

Figure 311 - Question 25 of 26 - Weldment current status

MIRRORING BODIES IN A PART

Click Mirror (Features toolbar) or Insert > Pattern/Mirror > Mirror. For Mirror Face/Plane, select a face in the graphics area as shown in the following image and under Bodies to Mirror, select the newly create Boss-Extrude by clicking it in the graphics area or in the Feature Manager Design Tree inside the _Cut List Folder_.

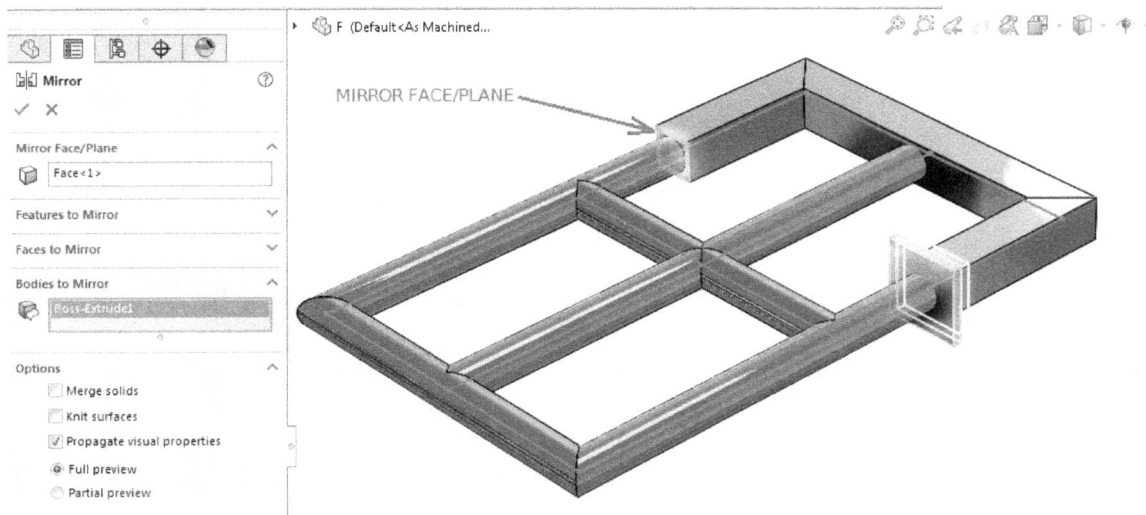

Figure 312 - Question 25 of 26 - Mirroring bodies in a part.

Click OK to close the Mirror Property Manager. Your part should now look as shown in the following image.

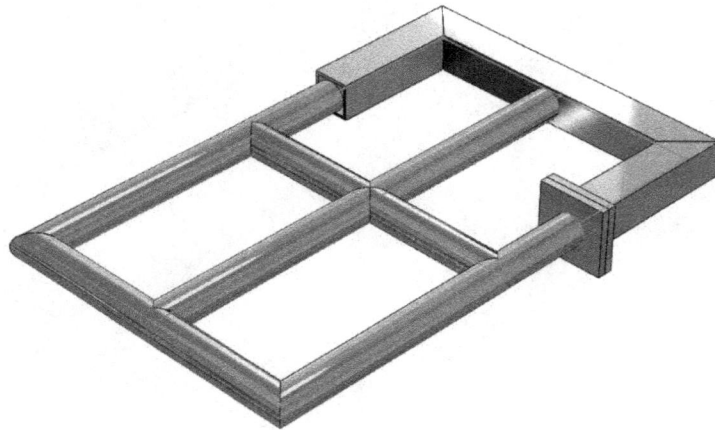

Figure 313 - Question 25 of 26 - Weldment current status

Save your part.

LINEAR PATTERN

Now, to create copies of the two boss extrude bodes on the other side we can either use the line pattern as shown in the following image or create a plane on the center of the weldment and then mirror the two bodies to the other side.

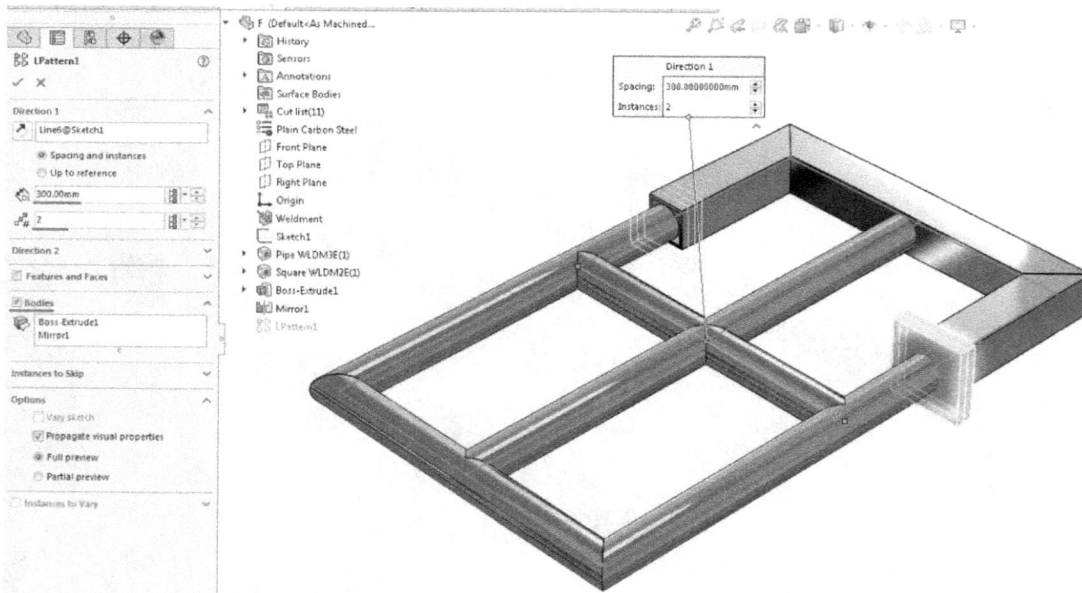

Figure 314 - Question 25 of 26 - Linear pattern

Click OK and your part should now look as shown in the following image.

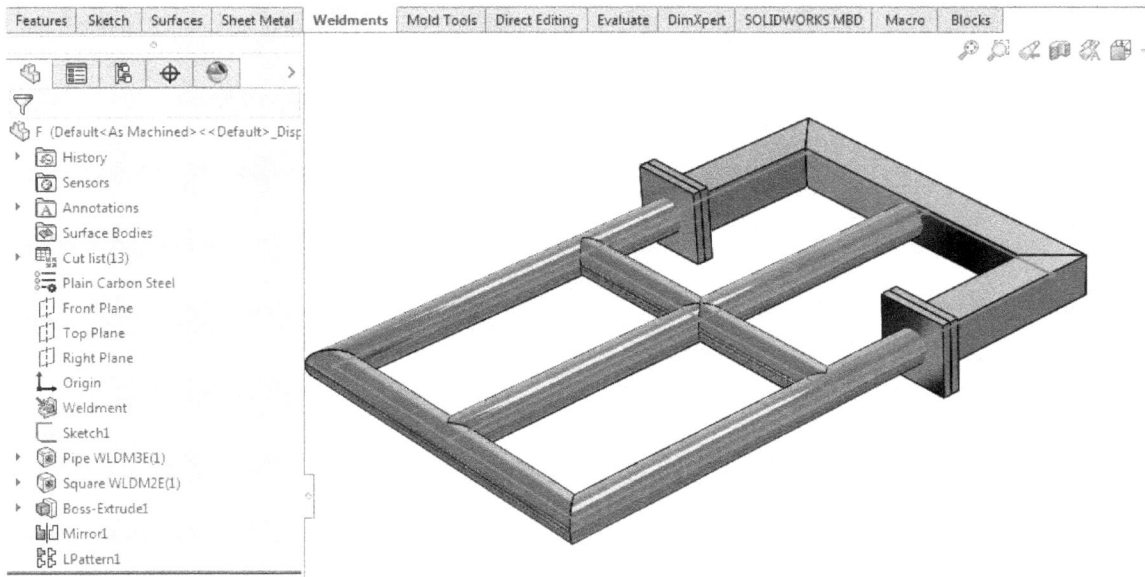

Figure 315 - Question 25 of 26 - Weldment Current Status

TRIM / EXTEND

Click Trim/Extend (Weldments toolbar) or Insert > Weldments > Trim/Extend and select options as shown in the following image.

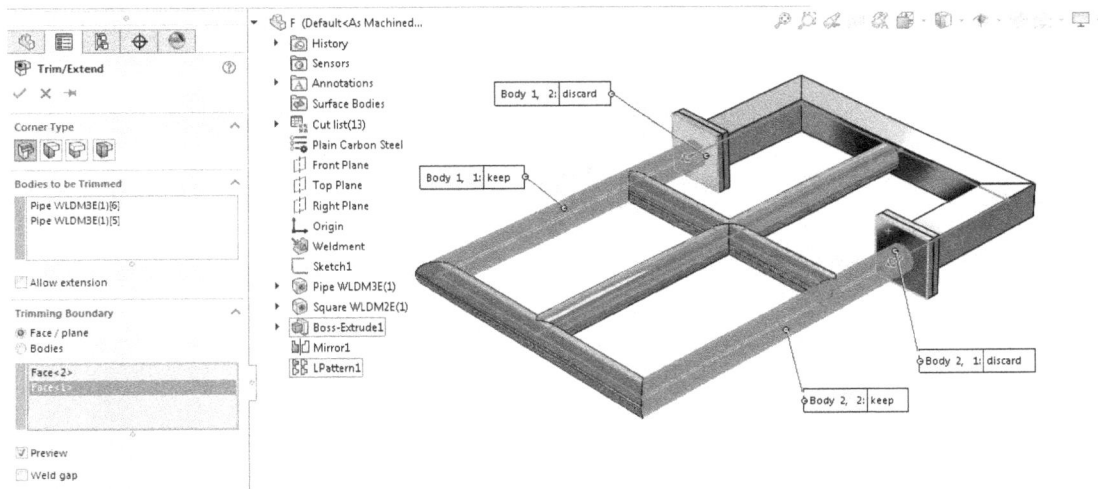

Figure 316 - Question 25 of 26 - Trim / Extend

Click OK.

Click Trim/Extend (Weldments toolbar) or Insert > Weldments > Trim/Extend and select options as shown in the following image.

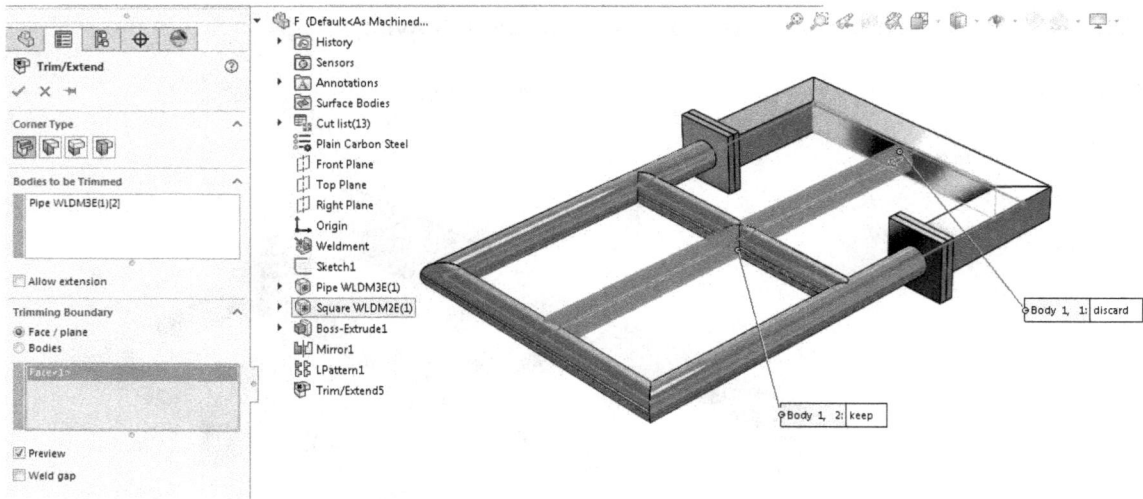

Figure 317 - Question 25 of 26 - Trim / Extend

Click OK.

Click Trim/Extend (Weldments toolbar) or Insert > Weldments > Trim/Extend and select options as shown in the following image.

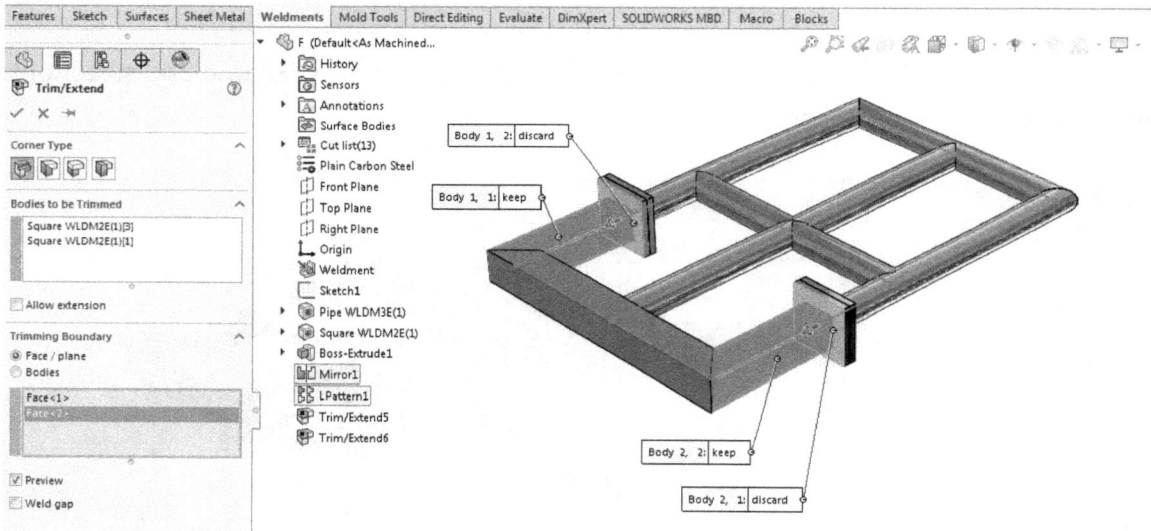

Figure 318 - Question 25 of 26 - Trim / Extend

Click OK.

Your part should now look as shown in the following image.

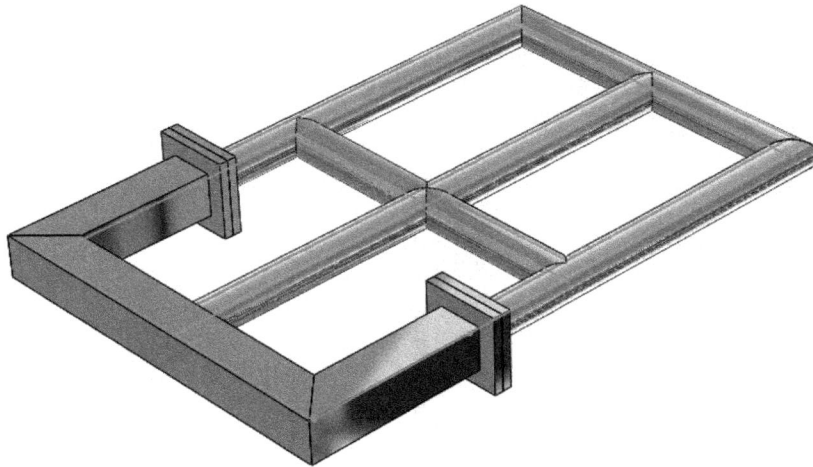

Figure 319 - Question 25 of 26 - Weldment Current Status

CREATING A SECTION VIEW IN A MODEL

Click Section View (View toolbar) or View > Display > Section View. In the Section View PropertyManager, select options as shown in the following image.

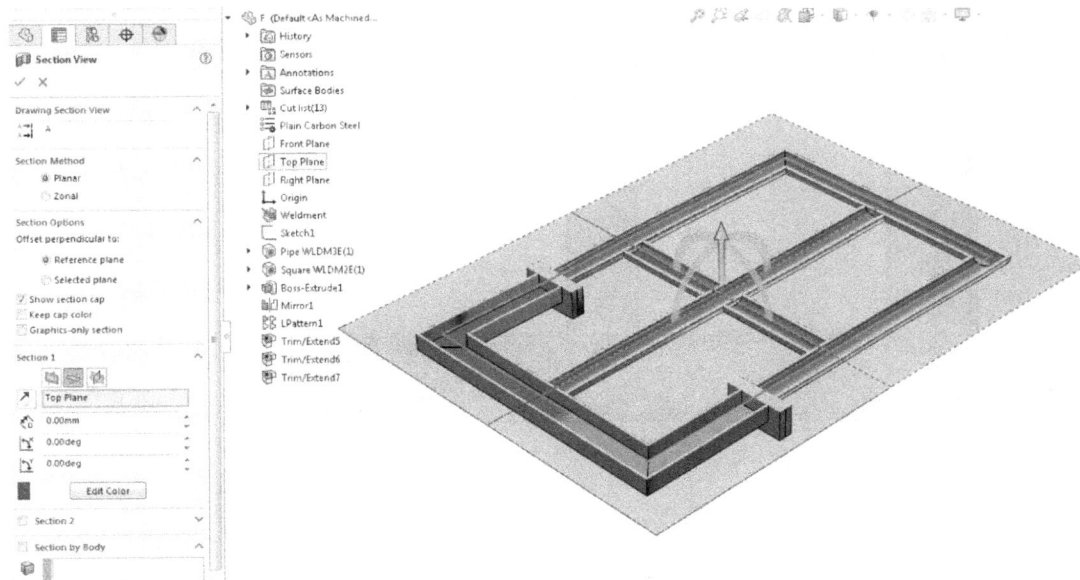

Figure 320 - Question 25 of 26 - Creating a Sectional View

Click OK and your part should now look as shown in the following image from the Top View.

Figure 320 - Question 25 of 26 - Sectional View from the Top View

Click Section View (View toolbar) or View > Display > Section View again to return the model to full view. Rebuild (Click Edit > Rebuild) and Save your part.

DISPLAYING THE CENTER OF MASS

Click Mass Properties (Tools toolbar) or Tools > Evaluate > Mass Properties. The calculated mass properties appear in the dialog box including the center of mass as shown in the following image - center of mass in underlined in red - thus X = 150.00, Y = 0.00 and Z = -226.00.

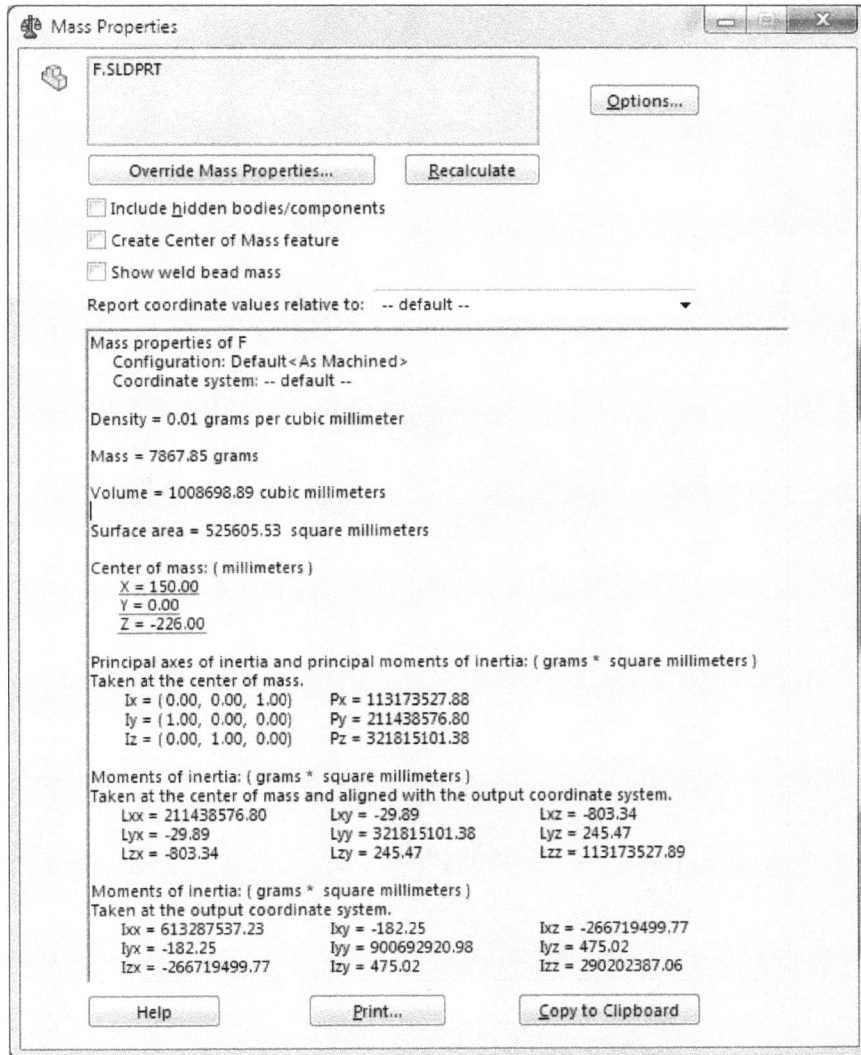

Figure 322 - Question 25 of 26 - Mass Properties

Click save to save the Part.

CHAPTER 26

QUESTION 26 EXAM SCREEN CAPTURES

QUESTION 26 EXAM SCREEN CAPTURE 1

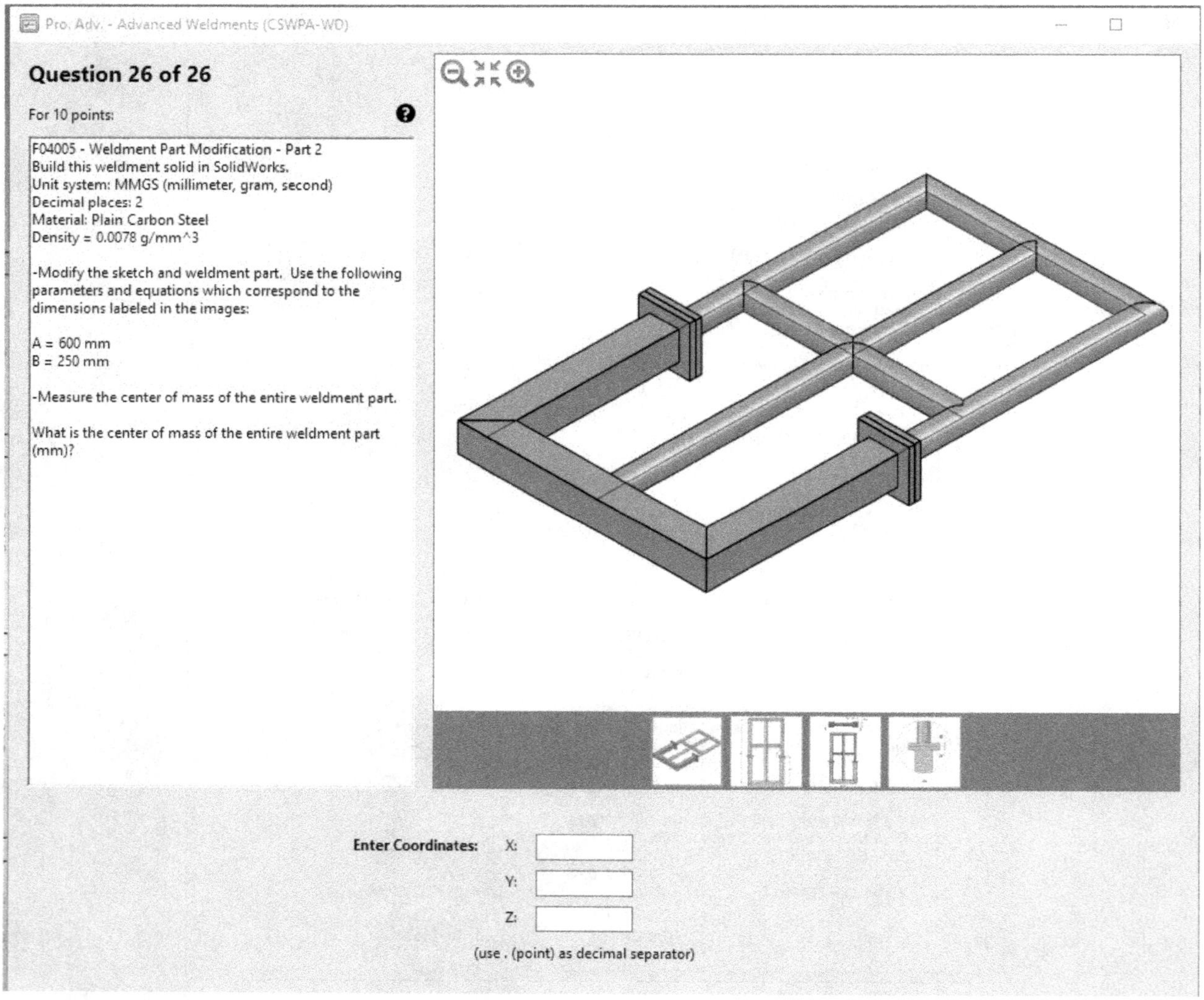

Figure 323 - Question 26 of 26 - Exam Screen Capture 1

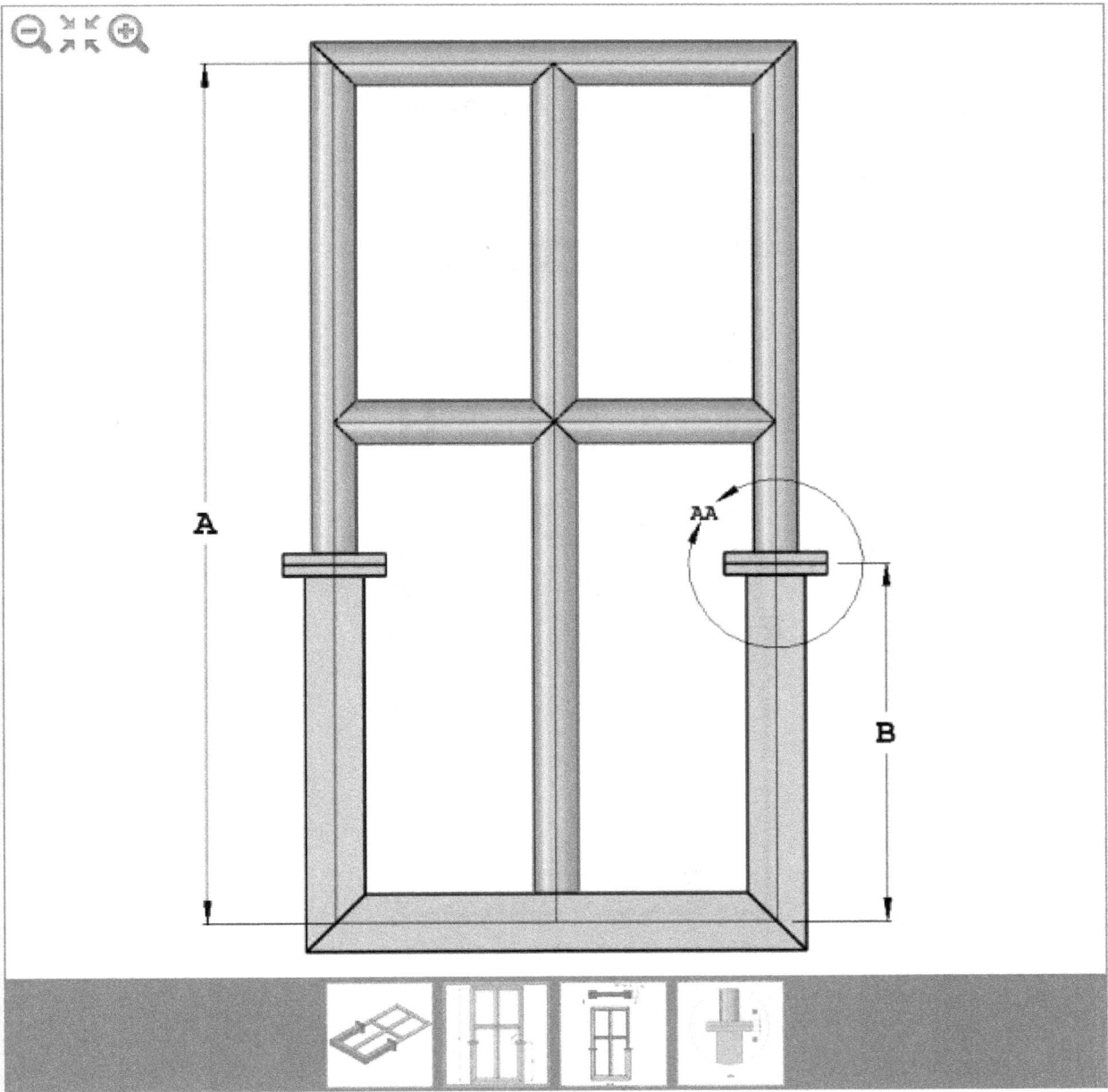

Figure 324 - Question 26 of 26 - Exam Screen Capture 2

Figure 325 - Question 26 of 26 - Exam Screen Capture 3

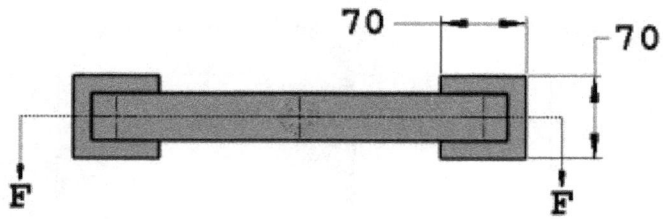

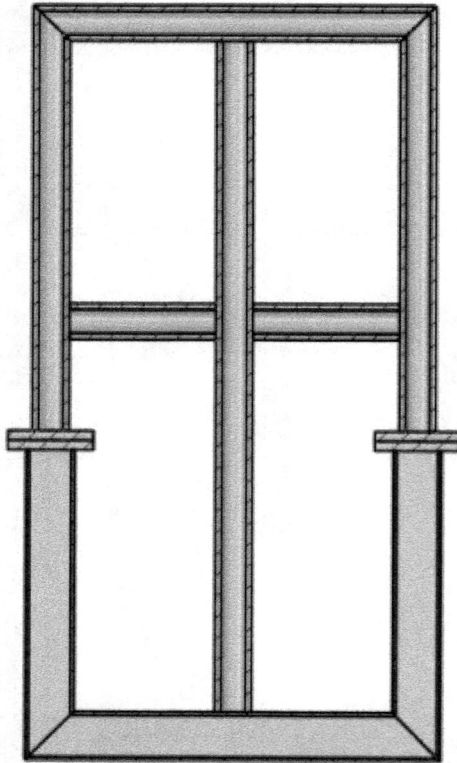

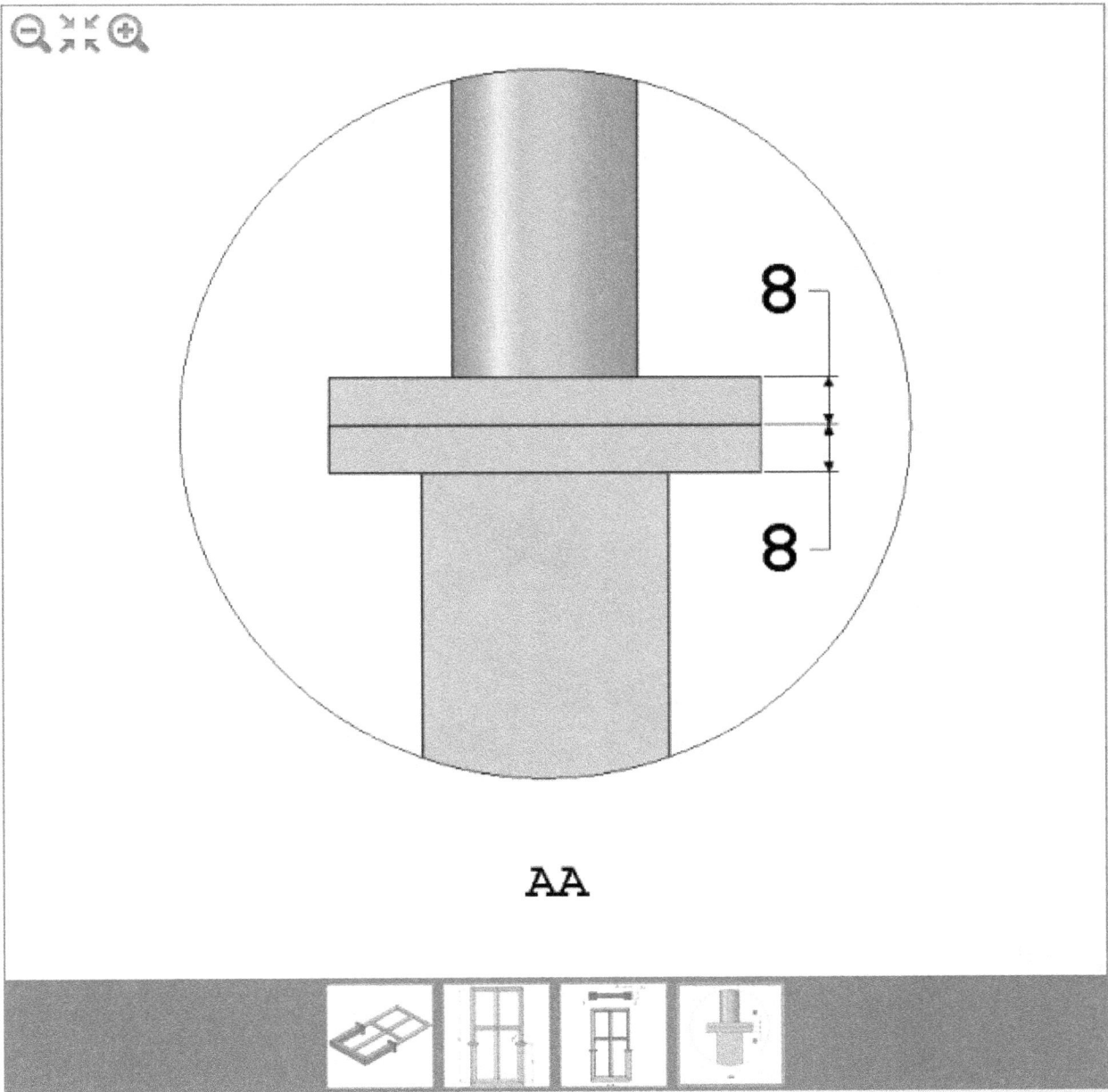

Figure 326 - Question 26 of 26 - Exam Screen Capture 4

EDITING AN EXISTING SKETCH

In the Feature Manager Design Tree, click or right click on Sketch1 and select Edit Sketch as shown in the following image. You may also click Sketch on the Sketch toolbar, or click Insert > Sketch then select an existing sketch to edit.

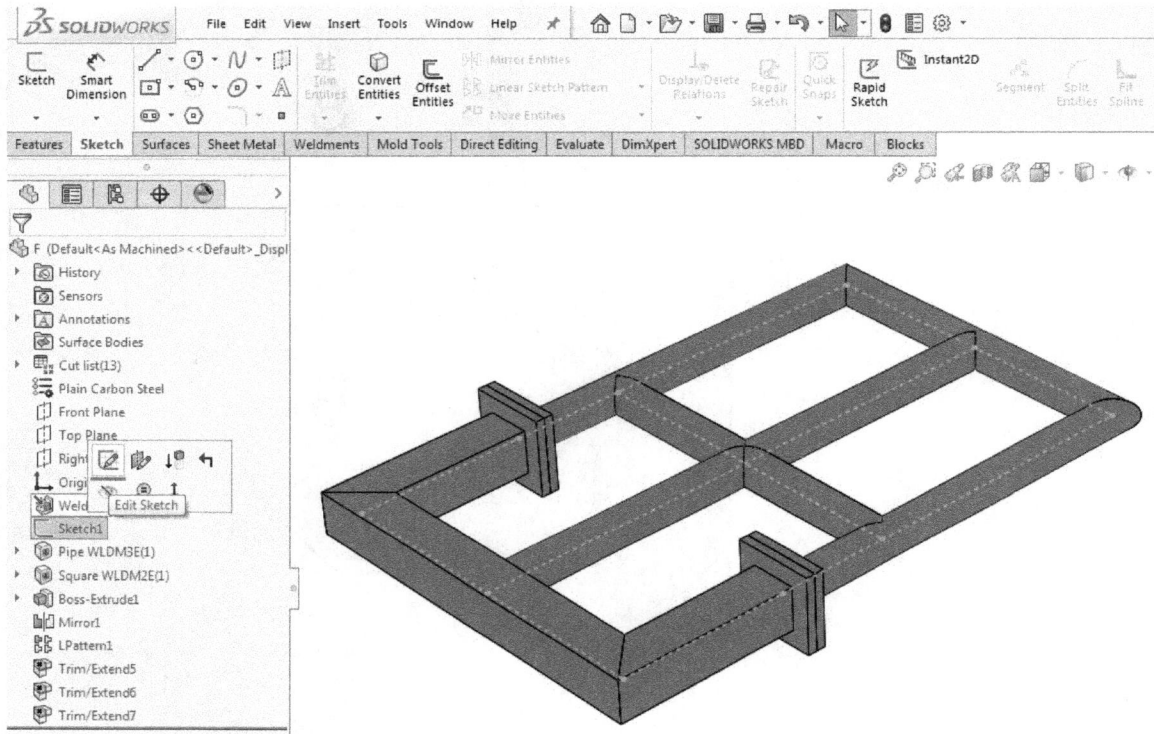

Figure 327 - Question 26 of 26 - Editing an existing sketch

Your part should now look as shown in the following image.

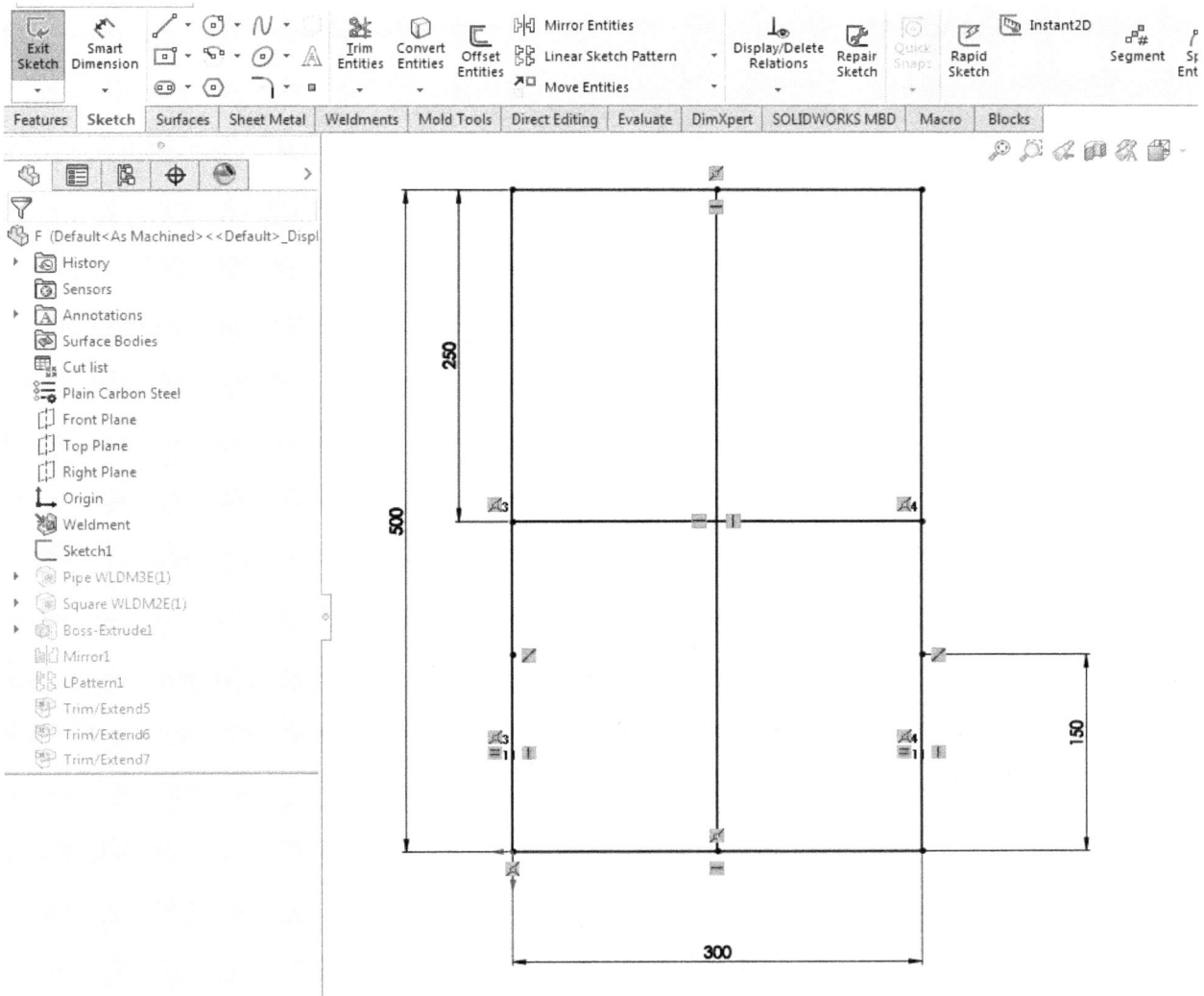

Figure 328 - Question 26 of 26 - Editing an existing sketch

CHANGING DIMENSIONS

Click on the 500mm dimension in the Graphics Area and in the Dimension Property Manager change the dimension to 600mm.

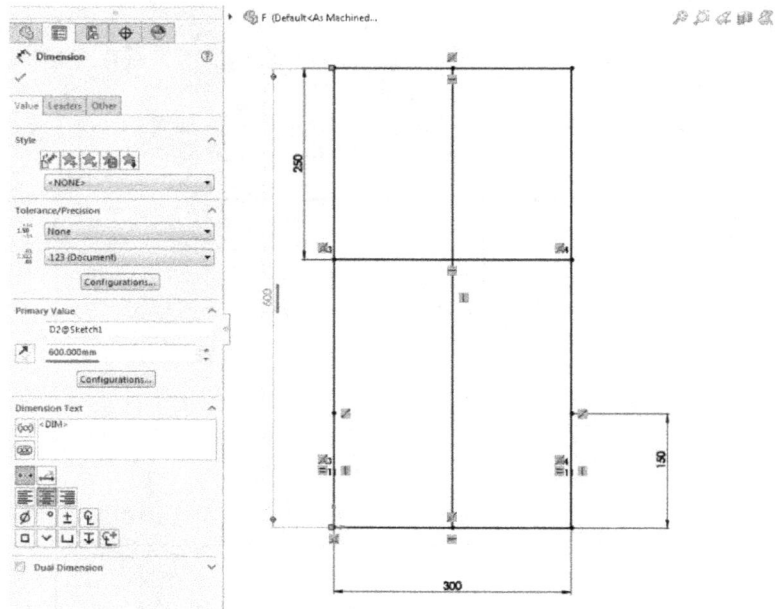

Figure 329 - Question 26 of 26 - Changing Dimensions

You may also Double click on the 500mm and the Modify Dialog box appears as shown in the following image. Change the dimension to 600mm then click the Rebuild Button underlined in red in the following image. Click Ok to close the Modify Dialog Box.

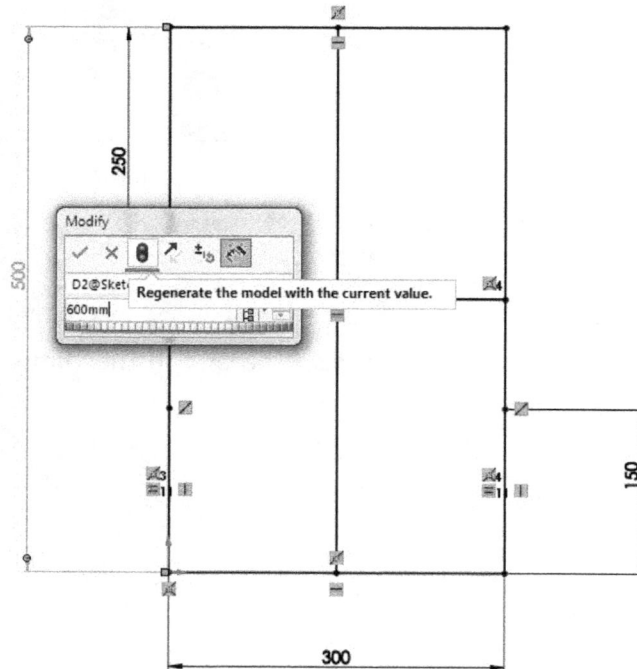

Figure 330 - Question 26 of 26 - Using the Modify Dialogue Box to change Dimensions

TIP : If the Modify Dialog box does not appear you need to change some system options by clicking on Options (Standard Toolbar). On the System Options tab, on the left pane, select General. On the right pane, select Input dimension value then Click OK. Now the Modify dialog box appears when you insert a dimension. To access the Modify Dialog box for an existing dimension, double-click the dimension.

Your part should now look as shown in the following image.

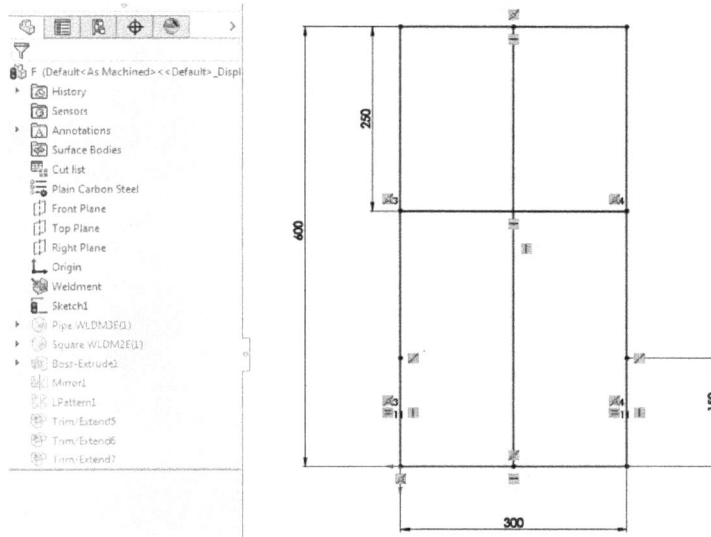

Figure 331 - Question 26 of 26 - Editing an existing sketch

While still in Edit Sketch Mode, repeat the same process to change the 150mm dimension to 250mm as shown in the following image.

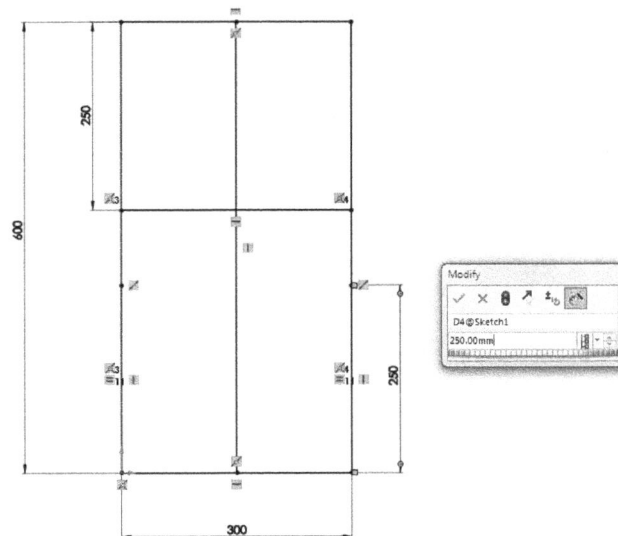

Figure 332 - Question 26 of 26 - Editing an existing sketch

239

Exit the sketch edit mode, save and rebuild you part. Your part should now look as shown in the following image.

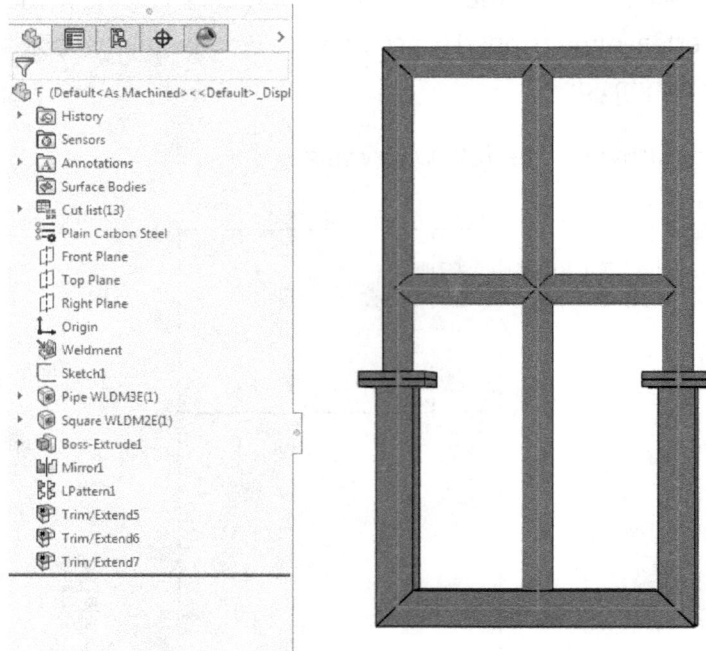

Figure 333 - Question 26 of 26 - Weldment Current Status

Take a cross sectional view as shown in the following image to double check the part and make sure or the segments are still trimmed correctly.

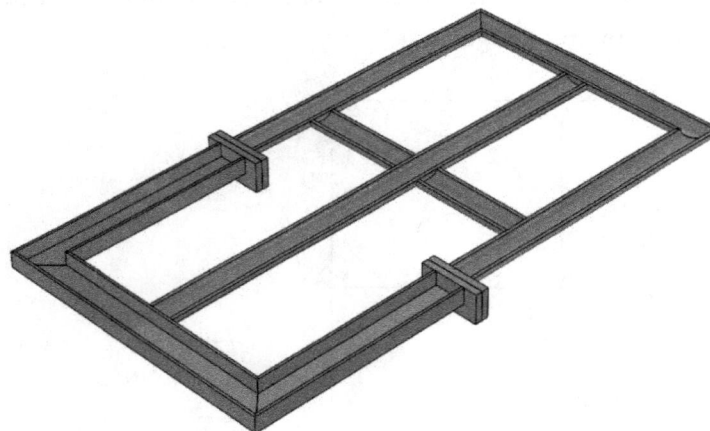

Figure 334 - Question 26 of 26 - Weldment Cross sectional View

Click Section View (View toolbar) or View > Display > Section View again to return the model to full view. Rebuild (Click Edit > Rebuild) and Save your part.

DISPLAYING THE CENTER OF MASS

Click Mass Properties (Tools toolbar) or Tools > Evaluate > Mass Properties. The calculated mass properties appear in the dialog box including the center of mass as shown in the following image - center of mass in underlined in red - thus X = 150.00, Y = 0.00 and Z = -285.45.

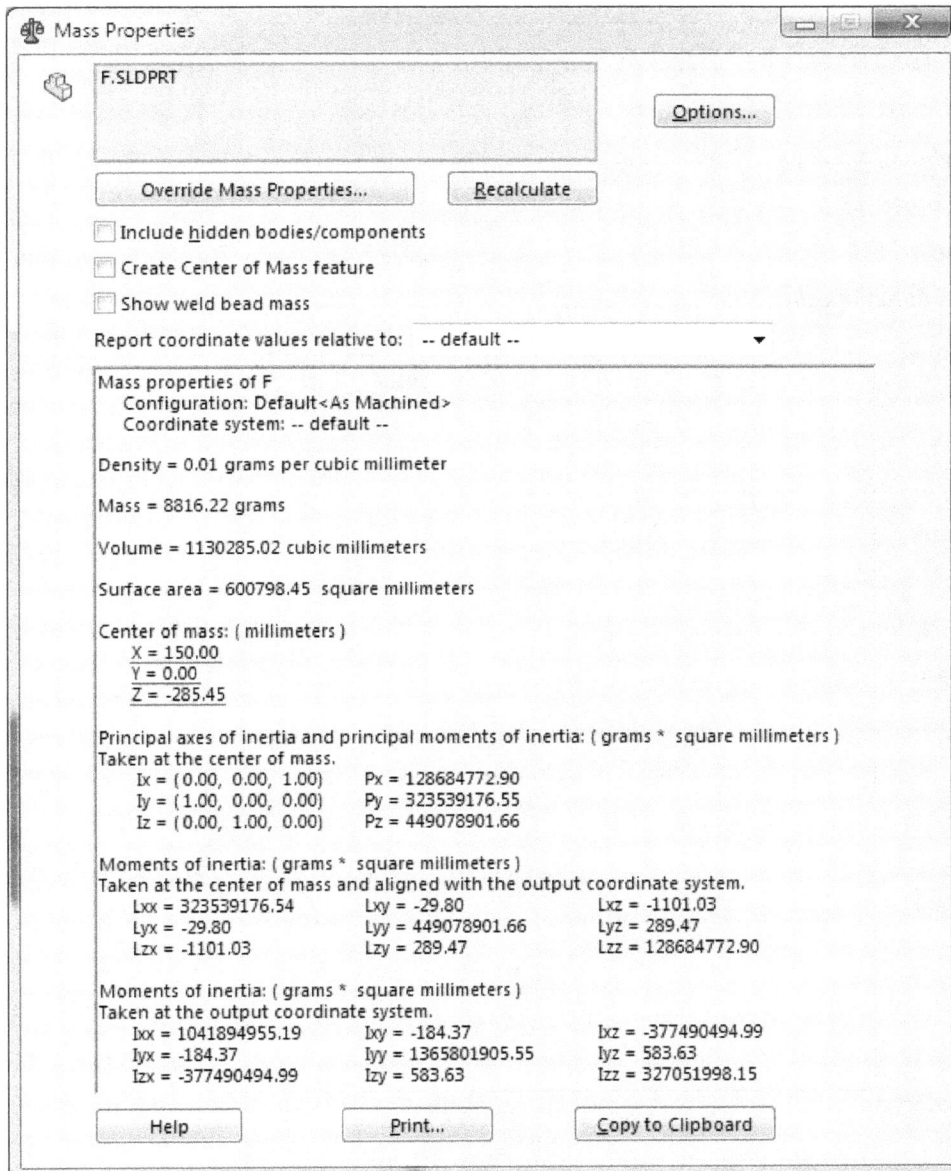

Mass Properties

F.SLDPRT

Options...

Override Mass Properties...   Recalculate

☐ Include hidden bodies/components

☐ Create Center of Mass feature

☐ Show weld bead mass

Report coordinate values relative to: -- default --   ▼

Mass properties of F
    Configuration: Default<As Machined>
    Coordinate system: -- default --

Density = 0.01 grams per cubic millimeter

Mass = 8816.22 grams

Volume = 1130285.02 cubic millimeters

Surface area = 600798.45 square millimeters

Center of mass: (millimeters)
    X = 150.00
    Y = 0.00
    Z = -285.45

Principal axes of inertia and principal moments of inertia: (grams * square millimeters)
Taken at the center of mass.
    Ix = (0.00, 0.00, 1.00)   Px = 128684772.90
    Iy = (1.00, 0.00, 0.00)   Py = 323539176.55
    Iz = (0.00, 1.00, 0.00)   Pz = 449078901.66

Moments of inertia: (grams * square millimeters)
Taken at the center of mass and aligned with the output coordinate system.
    Lxx = 323539176.54   Lxy = -29.80   Lxz = -1101.03
    Lyx = -29.80   Lyy = 449078901.66   Lyz = 289.47
    Lzx = -1101.03   Lzy = 289.47   Lzz = 128684772.90

Moments of inertia: (grams * square millimeters)
Taken at the output coordinate system.
    Ixx = 1041894955.19   Ixy = -184.37   Ixz = -377490494.99
    Iyx = -184.37   Iyy = 1365801905.55   Iyz = 583.63
    Izx = -377490494.99   Izy = 583.63   Izz = 327051998.15

Help   Print...   Copy to Clipboard

Figure 335 - Question 26 of 26 - Mass Properties

Click save to save the Part. Congratulations and wish you all the best.

INDEX

242

9 780620 868259